Work Activity Studies Within the Framework of Ergonomics, Psychology, and Economics

Human Activity: Efficiency, Safety, Complexity, and Reliability of Performance

Series Editor
Gregory Z. Bedny
Research Associate at Evolute, Inc.
Louisville, KY

Work Activity Studies Within the Framework of Ergonomics, Psychology and Economics
Gregory Z. Bedny and Inna S. Bedny

Work Activity Studies Within the Framework of Ergonomics, Psychology, and Economics

Gregory Z. Bedny and Inna S. Bedny

CRC Press
Taylor & Francis Group
Boca Raton London New York

CRC Press is an imprint of the
Taylor & Francis Group, an **informa** business

CRC Press
Taylor & Francis Group
6000 Broken Sound Parkway NW, Suite 300
Boca Raton, FL 33487-2742

First issued in paperback 2023

ISBN: 978-1-03-257027-3 (pbk)
ISBN: 978-0-8153-5710-0 (hbk)

DOI: 10.1201/9781351125000

Library of Congress Cataloging-in-Publication Data

Names: Bedny, Gregory Z., author. | Bedny, Inna, author.
Title: Work activity studies within the framework of ergonomics, psychology, and economics / authored by Gregory Z. Bedny and Inna S. Bedny.
Description: Boca Raton : Taylor & Francis, a CRC title, part of the Taylor & Francis imprint, a member of the Taylor & Francis Group, the academic division of T&F Informa, plc, [2019] | Includes bibliographical references and index.
Identifiers: LCCN 2018022126| ISBN 9780815357100 (hardback : acid-free paper) | ISBN 9781351125000 (ebook)
Subjects: LCSH: Work design.
Classification: LCC T60.8 .B43 2019 | DDC 620.8--dc23
LC record available at https://lccn.loc.gov/2018022126

Visit the Taylor & Francis Web site at
http://www.taylorandfrancis.com

and the CRC Press Web site at
http://www.crcpress.com

This one is for Marina and Maya, our two wonderful girls.

Contents

Preface

Businesses today face a constant need for innovation, but how can they determine if an innovation is going to work or if it is worth the effort required to implement it?

How can a company improve or perfect a user's online experience? How can software be made user-friendly or easy to use by customers?

This book strives to offer a number of methods that can help to answer these and a lot of other questions.

With increased cognitive demands on task performance, the psychological method of study of human activity plays an important role. Systemic-structural activity theory (SSAT) provides unified qualitative, analytical, and quantitative methods of task analysis from a psychological perspective.

This book presents advanced methods of morphological and quantitative analysis of computer-based and computerized tasks that involve multiple decisions. Such tasks are extremely variable and have complex probabilistic structures. Even instruction-based tasks include multiple decisions and therefore the logical sequence of cognitive and behavioral actions can change depending on the specifics of such decisions.

In this book, we continue our development of SSAT, which is a unified approach to the ergonomic design that goes beyond purely experimental procedures and includes analytical methods. This approach enables ergonomists to be involved in the design of new man-machine and human-computer systems at earlier stages. SSAT as a psychological theory not only provides a unified framework for the study of human work in ergonomics and work psychology but also can be utilized in economics and specifically in behavioral and labor economics. It can also be utilized by engineers working in such fields as efficiency of human performance, equipment and interface design, safety, and training. So in this book we consider the impact of SSAT not only in ergonomics but also in economics. An important area of the interaction of economics and ergonomics is time study and, associated with it, work complexity and the evaluation of the efficiency of ergonomic innovations and wage rates. There are no advanced methods of time study for highly variable tasks that include a lot of cognitive components. However, time study is critically important because activity unfolds in time as a process. In ergonomics, this process is considered as a linear sequence of events, where elements of activity follow one another. In cognitive psychology, time study is reduced to the analysis of various reactions that are performed with maximum speed. Cognitive processes such as perception, memory, decision-making, and so on are viewed as relatively isolated elements of activity. Time study in ergonomics is usually reduced to the techniques that are known as timeline analysis, which is just a simplified method of chronometrical study that utilizes task elements and their durations to determine the time of task performance. According to SSAT, this method employs technological units of analysis describing elements of tasks using common language. Such description does not clearly determine what really happens when an operator performs contemporary complex computerized tasks because the time structure of activity, where its basic elements are cognitive and behavioral actions, is not presented. These cognitive and behavioral actions have complex logical organization and occur in the activity structure with various probabilities. SSAT offers the new principles of time study for variable tasks with a high level of mental demands. Integration of ergonomics and economics is important when studying the efficiency of performance of such tasks.

SSAT introduced qualitative and quantitative stages of activity analyses during task performance. The most important one is functional analysis that considers activity as a complex self-regulative system. The SSAT model of self-regulation of orienting activity and its application to task analysis are presented in this book. Specific attention is given to the analysis of a pilot's task performance in emergency situations. Comparative analysis of concepts of situation awareness and orienting activity in a pilot's performance are considered.

There is no concept of activity structure in ergonomics, work psychology, and economics. This is the case despite the fact that the concept is critical for the design of equipment and human-computer interface, for the methods of improving task performance, and for evaluating usability.

In this book, the basic units of activity analysis are described in a detailed manner. The criteria that allow to extract cognitive and behavioral actions from activity process and to determine how these elements are combined in time during task performance are theoretically justified. Particular attention is paid to determining their logical organization and the probability of their appearance in the task structure. It is demonstrated that such a structure and the method of task performance depend on equipment configuration and/or the human-computer interface. Analysis of activity structure plays a critical role in design of equipment and computer interfaces. Understanding the structure of the operator's activity is important for determining the usability of equipment and in developing the training methods. This is because the structure of activity and methods of task performance depend on configuration of equipment and/or interface. Without understanding the structure of activity, it is impossible to determine the usability of the system.

Experiments dominate the study of human work, leaving to theory the modest role of data interpretation. In the field of engineering, design is understood as creation of models of an artificial object in accordance with previously set properties and characteristics with the purpose of creating material objects according to developed models. Thus, analytical methods play a leading role in this design. In ergonomics, there are no real analytical principles of design except for ones offered by SSAT.

There is a need for a powerful unified theory that would offer a standard-ized terminology, procedures, and qualitative and quantitative methods of analyses that can be used for the creation of models of designed systems.

In this book, we present a unified approach to design of man-machine and human–computer interaction (HCI) systems by using general principles of design. The basic principle of such design is development of models of human activity that describe its structure. Systemic description of the struc-ture of activity and evaluation of its efficiency are achieved by morphological analysis of activity. Morphological analysis and especially quantitative eval-uation of task complexity and reliability of task performance allow develop-ment of quantitative analytical methods of analysis.

The second section of the book considers contemporary complex tasks that include decisions with more than two outcomes that appear with vari-ous probabilities. Here we present a new method of algorithmic description of task performance when interacting with equipment and/or interfaces. Various decisions that are involved in task performance, the possible out-comes, and the probabilities of those decisions can be determined based on the suggested methods of analysis. The performance time of cognitive and motor actions is determined, which allows development of the temporal structure of activity. The method of morphological analysis of such complex variable tasks with the probabilistic structure is suggested for the first time.

Efficiency and usability of designed equipment, computer interface, and training methods can be evaluated based on the analysis of human activity structure because its safety, efficiency, complexity, and reliability depend on such design.

The description of an activity structure via morphological analysis is a prerequisite for development of various quantitative methods of task analy-sis. The results obtained allow determination of the probabilistic structure of a task, the performance time of the various versions of the variable task, the complexity of the versions, and the accuracy and reliability of the per-formance with high precision. In this book, the evaluation of complexity and reliability of computerized tasks is demonstrated utilizing analysis of their logical and probabilistic structures. The analysis of the probabilistic structure of activity is specifically relevant for a reliability analysis of task performance. A new method of reliability assessment is suggested in this book. In the final chapter of the book, the relationship between complexity and reliability is discussed.

Analytical methods of design of computer-based and computerized tasks with complex logical and probabilistic structure that have been developed in SSAT allow uncovering overlooked enhancements to human-computer interface systems at the design stage, putting ergonomic professionals ahead of user requests. Never-ending redesign of existing systems is a very costly process that can be avoided. This book presents methods of predicting vari-ous user requests which can reduce the number of testing and implementa-tion cycles and save money and resources. Being a step ahead of the game speeds up process enhancements that lead to reducing task complexity and failures of performance and increase efficiency and productivity without making the human performance more complicated.

Traditional methods of task analysis utilize methods of study not sufficiently relevant and theoretically grounded, where the main tool of analysis is conducting an experiment. It is combined with observation, surveying, interviews, and so on. The only analytical method that is successfully utilized in ergonomics is the anthropometric method of study.

SSAT offers a framework that allows ergonomics, work psychology, and economics to study human work, overcoming traditional separation of cognition, external observable behavior, and motivation. This framework presents new methods for the description of activity structure. This approach can be used at the design stage, when observation and experiment are not applicable. The economic efficiency of various innovations can be evaluated by using the described methods of study. The data obtained also can be useful for development of training methods.

Computer professionals utilize new methods and tools to improve the quality of the user interface.

The usage of these tools for user interface design enhances the produced results, and such results should still be evaluated through the SSAT approach to the analysis of human activity.

Acknowledgments

We are grateful to Executive Editor Cindy Carelli, who provided professional guidance and invaluable moral support for this book and for all our other publishing initiatives over the years.

We would also like to say thanks to our family members and friends whose love and affection were so important to us during this effort.

Gregory Z. Bedny and Inna S. Bedny

Authors

Dr. Gregory Z. Bedny has taught psychology at Essex County College and is a research associate at Evolute, Inc. Dr. G. Bedny earned his PhD in industrial organizational and a post-doctorate degree (ScD) in experimental psychology. He is also a board-certified ergonomist. He has authored and coauthored six scholarly books in English published between 1997 and 2015 by Lawrence Erlbaum Associates, Inc., Taylor & Francis Group, and CRC Press. Dr. G. Bedny has developed systemic-structural activity theory, which is the high-level general theory that offers unified and standardized methods of the study of human work. He has applied his theoretical studies in the fields of human-computer interaction (HCI), manufacturing, merchant marine, aviation and transportation, robotic systems, work motivation, training, and reducing fatigue.

Dr. Inna S. Bedny is a computer professional with a PhD in experimental psychology. She is doing research applying SSAT to HCI. She is the author or coauthor of a number of scientific publications in the field of HCI. In 2015, she coauthored the book *Applying Systemic-Structural Activity Theory to Design of Human-Computer Interaction Systems* published by CRC Press. She also has experience teaching math and physics at a high school level and training computer professionals. Dr. I. Bedny is also a research associate at Evolute, Inc.

Part I

Human Performance in Ergonomics, Psychology, and Economics

1

The Role of Ergonomics, Psychology, and Economics in Studying Efficiency of Human Performance

1.1 Brief History of Human Performance Analysis

Currently, in cognitive task analysis there is a tendency to excessively distinguish between methods of analysis of routine manufacturing tasks, or similar tasks in other types of industries, and contemporary task analysis in automated man-machine and human-computer interaction systems. However, the distinction between manual, mechanical (semiautomatic), and automated systems is not always a clear-cut one. In fact, within any given system different components can vary in the degree of their manual features versus automatic features (McCormick and Sanders, 1982). Of course, there are some differences in the tasks that are performed in manual, mechanical, and automated systems. At the same time, there are also general principles of studying human performance in any type of a system. These general principles derive from analysis of work activity. For example, in any manual system, subjects operate with some hand tools, which allow them to control their performance. In these kinds of operations, subjects use their own physical (or muscular) energy as a power source. Such types of activity also include simple cognitive components. Moreover, a subject should be motivated to perform his/her task. So, manual tasks include cognitive, behavioral, and motivational components that are integrated into a system and are directed to achieve some goal. Such a system is an example of human activity. Evolution of task analysis and therefore of human activity can be better understood based on an overview of the history of manual work studies and its relation to the cognitive aspects of task analysis. When a worker operates with a manual tool and interacts with equipment by using his/her physical actions, the results of such interaction or its output serves as feedback provided by various sense organs. Based on the feedback, a worker can correct her/his own actions. At the stage of mechanization, the power is typically facilitated by the machines, whereas a worker controls them. The term *machine* includes equipment, devices, tools, and so on that are used to carry out an intended work. In automated systems, the machine provides both the sensing and the controlling operating procedures. A worker, who is called an operator, functions as a monitor entering in the control loop to override the technical components of the system and to correct the system functions.

The last stage of work evolution involves computerization. According to the Activity Theory (AT) approach, in all man-machine systems an operator receives information, evaluates it, performs some cognitive and behavioral actions, evaluates results, corrects actions, and so on. Here, actions are understood as elements of work activity. In any type of activity, a subject strives to obtain results according to a goal of activity. The basis for the study of all kinds of labor is the analysis of human work activity. Although there are some specifics for conducting activity analysis at any stage of work evolution, there are also some general principles that are applicable in such studies. Therefore, we want to demonstrate that different methods of analysis of human activity are interrelated and influence each other. Previously developed research methods of studying human performance cannot be directly allied to new types of tasks that have different characteristics. However, they have influenced the development of modern methods of research. Moreover, some such methods can be modified and adapted to the contemporary studies of human performance. This is particularly relevant for the study of manual components of work. The idea that these components are going to be completely eliminated is not justified. Even in highly automated systems, elements of manual work will still be present. In our book, we will show that full automation often cannot guarantee complete system reliability. In an automated system, an operator functions as a monitor entering into an automatically functioning system in order to prevent it from failure. During the transition from automatic to manual control, in a case of a system malfunctioning, an operator can perform unnecessary and sometimes erroneous actions. For increasing reliability of an operator's cognitive functions, some manual elements of work should be preserved.

In the twentieth century, a body of knowledge directed toward increasing productivity and optimization of human performance has evolved (Kedrick, 1977). Originally, specialists such as industrial and process engineers, work psychologists, and physiologists were involved in this field. Later on, the range of professionals involved in the study of human work expanded significantly. Today these professionals are known as human factor specialists. They became involved in two interdependent areas of study. With the mechanization that occurred in the first half of the twentieth century one major area of study had to do with the analysis and design of efficient methods of work with existing technical equipment. Machines were designed by engineers and were an unchangeable factor which could be adapted to human needs. Workers had to adapt to constructive features of machines no matter how usable these machines were. Technical components of the workplace could be modified to a small extent. For example, a specialist could focus on the reduction of heavy physical work, or on elimination of unnecessary movements, on making changes in work organizational processes, and on introducing some additional devices in the workplace. However, the main equipment remained unchanged. The main object of study at that time was the blue-collar workers' performance. The basic concept, presently known as "a man-machine system," had not been formulated yet. The principle of integrating and incorporating human and machine functions into a unitary system had not been well recognized yet. At the next step, experts in the field of human performance began to pay attention to indirect company labor such as specialists in various offices, banks, hospitals, supermarkets, department stores, and so on. This area of research has grown significantly. To study these new areas of activity, experts developed appropriate specialized research methods.

According to Rasmussen and Goodstein (1988) and Vicente (1999), this period of study of human performance coincides with the phase of industrial development which is called industrial mechanization. This phase is usually associated with the names of Taylor (1911) and John and Lillian Gilbreth (1911, 1920). Taylor concentrated his attention mostly on studying physically demanding manual work. In contrast, the Gilbreths studied manual work which (in most cases) required small or moderate physical effort. The main focus of their study was production operations performed on production lines and conveyors. Performance time of such operations was a critical factor because all individual conveyor operations should be executed in about the same time, and their completion has to be coordinated with the rhythm of the conveyor movement. Typically, these operations are performed in a specific sequence on the production lines or conveyor. Such operations often involve high-precision motions and require coordinated movements of both hands. This type of work requires high perceptual control and concentration of attention for the movement regulation. The pace of task performance is also an important factor in performing the considered motions.

There was also another type of worker at that stage of industrial development—the blue-collar workers—craftsmen such as the bench assembly workers who operated sophisticated equipment and machine operators who produced complex parts. They performed manufacturing operations that required not only precise movements but also complex cognitive operations, skills, and knowledge as well. At this stage of mechanization, the power typically had been provided by machines and on some production lines there were processes that combined manual and machine control.

In the second half of the twentieth century there was dramatic progress in technology. This period of industrial development involved introduction of automation and centralization and the emergence of computerization as the third stage of industrial development (Vicente, 1999). Even in such highly automated systems, some degree of manual control is preserved in order to increase the reliability of the systems. At this period of time, cognitive components of work significantly increased. It is not a coincidence that

Method-Time Measurement by Maynard, Stegemerten, and Schawab (1948) was published at that time, offering a full description of the MTM-1 system and its application rules. This system was utilized in the US and later on in other industrialized countries. Currently, there are MTM-1 associations in the US, Canada, the UK, and several other countries. Some small changes have been introduced to this system since its first publication. Maynard and his colleagues utilized the Gilbreths' idea that all motor behavior can be described by using basic universal motor components (microelements). However, the MTM-1 system is much more complex and significantly more precise than the Gilbreths' idea. This system is utilized even now for analysis of complex manual production operations. There are some other microelement systems that address mental components of work. The latter are not well developed, and we will not discuss them in the course of this book.

Thus, the Gilbreths' could be considered as the founders of this approach to the study of human performance. We believe that the critical comments about the Gilbreths' contribution, which became sort of a tradition, are unfounded. Their system of behavioral units (Therbligs) was a critically important step in developing the principles of studying human work and became an important new method for work studies. Currently this system is not utilized, but it provided the impetus for the development of a number of methods for studying human performance based on these principles. The MTM-1 system that has derived from Gilbreths' ideas is sufficiently complex. According to the UK MTM-1 association, the minimum course duration for acquiring the knowledge of this system is 105 hours of classroom study (UK MTM-1 manual, 2000). The Gilbreths developed other important methods of studying movements and work processes in general. For example, they developed cyclograph and chrono-cyclograph methods of movement study. They were also the first ones to introduce the symbolic method of work process description and analysis. They developed principles of classification of basic elements of work processes such as operation, transportation, inspection, delay, and storage. They designated them by special symbols (Figure 1.1).

These symbols were used for the development of process charts. Such charts graphically describe the steps that take place during the performance of a production process. Frank and Lillian Gilbreth (1920) were the first ones who understood the importance of the principle of standardized description of work and production process elements and introduced standardized descriptions of each symbol. For example, a circle means an operation which occurs when an object is intentionally changed in one or more of its characteristics; it represents a major step in the process and usually occurs at

FIGURE 1.1
Basic Gilbreths' symbols.

a machine or work station. With the use of the process chart, such steps as transportation, machine operations, assembly, and storage can be described. The chart can give a clear picture of every step in a production process and helps to eliminate unnecessary steps. This method can be modified for various purposes. Some principles of symbolic description of work studies introduced by Frank and Lillian Gilbreth are utilized in ergonomics. For example, operational sequence diagrams or functional flow charts are based on these principles. For development of such diagrams, it is important to determine what elements of the production process or the task under study should then be used as the unit of analysis.

The automation and centralization stage of industrial development is associated with the emergence of highly automated systems which made it necessary to study human behavior in man-machine systems. If in the past the main figures studying human work were industrial and production engineers, then now in new conditions the role of designers who were involved in the design of equipment has increased. Human factor specialists became involved in the design of man-machine systems. This design was the main focus of study, which encompasses electronic products, aerospace vehicles, nuclear power stations, industrial systems, transportation systems, and so on. Design of such systems should be carried out not just considering the technological aspects but also from the point of view of the efficiency of human performance. This area of study is a foundation for the complex field which is now known as human factors/ergonomics. This area of study is concerned with the design of man-made complex equipment and software which can be used effectively and safely in a suitable environment. If equipment is poorly designed, then operators would not be able to perform required tasks efficiently. During the design, such characteristics of the environment as noise, lighting, temperature, vibration, and so on should be taken into account with the purpose of creating an optimum environment. Engineering psychologists, physiologists, and biomechanical specialists are the main contributors in the process of designing the equipment. Any equipment design solutions that are not based on the analysis of human behavior should be considered as pure engineering solutions and therefore not in line with ergonomics or human factors.

Task analysis is an important step in the design of man-machine systems. The term task is used instead of the term production operation. At present, traditional production operations are called skill-based tasks. The first person to introduce the concept of task analysis was Miller (1953). Today it is a broad area that integrates a variety of techniques for analysis of human performance during interaction with equipment or software. Task analysis methods are the main area of human factors and ergonomics (Meister, 1985). Up to the 1960s, task analysis was based on a behavioral approach. At that time, the cognitive approach appeared as the strongest force in psychology, and the nature of work also changed toward increasing the cognitive components of task performance. This evolution of science and the nature of work are the main reasons why scientists started focusing their attention on the cognitive aspects of human performance. Hence, cognitive task analysis surfaced as a leading branch of work psychology and ergonomics. There is a lot of interesting data in this field of study, but separation of cognition from behavior and concentration of effort on cognitive aspects of work is not very productive. According to AT, cognition and behavior should be studied in unity. Unity of cognition and behavior is the main principle of AT.

Some scientists who are working in the field of cognitive task analysis criticize existing methods in this field (Chipman, Schraagen and Shailin, 2000). They assert that the large number of existing task analysis techniques are described repeatedly and that little is said about their ability to yield the complete analysis of a task. Software engineers complain that "many task analysis methods were developed by researchers with psychological background, that is why these methods and their outputs often do not integrate well with those of software engineering" (Diaper, 2004, p.30). However, later in this book it will be shown that similar comments could be also addressed to the software engineers.

We will describe here a unified and standardized approach to task analysis that takes its roots in systemic-structural activity theory (SSAT).

Thus, a brief analysis of work studies demonstrates that there are two interdependent and interconnected fields. The first one involved the design of efficient methods of human performance of production operations or skill-based tasks. At the time of Taylor and the Gilbreths, there was no chance to redesign basic equipment. Only some elements of such equipment could be slightly modified. The efficiency of performance of various movements, their coordination in time, and performance with required precision was the main purpose of study.

The second field involved the efficient design of new complex equipment, its modification, and creation of adequate environmental components of work. Both approaches were utilized as key methods of analysis of production operations and of task performance. The two areas of design just described are closely interrelated, but they evolved relatively independently. Specialists in one of the areas were not familiar well enough with in other area.

We would like to emphasize that we focus on the design issues. Our interest is also in such areas of research as training and professional selection. All these areas of study are interrelated. Without solving the design issues, the training issues cannot be addressed, either. All of the above demonstrates that experience obtained in the study of production operations and designing of efficient method(s) of performance can be very useful in the study of any manual components of work, including contemporary work of operators in man-machine and computerized systems.

Manipulation with various controls like switches, levers, pedals, buttons, knobs, and so on can be described by using the contemporary method of time and motion studies. In human factors/ergonomics, this is not discussed. Usually, specialists use data related to the study of reaction or response time when analyzing discrete movements. Another important source in analysis of discrete movements is the study of positioning movements, when a subject hits two targets is Fitts' law (Fitts, 1954). However, humans should not be considered as a reactive system that performs independent movements at a maximum speed. Hand movements from one position to another during task performance cannot be executed with maximum speed as it is done in separate reactions demonstrated in Fitts's studies (1954) because movements are interrelated and influence each other. The pace of movements when performing a task does not match the speed of execution of isolated reactions or tapping on the two targets (Bedny and Karwowski, 2013). Motor activity is also important when performing tracking tasks. Moreover, existing methods of research in this area are not always adequate. The operators are not just trying to reduce tracking errors, but sometimes they even provoke them and evaluate the obtained results (Zabrodin and Chernishov, 1981). We will demonstrate that in such circumstances, modern research in the area of time and motion study can be very useful.

We would like to point out that the constant criticism of the Gilbreths' and their Therblings systems is undeserved, because these systems have historical importance in the development of time and motion study.

In the framework of SSAT, critical analysis of the MTM-1 was conducted. The strengths and weaknesses of the suggested system in contemporary work analysis are considered in this book. As will be demonstrated, the MTM-1 system is not adapted for the analysis of contemporary tasks that include cognitive components to a large degree. The task components have logical organization and the time of task performance is **a** variable parameter. Thanks to the method developed in SSAT of algorithmic description of the tasks, it is possible to use the MTM-1 system for analysis of manual components of activity in complex variable types of tasks, where cognition plays a significant role. In the MTM-1 system, units of analysis are human motions. These motions or movements have the standardized description; beginning and end points of movements are precisely defined. Information about such characteristics as distance, level of concentration of attention, and the ability to perform movements in sequence or simultaneously with other movements is also presented in the MTM-1 system. The description of motions is accompanied by a number of figures which demonstrate the most representative versions of motion performance. The standard pace of motions' performance is described. This allows professionals to uniquely interpret the description of motions. However, it should be noted that human activity consists of hierarchically organized units. This means that the analysis of manual work cannot be reduced to the only the description of motions.

In contrast to the MTM-1 system, SSAT utilizes a hierarchical system of units of analysis, the description of which is standardized to a significant degree. The main units of analysis are motor actions that in turn consist of motions. A goal of a motor action is an integrative mechanism that combines movements in a coherent system. Cognitive actions consist of mental operations. Several motor actions and similarly several cognitive actions can be integrated into the modes of performance by the high-order goal. In the first case, this is a goal of integrated motor actions, and in the second one it is a goal of integrated cognitive actions. The task can be viewed as a logically organized system of cognitive and motor actions that are directed toward achieving the goal of a specific task. As it can be seen from the brief analysis of the material presented, SSAT has developed methods of study that have adapted the MTM-1 system for contemporary task analysis of manual components of work. In SSAT, attention is paid not only to motor components of activity but also to its cognitive components. Cognitive and behavioral components of activity should be studied in close unity. In this book, we demonstrate interdependent motor and cognitive components of human activity. This interdependency has great importance for the study of computer-based and computerized tasks.

Another aspect of the historical analysis of the study of human performance is analysis of the relationship between psychology and economics. We will demonstrate the importance of psychology for labor economics. The interaction of economics and psychology has a long tradition. For example, psychology plays a significant role in the development of job evaluation methods, the main purpose of which is determining the value/worth of a job in relation to other jobs in an organization. Job evaluation strives to make a systemic comparison between jobs to assess their relative worth for the purpose of establishing a rational pay structure. We will show the role of SSAT in economic studies. Assessment of innovations and enhancements from an economic perspective is also an important aspect of studying human performance.

The purpose of this book is also to demonstrate that the study of human performance cannot be reduced to the application of multiple independent techniques which do not have a general theoretical background. This book presents a unified framework for the study of any type of human performance. Developed in SSAT, contemporary methods of study derive from the advanced studies of manual work and from the data obtained in cognitive psychology and in AT. SSAT demonstrates the successful integration of different directions in psychology to further their application (Chebykin, Bedny and Karwowski, 2008).

1.2 Study of Human Performance in Ergonomics, Psychology, and Economics

Contemporary ergonomics, economics, and work psychology are tightly interconnected because the main purpose of these fields is the study of human performance and productivity. There are two basic areas of study of human performance. The first area is concerned with the study of physically demanding work. The second area is associated with studying mentally demanding task performance where physical components of work are significantly reduced. The second area plays a dominant role in contemporary studies. Thus, ergonomics focuses, first of all, on studying cognitive and physiological responses to physical demands of human performance. Cognitive demands are currently the most important factor in ergonomic analysis. This makes work psychologists central figures among professionals that are involved in ergonomics. Psychologists focus their attention on human cognition and external behavior when they interact with equipment and computer interfaces in various environmental conditions. There are also engineering and human-computer interaction (HCI) approaches in ergonomics. Therefore, engineers with various specializations and computer specialists are involved in the field of ergonomics. HCI is cognitively demanding, and with the ever-increasing use of various computer gadgets, the role of psychology in ergonomics becomes even more important.

One critically important purpose of ergonomics is increasing human productivity or labor productivity, which is concerned with the quality and quantity of products or services per units of time. Increasing productivity makes it possible to sell products and provide services at lower prices, pay good wages to employees, and so on. Time study is an important method of measuring productivity (Bedny, 1979). Developing methods of task analysis in which not only motions but also cognitive processes are taken into consideration is a critical factor in increasing productivity. Hence, psychological aspects of studying productivity are important in the field of ergonomics. Psychological aspects of studying productivity are also important in labor economics. Labor economics is a branch of economics concerned with productivity, wage determination, labor markets, and similar questions concerning labor as a factor of production (Ammer and Ammer, 1984).

Justification of psychological and ergonomic innovations should normally be based on economic criteria. The evaluation of the relationship between the cost of innovations and the profit obtained by the company as a result of these innovations is the basis for economic analysis. The salaries and wages paid to employees are important data for the economic justification of psychological and ergonomic innovations. Concepts such as "unskilled labor," "skilled labor," simple or complex work, "tariff scale," and so on can be important when studying the efficiency of innovations from economic perspectives. Unskilled labor is associated with simple work, and skilled labor is considered complex work. According to the labor theory, the price of a commodity depends on the amount of labor required to produce it. Simple and complex work are also compensated differently. The values for the different types of labor are determined by the complexity of the work and the skills needed to perform it. All of this demonstrates the interconnection between ergonomics, psychology, and economics.

Let us now consider the concept of productivity in labor economics in more detail. Productivity in economics is the measure of efficiency that summarizes the value of outputs relative to the value of the inputs that are used to create them. There are various forms of productivity. For example, total factor of productivity is an overall indicator of how efficiently an organization utilizes all existing resources, including capital, materials, energy, and labor, for the creation of products and services. The total productivity is not sufficiently informative because all the listed criteria are considered together. For work psychology and ergonomics, labor productivity is the most important factor. It can be calculated as follows:

$$\text{Labor Productivity} = \frac{\text{Outputs}}{\text{Direct labor}}$$

Direct labor is the employees who are directly involved in production of the company's end-product. The company with a higher productivity than its competitors would have more product or services to offer at lower prices and therefore generate greater profit. Higher productivity can result in better wages. Productivity depends on two main factors: technological innovations and introduction of efficient methods of task performance. Increasing productivity is an important factor that helps businesses be more profitable. It is seen as key factor that indicates efficiency of innovations. The successful introduction of various innovations such as new products and services, or changes in a production process, facilitates growth of output that exceeds the growth of input.

In this chapter, we consider the interaction of economics and psychology in the field known as *job evaluation* that is shared by psychologists and labor economists. For a while, this was the most representative area of study in economics that was tightly connected with psychology. Economists recognize that inaccurate job evaluations may result in a highly inadequate distribution of income between employees (Arnaut, et al., 2001). Job evaluation is a procedure that is used for determining the relative value of jobs in the organization. This information, in turn, is useful for determining the level of compensation (Muchinsky, 1990). Therefore, job evaluation requires job classification according to the developed criteria. There are various criteria for job evaluation, such as skills factor, responsibilities, complexity of job, mental and physical demands. After calculation of the total number of points assigned to each job according to the listed criteria, specialists can determine the job's relative position and therefore the worth of the job for the company. At the next step, the market price of the product can be determined. The purpose of this stage of analysis is determining the external value of jobs. This allows the establishment of a wage and salary structure and based on it, the pay rates that not only correspond to the internal company criteria but also are adequate in relation to the job market. The main purpose of job evaluation is evaluation of the entire job that consists of different tasks which are performed by various employees who belong to the same profession. Job evaluation was established in the 1930s and 1940s as a part of the general trend toward scientific management. Job evaluation is a systemic way of determining the value/worth of a job in relation to other jobs in the organization. Based on job evaluation, specialists make a systemic comparison between jobs to assess their relative worth for the purpose of establishing a rational payment system. Thus, job evaluation begins with job analysis and ends at the stage when the worth of a job is defined for achieving pay equity for a variety of jobs.

Various techniques of job evaluation have been developed. They can be divided into two groups: non-analytical and analytical. Non-analytical job evaluation is based on the comparison of one job considered as a holistic system with another existing job. Such a method is based on the integrative evaluation of job without consideration of its distinct factors and quantitative evaluation of their complexity. We can mention here such methods as job ranking, paired comparison, job classification, and so on. These methods have limited precision because many aspects of the job are not being considered. Analytical methods are more precise because here each job is broken down into elements based on a number of factors. Such factors as education, skills, responsibilities, decision-making, and dexterity are being taken into consideration. The basis for the analytical job evaluation procedures lies in the choice of factors and in their weight. When workers are paid according to the tariff rates, an inequality factor can be often detected (Neilson and Stowe, 2010). The principle of comparable worth is important in job evaluation. This principle refers to giving workers equal pay for performing comparable work (Mahoney, 1983). Job evaluation can be performed based on one or more criteria. The method that uses one of the total criteria is less accurate than the method that is based on a number of different criteria. The more complex or difficult a job is, the higher should be its ranking or grade and job pay ranking.

Classification based on one criterion is called ranking method. An analytical method of job evaluation in most cases is based on a number of different criteria. At the final stage of analysis, each factor, depending on its significance, gets a certain number of points, and the specific weight for each factor is determined. At the next step, scores are translated into monetary assessment. One such method is the determination of the reference rate for zero-complexity jobs. At the next stage, depending on the complexity of the other jobs, the cost of the job can increase. Based on it, tariff scales for existing jobs can be developed.

It should be noted that there are no rigorous scientific criteria for selecting a number of factors based on which the job evaluation can be performed. Usually, choosing the factors for job evaluation include training requirements as well as requirements for the mental and physical effort, the impact of working conditions, the responsibility for the job of others, and so on. Hennecke (1960) compared the factors used in the various systems of analytical job evaluation in Germany. He showed that different systems utilize different factors, and these factors do not have a sufficiently clear and unambiguous description. The multiplicity of factors used in the various systems of job evaluation, different content put in the same factors, and factors that have similar names make it difficult to evaluate jobs objectively.

There are some limitations in job evaluation methods (Kedrick, 1977). For a long time in economics the differentiation of compensation for various types of work was based on the analysis of the position of the considered job in the scale between simple and complex labor. The complexity of the job was determined according to the employee's qualifications. Unskilled labor was associated with simple labor and skilled labor was considered as complex labor. According to the labor theory, the value of a commodity depends on the amount of labor required to produce it. Simple and complex jobs are compensated differently. Values for different types of labor are determined by the complexity of the job that depends on workers' skills. Economists developed the tariff scales based on the difference in their complexity. The tariff rate is a measure of the price of labor of a certain complexity. Managers set tariff rates for employees of different skills from the tariff scale. The basis of classification of employees in a particular skill group is the level of their professional education measured mainly by the time that is required for professional training (Barness, 1980).

However, such an assessment of complexity of work is no longer considered as an appropriate and sufficiently precise one. At the present stage of industry development, the diversity of jobs based on their content significantly increased. Jobs differ not only on the basis of the physical, but also emotional and mental effort, the degree of danger and responsibility, and so on. Determining the comparable worth of human work is a critically important factor in economics (Mohoney, 1983). In the 1960s the Department of Labor had already drawn attention to the fact that skilled workers are paid in accordance with what they are potentially capable of doing, and semi-skilled and unskilled workers are paid based on what they actually do (International Labor Office. "Job Evaluation," Geneva, 1960, p.7). As a result, the reason for different wages was reduced exclusively to the need for reimbursement of the costs incurred for the preparation of the labor workforce in the past. All this testifies to the fact that this approach to job evaluation is not very effective. Without taking into account the qualitative differences between simple and complex labor, it is impossible to understand why two types of labor create different values of the commodity working the same amount of time. Economists have come to the conclusion that the time of the preparation of the labor force may not be enough for identifying the differences between complex and simple work. Just the costs of education and training for more complex work cannot explain the fact that simple and complex labor create different values in the same intervals of time.

The economic studies started to introduce some additional factors that are causally related to the complexity of labor costs. These factors include concentration of attention, eye tension, muscular and mental effort, environmental factors, and so on. The cost of education and training became only one out of a number of other factors. This method depends on the correctness of the selection factors and their evaluation. The cost of education and training is the dominating one in the field of economics and is used for determining the relative value of jobs in the organization. This analysis, in turn, helps to determine the level of compensation depending on the work complexity as it is considered in economics. There is no one fixed set of compensable criteria. Different companies select different compensable criteria and therefore different methods of analytical evaluation of the complexity of work. We have to point out that there are some inaccuracies in understanding such concepts as complexity and difficulty of work.

Presently most economists are in agreement that concepts of *complex and simple work* have an important meaning for job evaluation (Moshensky, 1971). Therefore, a theoretical basis of any method for job evaluation that is based on the concept of *simple and complex work* should be evaluated adequately (Armstrong, Cummins and Hastings, 2003). In fact, any method of job ranking is performed based on such factors as complexity of work, which is considered as one of the most important criteria for evaluation of job worth. However, methods of work complexity evaluation are often not precise and are even incorrect. It should be pointed out that for a long time, one of the most important criteria in evaluating work complexity was the level of the workers' qualification for a particular job. As a result, other factors were not being sufficiently taken into consideration. Overemphasis on the qualification factor can lead to the situation where workers with a high level of qualification are paid in accordance with what they potentially can do. At the same time, workers with a lower level of qualification are paid based on what they really do (Veikher, 1978).

Complex work is a function of a qualified work force which requires more mental effort than simple work. According to many specialists in psychology and economics, complex work requires more concentration of attention, more precise motions, and more mental effort in general. Complex work results in changes of the value of time units of work. Such units become more significant components of work in comparison with time units in simple work. All of the above-discussed is evidence that in any method of job evaluation the factor of work complexity must be taken into account. Job evaluation which is based on the more precise work complexity assessment permits elimination of the job evaluation where the dominant principle is the expenditures associated with training and education. Without understanding the differences between complex and simple work, it is difficult to explain why expenditures of mental effort for these two kinds of work during the same time period are different. Development of principles of work complexity evaluation is a theoretical basis of job evaluation. Economics as a science cannot evaluate complexity of work with sufficient precision without utilizing psychological methods of work evaluation.

Analysis of efficiency of various innovations is important for ergonomics, psychology, and economics. Evaluation of the relationship between the cost of innovations and profit obtained by company as a result of their implementation is the basis of economic analysis. The salaries and wages paid to workers are important data for economic justification of psychological and ergonomic innovations.

However, these issues face certain difficulties due to the fact that in the economics there is no clear-cut demarcation between complex and simple labor. This, in turn, gives rise to some difficulties in implementing analytical methods of job evaluation in economic studies.

Simon (1999) postulates that complexity is the basic property of a system. Human activity can also be considered as a system. However, the complexity of human cognition or behavior has not been significantly studied in cognitive psychology. Here, we briefly consider psychological aspects of complexity and its relation to the concept of complex work. Complex work requires greater concentration of attention, more precise motions, and mental effort than the simple one (Veikher, 1978). Such understanding gives economists and ergonomists the psychological perspective of work complexity. There are multiple factors that determine the complexity of a task (Payne, 1976; Gallwey and Drury, 1986; Bedny and Meister, 1997; Bedny, Seglin, and Meister, 2000; Bedny, 2015; Bedny, Karwowski, I. Bedny, 2015).

Here we will shortly discuss task complexity. The more complex the task is, the more mental or cognitive effort is required for its performance. Complexity of task depends on the quantity of task elements, specificity of their interaction, and on the number of static and dynamic elements. A degree of uncertainty or unpredictability of the task is also an important component of task complexity. An increase in the information processing speed under the time-restrained conditions is one possible factor of the task's complexity. According to cognitive psychology, the main causes of the complexity of a problem-solving task are the complexity of the rules for its solving, ease of developing such rules and their application, and memory workload during utilization of these rules. Release of working memory from storing information about the current state of the task solution is also an important factor in reducing the difficulty of a problem-solving task (Kotovskya and Simon, 1990). These are the most general characteristics of task complexity.

SSAT presents various factors that can influence task complexity. For example, a number of contradicting solutions influences the complexity of the task. The decision-making process is more complicated when it involves extracting information from the memory. On the other hand, decision-making processes are easier to perform when they are predominantly determined by external stimuli or information provided by external sources. It was discovered that the more complex the task is, the more attention and concentration are required. This criterion can be used for evaluation of the complexity of motor components of work. The concept of complexity is also associated with the emotional-motivational component of activity. Emotional tension and motivational forces increase as the task complexity increases. Specificity of the combination of elements of activity is another factor of complexity.

Complexity is an objective characteristic of a task. Task difficulty is another characteristic of a task that depends on its complexity. While complexity is an objective characteristic of the task, its difficulty is the subjective evaluation of the task complexity. Therefore, complexity and difficulty cannot be considered as synonymous. The same task can be subjectively perceived as being of different level of difficulty by subjects with different past experience and different individual features. An increase in the complexity of a task increases the probability of requiring more cognitive effort for its performance. Complexity itself does not have a subjective component. A subject cannot directly experience complexity of the task by itself but rather perceives its subjective difficulty.

Analysis of psychological, ergonomics, and economics literature demonstrates that there are four basic characteristics of tasks or, more precisely, human activity during task performance: *heaviness of work (physical demands), intensity of work, complexity, and difficulty of work*. It should be noted that these terms are not precisely defined outside of SSAT and are often used synonymously. For example, heavy work is considered as synonymous with difficult work (S. Rodgers, ed., 1986, V. 2). We consider these characteristics to be totally different.

Some economists are concerned with the effect of the energetic components of work when considering such term as "heavy work." Physical demands of work can be *extremely heavy, moderate, or light*. If physical effort is minimal, we can neglect this characteristic altogether. Usually, though, the degree of physical effort is an important characteristic for manual tasks. Physical effort has subjective components, such as the feeling of physical stress that is affected by the individual's physical condition. Physical and mental efforts can influence each other. For example, coordination of actions (mental effort) is much more difficult for a subject when physical effort increases. Physical effort or "heavy work" can be measured by utilizing such indexes as energy expenditure, heart rate, and so on.

Another important characteristic of task in ergonomics and psychology is the *intensity of work*. It is also important when studying human performance in economics. (Boisard, et al., 2003; Gal'sev, 1973; Veikher, 1978). The work intensity depends on the speed or pace of performance. Studies demonstrate that workers reported an increase in work intensity when they were exposed to a higher pace of work or had insufficient time to complete the job. In recent years, work in general has become more and more intense. Economists sometimes utilize the term "work tension" instead of the term "work intensity." Tension of work includes such characteristics as pace of performance and amount of mental and physical effort per unit of time (Kholodnaya, 1978). Russian psychologist Nayanko (1976) distinguished between what he called operational and emotional tension. Operational tension is determined by a combination of task difficulty and lack of available task performance time. Emotional tension is determined by the personal significance of activity to a subject. These two types of tension are closely related and under certain conditions one type of tension causes the other. Analysis of such concepts as intensity and tension demonstrates that these two concepts are very similar and are contaminated by such concepts as heavy work, complexity, and difficulty. According to our analysis, *heaviness of work* (degree of physical effort), *task complexity*, and *task difficulty* are the most productive characteristics of task. Intensity of work can be used as an additional characteristic of work that is associated with the lack of available time for a task performance. If mental aspects of work are a dominating factor, then the intensity of such work can emerge as a component of task complexity. However, if the work requires physical effort, its intensity should be characterized as an aspect of the work heaviness (heavy work). The *time study* is used not only for salaries and wages analysis but also for organization of work in general.

Knowing the salaries and wages paid to employees with different skill sets, and the standardized method and performance time of various specific tasks, we can provide economic justification of psychological and ergonomic innovations in human performance. With the increase in mechanization and automation, the role of indirect labor has increased. Similar economic criteria for evaluating efficiency of innovations should be applied to indirect labor. Time study in this area of work is also important. Therefore, the economic analysis of benefits of psychological and ergonomic innovations is an important aspect of studying human work in general. However, there are certain difficulties in this area. First of all, there is no clear concept of what complex and simple labor is in economics. This, in turn, gives rise to some difficulties in implementing analytical methods of job evaluation in economic studies.

Economics play an important role in application of work psychology and ergonomics. Management is often not interested in carrying out ergonomic or/and psychological studies that are accompanied by some expense without evidence that they would result in increased productivity or other justification via tangible benefits of such studies. Thus, the purpose of this section is to demonstrate the importance of interaction of economics and psychology in ergonomic studies and in studies of work activity in general. The salaries and wages are important for economic justification of psychological and ergonomic innovations. The material presented demonstrates that such concepts as "unskilled labor," "skilled labor," simple or complex work, tariff scale, and so on. are important in studying the efficiency of innovations from an economic standpoint.

Another area of economics that was developed under the influence of psychology is behavioral economics. The main contribution to the development of this direction of science was made by such scholars as Kahneman and Tversky (1973, 1984) and Kahneman (1991). They stated that human beings rely on a set of heuristics for their decision-making and that these heuristics often lead to systemic deviation from the normatively correct decisions. According to this approach in economics, a subject is not a rational decision maker. The data obtained in this area of study made an important contribution in the development of behavioral economics. The strategies of choosing an alternative that are based on a formalized method of study have not been sufficiently developed for analysis of human decision-making. The actual behavior of a subject in a decision-making situation and rational choice are not the same. This theory focused on financial decision and how people behave in an environment driven by the free market.

Rational decision-making theory identifies information that is relevant to a decision and considers how this information can be used for decision-making. The purpose of such analysis is optimization of a decision. People have to choose the alternative with the greatest value. This strategy has the purpose of maximizing the gain and minimizing the losses. However, this strategy of choosing an alternative that is based on formalized methods of study has not been sufficiently developed. This is due to the fact that different subjects evaluate gains and losses differently depending on the factors of positive and negative significance of these events for a subject in a particular situation. The concept of significance plays an important role in analysis of human performance in SSAT.

Outside of SSAT, the key point in studying rational decision-making is to determine how subjects evaluate values and cost when making a decision. The psychological worth of subjects' alternatives depends on their subjective *utility*. The psychological worth of money is an interesting area of studying utility. It was discovered that for different subjects the value of money is not the same. Utility can be different for people with different wealth and personality features. The basic conclusion is that the psychological value of alternatives or different items for various subjects does not increase in direct proportion to their objective value.

Decisions can be of different complexity. For example, they can have a number of dimensions or have to be performed in uncertain situations. Examples of such decisions are risky decisions. When gambling, a subject can win or lose money. For risky decision-making, when outcomes are uncertain, both chance and utility should be considered. Such concepts as probability and expected value should also be considered. Analysis of material presented in this chapter demonstrates that there are various aspects of the relationship of economics, psychology, and ergonomics. In our later discussion, we will consider this interaction from the perspective of the SSAT framework.

Developed in SSAT psychological aspects of time study, morphological analysis of activity and principles of task complexity evaluation are important for integration of ergonomics, psychology, and economics in the study of human performance (Bedny, 2015). Such data can be useful in evaluation of the efficiency of various innovations that can be introduced in a production process.

1.3 Examples of Evaluation of Economic Efficiency of Innovations

Ergonomists have to provide an economic justification of ergonomic interventions. Hence, the purpose of this chapter is to demonstrate some examples of economic efficiency of ergonomic innovations. Prediction of economic the efficiency of innovation can be performed based on increased productivity and decreases in compensation cost for injuries, turnover rates, absenteeism, excessive downtime during production process, and so on. Prediction of the potential economic impact of innovation based on the factors listed above and sometimes vague criterion is not easy to achieve. Thus, in this chapter we present some examples that demonstrate our method of estimating the economic efficiency of some ergonomic innovations.

When we analyze the economic effectiveness of various innovations in production environment, based on an analysis of safety, fatigue, absenteeism, and so on, then we are often talking about the efficiency of corrective ergonomic design. This approach to design in applied activity theory (AAT) usually involves such methods as observation, survey, discussion, measurement of certain characteristics of equipment, and so on, because the analytical methods are not sufficiently developed in AAT and in ergonomics in general. We will show later that analytical methods of study that are efficiently utilized in SSAT are often more effective. One of the important requirements of ergonomic research in conditions of a current production process is compliance with safety regulations. Another requirement is that such research and innovation should not be expensive. Failure to comply with these requirements results in not implementing the ergonomic recommendations in production. Innovations normally consist of two stages. The first one involves uncovering deficiencies in the existing production process, and the second one includes development and application of innovations. As was already mentioned the first stage can be more complex than the second one. Let us discuss a well-known example that demonstrates the importance of timely detection of deficiencies in human performance and prediction of the efficiency of innovations for the development of new software.

The IRS conceded that it had spent $4 billion developing the modern computer system that a top official said, "does not work in the real world," and proposed contracting out the processing of paper tax returns filed by individuals. Customer service representatives had to use as many as nine different computer screens, each of which were connected to several different databases, to resolve each problem. The cost of shutting down the project was astronomical. The IRS had 12 other systems under review to determine if they also should be killed. Although this took place 20 years ago, abandoning projects and losing millions of taxpayer dollars is still happening in government agencies and private businesses.

The tasks that will be performed utilizing the designed computer systems should be analyzed at the design stage. Their poor usability often becomes obvious only when millions of dollars are spent developing them. Ergonomic design cannot be efficient without analytical methods of study. The basic principle of design is creation of analytical models of designed objects, creation of their physical models or prototypes, and their subsequent experimental evaluation. This principle should be used not only for design of radically new systems but also for redesign of existing systems. These principles of design are used in engineering.

In this chapter, we describe as examples some innovations which were developed by one of the authors of this book at the beginning of his professional career. At that time, he worked in the field presently known as AAT. His first position was called a psychologist-physiologist (now called ergonomist) at a factory that produced equipment for the food industry. The study considered was carried out in the factory by utilizing relatively simple and inexpensive research methods. Based on observations, interviews of workers, timing, analysis of work postures, and anthropometric analysis, a number of shortcomings in the design of the equipment and methods of work performance of the nibbling shear operation were revealed. Analysis of medical records and interviews showed that one worker had to quit his job because he had symptoms of vibration-induced illness (VWF); other workers complained that they are too tired, and this was proved by measuring their fatigue. Fatigue was considered to be the critical factor in existing work conditions because it could cause disability. Hence, at the first stage of our study, we measured workers' fatigue (Bedny, 1967). The first stage of the study consisted of observation and interviews of workers. The methods of evaluation of fatigue had been selected based on the data obtained. The term fatigue is tightly connected with the term work capacity. Work capacity and fatigue refer to a person's dynamic state that may change during the shift and that represent psychological and physiological functions which determine the ability of a subject to perform his/her mental or physical work (Bedny and Meister, 1997).

Changes in the worker's functional state can be detected based on psychological and physiological indexes, such as heart rate, blood pressure, skin electrical resistance (GSR), fingers' tremor, brain activity, reaction time, and so on. Selected psychological and physiological indices are measured during the shift, in the middle of the shift, and at the end of the shift. Discovered changes are used as information about the level of fatigue. Measurement of mental and physical fatigue have their specifics. The greater the fatigue is, the less is the work capacity. The result of the measurements presented the information about workers' functional states. Selected methods of measuring fatigue depend on the specifics of work under consideration. Incorrectly selected methods are not informative. The quality of performance and productivity might change as a result of increasing fatigue. Hence, productivity changes during the shift can also be useful for measuring the index of fatigue. Mental and physical fatigue are interdependent and can influence each other. Let us present the general description of the job under consideration before considering the selected measurement methods.

Five workers performed the nibbling shear production operation. During their work using the nibbling shear equipment, the workers fed and supported sheet metal with their hands. While performing this task, they were affected by vibration. The vibration amplitude depends on the characteristics of the sheet metal that a worker holds and pushes through. The vibration of the sheet metal was measured by a special device using several versions of metal sheets. Analysis of vibration during cutting these sheets showed that the amplitude of vibration was 0.7–0.8 mm when the frequency range was 50 Hz for sheet metal width ranging from 1 to 3 mm. When the width of the sheet metal was 4 mm, the amplitude of vibration was approximately 1.5 mm with the frequency range of 25 Hz. Physiological studies have shown that the low vibration (25–50 Hz) causes involuntary synchronic activation in motor neurons which can eventually cause the vibration illness. Long-term exposure to vibrations in the range of 25–50 Hz can exert the most severe consequences and vibration syndrome. The floor on which the workers stand when performing various tasks also vibrates. When cutting out sheet metal parts, workers carefully monitor the lines drawn on the metal for the cutting to be carried out exactly along the lines. This requires considerable visual control, concentration, and nervous tension. Visual control is complicated by the vibration of the sheet metal surface. In order to motivate the workers, in addition to high wages, the management used the low requirements for the time of tasks performance. Workers could complete their daily job in 60% of the shift time. Hence, there was excessive downtime that was the most important factor affecting productivity of these workers. In our discussion with the management, it had been discovered that the excessive downtime had been intentionally introduced by the management to motivate the workers to stay. Another motivational factor was good compensation despite excessive downtime. The reason for using such motivational factors was the realization that this job was accompanied by the impact of harmful vibration. It was an attempt to compensate for hazardous working conditions, to reduce fatigue, and to give an opportunity to earn more money.

The purpose of our follow-up study was to demonstrate that low time standards and excessive downtime cannot reduce fatigue and provide protection from vibration illness. We will demonstrate that redesign of the equipment was the main factor in preventing vibration illness.

Thus, at the first stage of the research study, we measured fatigue in the existing work conditions. At the second stage we measured fatigue after the redesign of the equipment and a reduction of a certain amount of time for rest period. Hence, the purpose of our study was to demonstrate that correct redesign of equipment in combination with an optimal rest period was the main factor in preventing vibration illness and reducing fatigue.

Physiological and psychological functions may change at different rates. The most significant changes occur in the systems of the human body that are directly involved in work performance. Changes in functions that are not directly involved in work performance cannot be observed. This emphasizes the importance of correct selection of test procedures for measurement of fatigue. For example, the requirement for military pilots to fly a helicopter at a precise altitude within a ± 5 m margin for 50 minutes could be achieved. However, the study shows that with increasing flight duration, such a task is accompanied by a heart rate increase, as well as an increase in the number of control manipulations (Zarakovsky and Pavlov, 1987).

In our study, we utilized the following measures: changes in the muscular strength of the right and left hands; changes in the static muscle strength of the dominant hand; dominant hand tremor during its movement; and mental work capacity. We used special test procedures to take these measurements. The changes in muscle strength were measured using two dynamometers. The worker simultaneously clamped two wrist dynamometers. The test was repeated three times. The highest result was taken into account. The static muscle strength was measured by using the following method. At the preliminary stage of our study, the maximum strength of the dominant hand was measured. At the main stage of our study, workers had to squeeze the designed dynamometer with 75% of the maximum strength of their dominant hand. The endurance time of maintaining the required effort was measured throughout the shift.

Static muscle strength was determined with the help of a specially designed dynamometer.

An additional scale was developed for this dynamometer by Bedny, and he received a patent for original design of this device (Bedny, 1971). The need to use the dynamometer of our own design was due to the fact that the existing dynamometer was not adapted to production conditions. The device with the old design was produced by the medical industry. It contained glass elements and was not stable when it was installed on a horizontal surface. As a result, it could easily be broken in production conditions and required a lot of care for its use.

When studying the hand tremor, the worker was asked to move the metal stick in a zigzag slot without touching the edges. If the stick touched the edge of the slot, the light bulb went on and the number of touches was recorded by a special counter. Movement of the hand was carried out for 30 seconds. The workers did not know what touches were taken into account in the test.

The mental fatigue was measured using two special tests. First, the workers were given tables with two-digit numbers which were to be alternately added or subtracted in time-restricted conditions. In the second test, the workers had to write down from memory ten two-digit numbers that the experimenter had previously read aloud to them. The workers had been trained to perform all tests. The test results were reported to the workers only during training. It should be noted that all tests and measurements were carried out at the workplace.

We will not discuss here all the test results but just point out that the analysis of the obtained results showed a deterioration of all test scores during the shift. It was also found out that a significant deterioration in performance was observed in the middle of the working shift. Further deterioration of the results was less significant. This means that *excessive downtime did not protect workers from fatigue.* This psycho-physiological study discovered that although workers involved in the production process only 60% of the work shift, they still had experienced significant fatigue. Based on physiological measurements, we proposed a more efficient work schedule. For further reduction of fatigue, we proposed ten-minute breaks after every 50 minutes of continuous work on the nibbling shear equipment.

Also, based on the analysis conducted, a number of recommendations relating to changes in the working methods and the *redesign* of the existing equipment were made. As we have mentioned earlier, this work required intense visual control during the metal cutting operation when the workers had to guide the movement of the tool along the preliminarily drawn lines. Surface vibration and glare made this visual control very complicated and uncomfortable.

Observations showed that workers often had to exert great effort to move sheet metal, which increased the impact of vibration. The study, described previously, brought us to the conclusion that in order to reduce fatigue and influence of vibration, it would be necessary to change the equipment design. Figure 1.2 depicts the old version of considered equipment.

The new design that decreased vibration is shown in Figure 1.2.

In this section, we describe workplace and equipment analysis. The considered equipment was used to cut the metal sheets to various complex configurations. There were two cutting tools in a vertical position. A lower cutting tool was mounted gradually. A top cutting tool performed reciprocating

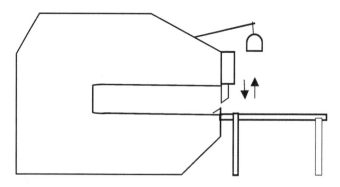

FIGURE 1.2
Schema of old version of the nibbling shear equipment.

motion at a high rate. Moving metal sheets between various cutters, one could cut holes in the sheet metal. Punching could not be used here because the metal forms are produced in small batches. Therefore, equipment at this workplace required ergonomic improvement.

The table was connected to the nibbling shear as depicted in Figure 1.2. As a result, vibration was transmitted from the nibbling shear to the table which supported the sheet, because the table and the nibbling shear were tightly connected. This, in turn, increased the impact of vibration on the workers. The anthropometric data should be considered in the context of the performed tasks. Various positions of a worker's body while working with sheet metal of different weights and sizes have been analyzed. Considering anthropometric characteristics of the workplace we took into account subjective opinions of workers, their critical comments about discomfort, and the spatial characteristics of the workplace.

Based on the conducted analysis, an improved version of the nibbling shear was suggested (see Figure 1.3)

The new design allowed separation of the work table from the tool. This prevented transmission of vibration from the nibbling shear's bed to the table. The table could be adjusted in height and position depending on the task at hand. Preventing the flank of the cutting tool from hitting the surface of the sheet metal also allowed a decrease in the vibration. We also made the suggestion to install small metal balls along the surface of the table with springs under them. As a result, sliding friction was replaced by rolling friction. This innovation significantly reduced static and dynamic physical effort of the workers. We also recommended improving local illumination by replacing the direct type of light with a semi-indirect light which reduced the reflected glare from the sheet metal. Further, we recommended using a special clamp that had a roller with a rubber rim which diminished vibration of the sheet metal by pressing the metal sheet to the table. This clamp can be installed into a working position when needed by pressing the metal sheet to the table.

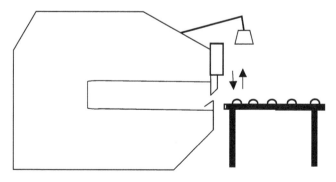

FIGURE 1.3
Schema of improved version of the nibbling shear equipment.

After redesign of the equipment, we repeated measurements of fatigue and discovered that it significantly reduced. It is also important to pay attention to the fact that excessive downtime that was intentionally introduced by the management for workers cannot reduce the impact of harmful vibration or reduce fatigue. In conclusion, we have to point out that prior to the implementation of the recommended improvements, musculoskeletal sickness absence was steadily at 7%–9% of the total production time. After implementation of the innovations, it had dropped to 3% in the following two years. There were also no symptoms of possible VWF among the workers. Productivity has increased approximately 15%. These findings have implications for the assessment of economic efficiency of considered innovations. Thus, this study demonstrates interdependence of work psychology, ergonomics, and economics.

The next example demonstrates the evaluation of economic efficiency of the redesign of an assembly line production task or production operation (Bedny, Karwowski and Known, 2001). In our previous studies, we did not consider this aspect of analysis of this production operation. This operation involves welding of two brackets to the neck of a milk jug or flask (Figure 1.4). The purpose of this study was totally different from the first one. Here, we demonstrate that the two considered production operations involved practically the same motor actions and motions, but their cognitive requirements were different. This critically important factor is ignored in contemporary time studies. SSAT offers a totally new psychologically grounded method of time study (Bedny, 2015). We now consider only one production operation from different perspectives. The purpose of our analysis is to demonstrate how we can evaluate the economic efficiency of innovation or redesign of the considered production operation.

Consideration of this manufacturing operation and its enhancements from the economic efficiency perspective has both theoretical and practical importance. There are the following stages of analysis of this production operation: qualitative analysis, morphological analysis that includes algorithmic analysis and temporal or time structure analysis, and quantitative complexity evaluation. For our discussion, it is sufficient to conduct an abbreviated

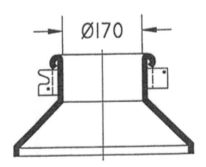

FIGURE 1.4
Neck with welding brackets.

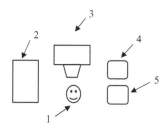

FIGURE 1.5
Plan of the workplace for operation of welding brackets. 1 – welder; 2 – bin with neck; 3 – welding apparatus; 4 – bin with rear brackets; 5 – bin with front brackets.

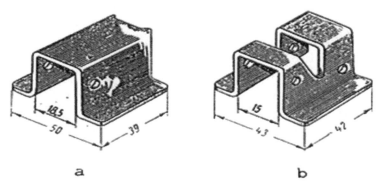

FIGURE 1.6
The rear (a) and the front (b) brackets.

qualitative analysis of this operation. Figure 1.5 depicts the plan of the workplace for the considered operation.

The rear and front brackets are presented on Figure 1.6.

Brackets are notched, which means that they should have defined orientation before welding. Brackets also have to be aligned with the axis on the neck. The jig that is presented in Figure 1.7 is used to align the brackets along the neck axis, guaranteeing that the brackets will be positioned strictly opposite one another. According to the technological requirements, the arms of the jig are different only in their width. So the narrower arm corresponds to the front bracket and the wider arm corresponds to the rear bracket (see Figure 1.7). Figure 1.7 shows that the jig's arm for the rear bracket has a width of 18.5 mm and for the front bracket it is 15 mm.

The operation of welding brackets to the neck of a milk jug consists of the following steps:

1. Take a neck from bin 1 with the left hand;
2. Place a jig on the neck to align the brackets;
3. Put the neck (with a jig) into the working position on a welding machine;

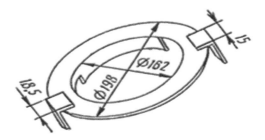

FIGURE 1.7
Jig for aligning the welding brackets.

4. Hold the neck (with a jig) by the left hand; take one of the two brackets (either the front or the rear) from bin 4 or 5 with the right hand;

5. Turn the bracket to the required position, weld the bracket to the neck with two sets of three spots on each flange of the bracket by operating a foot-switch;

6. Rotate the neck (with a jig still on it) 180° on the welding machine;

7. If the front bracket was welded in step 5, then take the rear bracket, put it into a working position, and weld it. If the rear bracket was welded in step 5, then take the front bracket, put it into a working position, and weld it;

8. Take the jig out and put it on the lap, and then take the neck with brackets and place it on the conveyer slide which drops it into a bin for the next worker (who undertakes the next operation "welding the handle to the neck").

Since the arms of a jig are of two different sizes, each call for the corresponding bracket. Thus, in the passage from one complete welding cycle to another, a jig's arm (at the position to be worked on) alternates from one size to the other. Due to this alternation of the jig's arm, the operator on one occasion has to begin the welding cycle process by taking a bracket from the front bracket bin, and on the next by taking a bracket from the bin containing the rear brackets (see Figure 1.6). Thus, at the beginning of each welding cycle, an operator has to recognize whether a jig's arm in the working position corresponds to the front or rear bracket. After that an operator makes a decision to select the corresponding bracket. Next decisions are related to installing each bracket in a specific position as it is presented on the Figure 1.4. Hence, the worker constantly makes various decisions during one cycle. Decisions can be made based on external visual information or extraction information from memory. In either case, an extraneous mental load is added to the task.

The six decision-making actions of an operator are as follows:

1. If the narrow jig arm is in the working position, then reach for the front bracket;

2. If the wide jig arm is in the work position, then reach for the rear bracket;

3. Decide how to orient the rear bracket to place it on the jig arm after grasping it;

4. Examine and decide upon the acceptability of the rear bracket's quality while orienting it in relation to the jig's arm;

5. Decide how to orient the front bracket in the appropriate position;

6. Examine and decide upon the acceptability of the front bracket's quality while orienting it in relation to the jig arm.

Decisions 4 and 5 are performed relatively seldom and do not cause significant mental workload. Hence, we do not consider them as a critical factor in task performance and concentrate attention only on the four remaining decisions. During the shift, workers usually perform 1250 production operations. This means that about 5000 decisions out of two alternatives should be performed by workers during each shift. Such a quantity of decisions is a negative factor that makes the task very difficult to perform. As the difference in widths between the arms of the jig is not highly salient, distractions or other sources of attention lapse typically result in a high frequency of mismatches between jig's arms and brackets, leading to a longer welding cycle time and a greater error rate. One factor confounding the workers' judgment is that the differences between the two arms of a jig are not at all salient, as they differ in their width only by about 3 mm. Thus, direct perception of the width of the jig's arm provides little inherent indication as to whether a welder should reach for a rear or a front bracket. The other difficulty is associated with the correct orientation of the brackets before they are installed on the arms. Such orientation is performed not based on the externally presented perceptual information but based on the instructions that are kept in memory.

Thus, the selection, examination, and orientation of the brackets in the limited time and quantity of decisions during the shift constituted the greatest difficulty for the welders. The workers "solved" this problem by simply cutting down the jig's wider arm (18.5 mm) to the same size as the narrower arm (15 mm). This "solution" allowed workers to begin each cycle by taking a bracket from the same bin. Although, this clearly eased the mental workload of the decision-making, it also introduced large variations in the post-weld position of the brackets, degrading the quality of the task performance.

There was another reason for the deterioration in the quality of this task performance. The length of the rear bracket was 39 mm, and the front bracket was 47 mm long (see Figure 1.6). The worker cut down the length of both arms of the jig and made them equal to approximately 25 mm. This contributed to the additional deviation of brackets from vertical position. Data on the type of errors and associated with them the types of defective necks is presented in Table 1.1 (see Table 1.1 before innovations).

TABLE 1.1

Type of Defected Necks

Errors and Associated with them Types of Defective Necks	Number of Cases	
	Before Innovations	After Innovations
Misaligning of the brackets	590	0
Welding two rear brackets	3	1
Welding two front brackets	2	0
Wrong installation of brackets	5	2
Installation of brackets with inadequate quality	7	4

In cases when the welders "simplified" the selection process by cutting the jig's arms to the same width, another distinguishing feature related to the sequence of alternation between rear and front brackets, requiring the welder to memorize which bracket was welded in the previous operation, was eliminated. The task is further complicated by the fact that if the front bracket was welded first on the previous neck, then the rear bracket must be welded first on the next neck, otherwise the welder has to turn a jig 180° manually. Typically, a welder controls the sequence in which the separate elements of the operation are performed, using a mental scheme of actions that primarily depends on memorized information. All of this causes mental fatigue.

Misaligning of the brackets means that the flasks' covers are difficult to open and close. A worker at the end of the assembly line has to force the covers with a mallet. This in turn means that flasks' covers are not tight enough to safely hold the liquid for which they are intended.

Qualitative analysis, further morphological and quantitative analysis that are not discussed here allowed the suggestion and implementation of workplace reconstruction. This reconstruction allowed the production operation to be performed without violating the technological requirements and instructions Redesign includes the following steps:

1. Redesign the body of the jig in such a way that it consists of two halves of unequal width, by increasing the width of one half by 3 mm and decreasing the other half by 3 mm, resulting in a 6 mm difference between halves that still preserves the overall weight of the jig. The wider section should be associated with the wider arm, and the narrower section with the narrower arm.

2. Cover the wider section of the jig ring with yellow plastic, and the narrower section with a dark green plastic. These color cues enhance the discriminative properties of the two jig ring sections: the yellow color tends to enhance the impression of largeness, while the darker color tends to create an impression of diminished size. So, these two

features intensify the intrinsic relationship between the jig arm and its associated bracket.

3. Cover the bins which contain the rear, wider brackets with yellow plastic, and the bin with the front, narrower brackets with dark green plastic. Correspondingly colored plastic coverings should also be applied to the table surfaces on which the bins rest (see Figure 1.8).

4. Reconfigure the shape of the jig's arms to produce a notch on one, and an aperture on the other, providing an additional cue as to the relationship between each arm and the orientation of the corresponding bracket.

5. Make the length of the arm approaching to the same as the length of the corresponding bracket.

6. Reduce the overall weight of the jig by perforating the metal part of the jig's ring.

Principles of reconfiguring the shape of the jig's arm are demonstrated in Figure 1.9.

When the changes just listed were actually implemented in the workplace, they substantially improved the salience of the discriminative features of the task, simplifying the workers' decision-making processes. Not only were the visual cues enhanced by using the color scheme but haptic cues were also introduced through the alteration of the jig's ring sections, reducing the visual workload. This production operation was the most critical one on the production line. So after its enhancement productivity increased by 4% and misalignments of the brackets were eliminated.

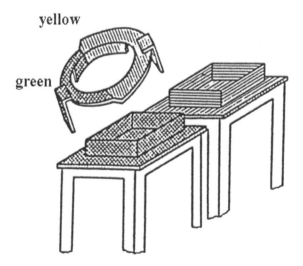

FIGURE 1.8
Demonstration of external association between a jig position and a position of bins with brackets.

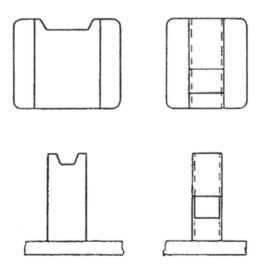

FIGURE 1.9
Association between jig's arms and position of the brackets.

This innovation is of interest from the economic efficiency point of view. From Table 1.1 we can see that after innovation, the quality of work significantly improved, and mental workload has not been increased. The important fact is that the errors, such as misalignments of the brackets, have been totally eliminated. This type of error before innovation was totally ignored by the management of the factory despite the fact that this type of error was registered in approximately 50% of the produced flasks. This can be explained only by *economic reasons*.

As we have shown earlier, a worker had to make about 5000 decisions during the shift, which was extremely stressful and difficult. The considered production operation was critical for the production process. The factory's management had a choice to *force workers to obey technological requirements*, or *ignore the fact that they have been violated*. If they chose the first alternative, that would create high turnover and perhaps even the inability to find the required workforce. If they chose the second alternative, it still would be possible to sell the product, because customers could not buy a better product. Despite the fact that the flasks were of poor quality (covers and locks quickly wear out, and the liquids stored in the flasks would not be adequately protected), the customers could not change the vendor. The administration did not know how to change technology or the method of task performance. So they simply *"closed their eyes"* to the poor quality of the product. The major conclusion is:

> Compliance with normative technological requirements to quantity and quality of the product often leads to economic losses and such contradiction might be difficult to overcome without introducing ergonomic innovation.

For example, just recently it was uncovered that Daimler, like Volkswagen before it, had equipped its cars with software designed to cheat testing cycles by running the engine more cleanly in lab conditions than it would normally on the road. This situation can be explained by the fact that the company could not find an appropriate technological solution to the problem. Ergonomics often can help to find such solutions.

Let us consider shortly one more example of the cost-effectiveness of ergonomic innovations. In this section, we introduce the method developed to establish the benefits of ergonomic intervention that utilizes heart-rate evaluation (Bedny and Seglin, 1997; Bedny, Karwowski and Seglin, 2001). Oxygen consumption in calories per minute or heart rate in beats per minute can be used for this purpose. The second method is much better because it is not as complex as the first one. Physiological studies demonstrate that 4.17 kcal/min or 100 beats/min are used as benchmarks for the boundary between acceptable and unacceptable strenuousness of work. The last criterion was justified in Rozenblat's study (1975). These criteria correspond to the boundary between low and high physical intensity of work. Rozenblat utilized 100 beats/min as a benchmark for introducing special allowances. The method requires determining an average pulse rate during the shift PR_w, and an average pulse rate during the break PA_{br}. The *average pulse rate during the shift PA$_{sh}$* is determined using these data.

PA_{sh} is the main criterion for evaluating the intensity of work and for estimating the break time. If the value PA_{sh} is less than 100 beats/min, one needs no additional time for rest. If PA_{sh} is more than 100 beats/min, then additional break time should be introduced (Rozenblat, 1975).

The average pulse rate PR_w during the shift can be determined using the following formula:

$$PR_w = P_1T_1 + P_2T_2 \cdots + P_nT_n, \qquad (1.1)$$

where $P_1, P_2, \ldots\ldots P_n$ are the pulse rates of the first, second, ..., n-s operation, and so on; $T_1, T_2, \ldots\ldots T_n$ performance time of these operations.

The average pulse rate during the break time PA_{br} is determined similarly. At the next step, the calculation of the pulse rate during the shift is performed.

$$PR_{sh} = \frac{\left(PR_wT_s + PA_{br}T_{br}\right)}{\left(T_s + T_{br}\right)} \qquad (1.2)$$

Then the break time BT_{cal} is determined as a percentage of the shift time. The criterion BT_{cal} is based on the average pulse rate during the shift being less than 100 beats/min and is calculated as follows:

$$BT_{cal} = \frac{100\left(PR_w - 100\right)}{\left(PR_w - PA_{br}\right)}\% \qquad (1.3)$$

Thus, in our further analysis, PR_w should not exceed 100 beats/min. We consider only the situation when PR_w is less than 95 beats/min (regular situation). Lunch break is not taken into account in these calculations.

Using the theoretical and the real break time (BT_{rl}), the required break time (BT_{rq}) is determined according to the formula:

$$BT_{rq} = \frac{(BT_{rl} + BT_{cal})}{2} \tag{1.4}$$

A real or standard break time is the length of the rest during the shift for various professions.

A detailed description of this method can be found in Bedny, Karwowski, (2007). Bedny suggested applying this method in order to evaluate the cost-effectiveness of various innovations. This method was utilized for evaluation of the economic efficiency of installing the air-conditioner into the cabin of a large excavator. This excavator did not have an air-conditioner because it has operated in moderate climate conditions. However, this excavator started being used in hot climate areas and the air temperature inside the cabin of the excavator during the shift on average was 40°C–41°C, with air speed 1 m/sec, and relative humidity 60%.

The president of the company that produced the excavator announced that the installation of an air-conditioner would sharply increase its cost. Therefore, without justifying the need for an air-conditioner, it can't be installed. For economic justification of installation of an air-conditioner, we conducted the following experiment. The excavator cabin was installed in the laboratory with two control levers inside the cabin that had similar physical resistance as real controls levers in the type of excavator being considered. Ten subjects, all trained in the use of the excavator imitator by moving controls according to the defined trajectory, worked in the cabin. The pace of movement, standard short rest period, errors, and so on have been controlled and registered. Before the task began, each subject spent 25 minutes in the cabin adapting to the climate. Each experimental trial without an air-conditioner and with it in the cabin lasted two hours.

The pulse rate (beats/min), breath rate (bits/min), and blood pressure were registered during work and rest periods. The pulse rate was determined by a photopleismograph (PPG). The pulse rate detector was fixed to the ear lobes of a subject. The breath rate was determined with a gauge fixed to the operator's nostrils. The blood pressure was measured during short breaks using the standard medical method. Work on the imitator was performed with the air-conditioner turned off and turned on. Here, without discussing physiological data in great detail, we only present calculations which determined duration of required rest without and with an air-conditioner. According to the following formula BT_{cal} is determined as:

$$BT_{cal} = \frac{100(105.1-100)}{(105.1-94.6)} = 48.57\% \tag{1.5}$$

Real break time $BT_r = t_{pp}$ according to the standard data is 17.88% [based on the handbook of Kantorer, (1977)]. The required proportion of rest time can be determined as:

$$BT_{rq} = \frac{(17.88+48.57)}{2} = 33.22\% \tag{1.6}$$

In other words, to obtain an average pulse rate PR_{sh} less than 100 beats/min, 33.22% break time should be allowed instead of the prescribed 17.88% of work time. This means that the break time should be increased by 15.34% of the overall time of the shift $T_{ex.w}$.

The work time on the excavator during the shift included breaks for maintenance and other technical reasons t_{tr}. According to the standard requirements, $t_{tr} = 39$ min* (Kantorer, 1977). In the case at hand, the work time ($T_{ex.w}$) when the excavator was operated during the shift was:

$$T_{ex.w} = T_{sh} - (t_{pp} + t_{tr}) = 492 - (88 + 39) = 365 \tag{1.7}$$

Thus, the required increase for work breaks BT_{rq} should be 15.34% of 365 min. So, according to physiological data in the absence of the air-conditioner, the additional 56 min (15.34%) break time per shift should be given. When the air-conditioner is turned on, the pulse rate is much lower than 100 beats/min. This reduces the need for additional breaks. In this example, the use of the air-conditioner saves 56 min of work time per shift. Thus, without the air-conditioner, an operator should have additional break time per shift in order to avoiding overloading the cardiovascular system. It is easy to demonstrate that one hour of excavator work time per shift provides enough saving during the year that makes the use of air-conditioner economically sound. Moreover, there might be legal consequences if management insists that an operator can work in the excavator without an air-conditioner.

This data just presented demonstrates that cost-effectiveness of the innovations that reduce physical workload and stress can be determined on the basis of the pulse rate method and can be accurately estimated using monetary measurements.

This approach allows to convince the management of the need to introduce innovations based on economic and ergonomic criteria. Violation of technological requirements by workers or management can be often explained by an inability to find an appropriate technological solution to the problem. Existing contradictions between productivity and technical requirements can be eliminated based on ergonomic criteria. In the later chapters of this book we will show how it can be achieved when considering computerized and computer-based tasks.

References

Ammer, C. and Ammer, D. S. (1984). *Dictionary of Business and Economics*. 2nd ed. New York, NY: The Free Press.

Armstrong, M., Cummins, A. and Hastings, S. (2003). *Job Evaluation: A Guide to Achieving Equal Pay*. London, UK: Kogan Page.

Arnaut, E. J., Gordon, G. M., Joines, D. H. and Philips, G. M. (2001). An experimental study of job evaluation and comparable worth. *Industrial and Labor Relations Review*, 54(4), 806–815.

Barness, P. M. (1980). *Motion and Time Study Design and Measurement of Work*. New York, NY: John Wiley and Sons.

Bedny, G. (1967). Psycho-physiological analysis efficiency of work. *Socialistic Labor Journal*, 77–80.

Bedny, G. Z. (1971). Committee for inventions at the Council of Ministers of the USSR. Patent 297876. Bulletin 10.

Bedny, G. Z. (1979). *Psychophysiological Aspects of a Time Study*. Moscow, Russia: Economics Publishers.

Bedny, G. Z. (2015). *Application of Systemic-Structural Activity Theory to Design and Training*. Boca Raton, FL and London, UK: CRC and Taylor & Francis.

Bedny, G. Z. and Karwowski, W. (2007). *A Systemic-Structural Theory of Activity. Application to human Performance and Work Design*. Boca Raton, FL and London, UK: CRC Press and Taylor & Francis.

Bedny, G. Z. and Karwowski, W. (2013). Analysis of strategies employed during upper extremity positioning actions. *Theoretical Issues in Ergonomics Science*, 14(2), 175–174.

Bedny, G. and Meister, D. (1997). *The Russian Theory of Activity: Current Application to Design and Learning*. Mahwah, NJ: Lawrence Erlbaum Associates, Publishers.

Bedny, G. and Seglin, M. (1997). The use of pulse rate to evaluate physical work load in Russian Ergonomics. *American Industrial Hygiene Association Journal*, 58, 335–379.

Bedny, G., Karwowski, W. and Known, Y.-G. (2001). A methodology for systemic-structural analysis and design of manual-based manufacturing operations. *Human Factors and Ergonomics in Manufacturing*, 11(3), 233–253.

Bedny, G., Karwowski, W. and Seglin, M. (2001). A hart evaluation approach to determine cost-effectiveness an ergonomics intervention. *International Journal of Occupational Safety*, 7(2), 121–133.

Bedny, G., Seglin, M., Meister, D. (2000). Activity Theory. History, Research and Application. *Theoretical Issues in Ergonomics Science*, Vol. 1, 2. pp. 165–206.

Boisard, P., Cartron, D., Gollac, M. and Valeyre, A. (2003). Time and work: Work intensity, European Foundation for the improvement of living and work conditions. Website: eurofound.eu.int.

Chebykin, O. Y., Bedny, G. Z. and Karwowski, W. (Eds.), (2008). *Ergonomics and Psychology. Development in Theory and Practice*. Boca Raton, FL and London, UK: CRC Press and Taylor and Francis.

Chipman, S. F., Schraagen, J. M. and Shailin, V. L. (2000). Introduction to cognitive task analysis. In J. M. Schraagen, S. F. Chipman and V. L. Shalin (Eds.), *Cognitive Task Analysis* (pp. 3–23). Mahwah, NJ: Lawrence Erlbaum Associates, Publishers.

Diaper, D. (2004). Understanding task analysis for human-computer interaction. In D. Diaper and N. Stanton (Eds.), *The Handbook of Task Analysis for Human-Computer Interaction* (p. 30). Mahwah, NJ: Lawrence Erlbaum Associates, Publishers.

Fitts, P. M. (1954). The information capacity of the human motor system in controlling the amplitude of movement. *Journal of Experimental Psychology, 47,* 381–391.

Gallwey, T. J. and Drury, G. G. (1986). Task complexity in visual inspection. *Human Factor, 28,* 596–606.

Gal'sev, A. D. (1973). *Time Study and Scientific Management of Work in Manufacturing.* Moscow, Russia: Manufacturing Publishers.

Gilbreth, F. V. (1911). *Motion Study.* Princeton, NJ: D. Van Nostrand Company.

Gilbreth, F. V. and Gilbreth, L. M. (1920). *Motion Study for Handicapped.* London, UK: George Routledge and Sons.

Hennecke (1960). International Labor Office. "Job Evaluation" Geneva, p.7.

Kahneman, D. (1991). Judgment and decision making: A personal view. *Psychological Science,* 2(3), 142–145.

Kahneman, D. and Tversky, A. (1973). On the psychology of prediction. *Psychological Review,* 102, 211–245.

Kahneman, D. and Tversky, A. (1984). Choices, values and frames. *American Psychologist,* 39, 341–350.

Kantorer, S. (1977). Calculation of the economic efficiency in the construction industry. In Kantorer, S. (Ed.), *Using Machine in Building Construction* (pp. 273–311). Moscow, Russia: Building Construction Publisher.

Kedrick, J. W. 1977. *Understanding productivity: An introduction to the dynamics of productivity change.* Baltimore: Johns Hopkins.

Kholodnaya, G. N. (1978). *Time Study in Manufacturing.* Moscow, Russia: Economic Publisher.

Kotovsky, K., Simon, H. A., (1990). Why are some problems really hard: explorations in the problem space of difficulty. *Cognitive Psychology,* 22, 143–183.

Lytle, C. W (1954). Job Evaluation Methods. New York, NY

Mahoney, T. A. (1983). Approach to the definition of comparable worth. *Academy of Management Review,* 8, 14–22.

Maynard, H. B., Stegemerten, G. J., and Schawab, J., L. (1948). *Method-Time Measurement,* New York, NY: McGraw-Hill.

McCormick, E. J., Sanders, M. S. (1982). *Human Factors in Engineering and Design.* New York, NY: McGraw-Hill.

Meister, D. (1985). *Behavioral Analysis and Measurement Methods.* New York, NY: John Wiley & Sons.

Miller, R. B. (1953). *A Method For Man-Machine Task Analysis* (Report 53-137). Wright-Patterson AFB, OH: Wright Air Research and Development Command.

Moshensky, M. G., (1971). *Time Study and Wages in the West and USA.* Moscow, Russia: Thinking Publisher.

Muchinsky, P. M., 1990. Psychology applied to work. *An Introduction to Industrial and Organizational Psychology,* Pacific Grove, CA: Brooks/Cole Publishing Company.

Nayenko, N.I. (1976). *Psychic Tension.* Moscow, Russia: Moscow University.

Neilson, W. S., Stowe, J., (2010). Piece rate contracts for other-regarding workers. *Economic Inquiry,* 48, 575–586.

Payne, John W. (1976), Task complexity and contingent processing in decision mak-
     ing: An information search and protocol analysis; *Organizational Behavior and
     Human Performance,* Vol. 16, Issue 2, pp. 366–387: Science Direct.
Rasmussen, J., Goodstein, L. P. (1988). Informational technology and work. In Helendar
     (Ed). *Handbook of Human-Computer Interaction* (pp. 175–201. Amsterdam, The
     Netherlands: Elsevier.
Rodgers S. H. and Eggleton E. M. (Eds) 1986. *Ergonomic Design for People at Work,* Vol.
     2. New York, NY: Van Nostrand Reinhold.
Rozenblat, V. V. (1975). Principle of physiological evaluation of hard labor based on
     pulse measurement procedures. In V. V. Rozenblat (Ed.) *Function of Organism in
     Work Process* (pp112–126 Moscow, Russia: Medicine.
Simon, H. A. (1999). *The Sciences of the Artificial* (3rd, rev. ed.). Cambridge, MA: MIT
     Press
Taylor, F. W. (1911). *The Principles of Scientific Management,* New York, NY: Harper and
     Brothers.
UK MTMA. MTM-1. Analyst Manual, London, UK: The UK MTM Association, 2000
Veikher, A. A., (1978). *Complex Labor,* Leningrad, Russia: Science Publisher.
Vicente, K. J. (1999). *Cognitive Work Analysis: Toward Safe, Productive, and Healthy
     Computer-Based Work.* Mahwah, NJ. Lawrence Erlbaum Associates, Publishers.
Zabrodin, Yu.M., Chernishov, A.P. (1981). On the loss of information in describing
     the activity of human-operator by the transfer function. In B.F. Lomov and
     V.F. Venda (Eds.), *Theoretical and Morphological Analysis in Engineering Psychology,
     Psychology of Work and Control.* Moscow, Russia: Science Publisher.
Zarakovsky, G. M., & Pavlov, V.V. (1987). *Laws of Functioning Man-Machine Systems.*
     Moscow, Russia: Soviet Radio.

2

Concept of Task in Systemic-Structural Activity Theory

2.1 Overview of Systemic-Structural Activity Theory

The analysis of publications in activity theory allows the identification of three areas in this field: general activity theory (AT), applied activity theory (AAT), and systemic-structural activity theory (SSAT). The word "activity" is not a precise translation of the Russian word *"deyatel'nost"*, which exclusively designates human behavior. Activity can be presented as a logically organized system of cognitive and behavioral actions. In AT, activity relates to actions as the whole relates to its parts. Human activity always includes consciousness, because it is directed toward achieving a conscious goal. Rubinshtein (1922, 1986, 1958) and Leont'ev (1977, 1978) were recognized as founders of the general activity theory. At the same time, Vygotsky's publications around 1930 formed the foundation of the social-cultural theory of higher mental functions. He introduced the notions of tools and signs that were used to explain the origins of consciousness and cognition in general (Vygotsky, 1971). The work of Vygotsky had a great influence on the development of activity theory.

General AT is not well adapted for the study of human work. This becomes obvious through the analysis of some examples when specialists in general AT apply it to specific work situations. For example, Leont'ev (1978) mixed the term "psychological operation" as an element of action with "technical operation" or production operation. He wrote that in some cases the operation of cutting metal is more adequate than the operation of sawing it, and the workers should know what tools to use in each case (the knife, the saw, etc.). However, in this example a worker performs multiple motor actions, which, in turn, include various motor operations. Motor operations are motions. For instance, the motor action "move a hand and grasp a control" is a motor action that includes motion "move a hand" (motor operation 1) and "grasp" (motor operation 2).

Strelkov (2007), who worked at Moscow State University, wrote in his study of pilot activity that some pilots' actions last 15–20 min. One of the actions he considered was a decrease in the required aircraft flight altitude. However, this is not a separate action but a complex task that includes numerous cognitive and behavioral actions. Actions are small elements of activity that have short duration. This author mixes the time of task performance with duration of much smaller elements of activity that are actions during task performance. A task includes a logically organized system of mental and behavioral actions. Zinchenko (1978) often used in his studies action and activity interchangeably. These examples clearly demonstrate that there is a big gap between theoretical studies in general AT and its applications in work psychology and ergonomics.

The development of cognitive psychology in the West coincided with the emergence of AAT in the former Soviet Union. However, nobody used the term AAT at that time. A lot of data from cognitive psychology has been adapted and accumulated in AAT. Recognition of this fact and introduction of the term *applied activity theory* was impossible at that time for ideological reasons. *"Such bourgeois science as cognitive psychology could not have significant impact on such Soviet ideologue science as activity theory."* Zinchenko developed some basic theoretical aspects of AAT by introducing in its content some data in cognitive psychology. However, further significant development of this field should be associated with names of such scientists as Kotik (1974, 1978); Konopkin (1980); Pushkin and Nersesyan (1972); Zarakovsky (1966), Zarakovsky, et al. (1974); Ponomarenko and Zavalova (1994); Platonov (1970); Landa (1984); Bedny (1981, 1987) and others. These scientists did not differentiate their work from the general activity theory (AT) that time because it would contradict with ideology according to which AT was the one and only best representative of Soviet psychology. AAT is not a unitary theory. The terminology used by different authors in this field is not the same, and therefore, it does not unify the theory. There are no standardizing units of analysis and standardized procedures of work activity analysis. This creates difficulties in translating the studies conducted in AAT into English.

During the last several decades, a new approach within AAT has been developed. This approach is called SSAT. The first book in this field has been published by Bedny (1987), who is the founder of this approach. Some aspects of SSAT were published in Bendy and Meister (1997). As an independent topic, SSAT has been detailed in Bedny and Karwowski (2007), Bedny, Karwowski and Bedny (2015) and Bedny (2015). In this chapter, we give only a brief overview of SSAT. An SSAT is a unique and independent approach to work activity analysis and is distinct from those that preceded it. This approach is derived from general and AAT. SSAT, as with AAT, accumulates some basic findings in cognitive psychology and uses them in its further development. In general, we can conclude that applied and systemic-structural activity theories can be efficiently utilized in the study of task performance. Useful information in this field can be found in such publications as Bedny (2004); Bedny and Karwowski (2004); Bedny and Harris (2005); Bedny (2006); Bedny and Karwowski (2008a,b).

The topic of systemic analysis of work activity has been very popular in Soviet psychology. Systemic analysis as an interdisciplinary field should not negate the need for the development of the proper systemic psychological methods of activity study. Implementing such an approach becomes possible when activity is described as a *complex structure evolving over time*.

Understanding activity as a structure that unfolds in time makes it different from general AT, AAT, and cognitive psychology which consider their object of study simply as a process (Chebykin, Bedny and Karwowski, 2008).

Creation of such an approach is possible only when we can offer methods of analysis of activity as a systemic-structural entity. Here, a description of activity as a system that consists of the elements that have specific relation and interact with other elements of activity is at the forefront of the research.

The transition from a general philosophical discussion about systemic analysis to one relating to its practical application is not an easy one. Existing methods of systemic analysis of activity outside of SSAT are important and useful from the theoretical and practical points of view. However, they are fragmented and cannot substitute for a unified and, to some extent, standardized approach to systemic analysis of work activity.

As per Schedrovitsky (1995), the nature of the systemic-structural method is determined by what methods and tools are used to describe an object as a system. This means that the same object can be presented as a system in different ways, depending on the methods used. Hence, we need to develop standardized methods, procedures, and operations to facilitate development of various types of models that described such an object as a system and as a structure.

Farther in this book we concentrate our efforts primarily on the application of the systemic-structural approach to an area of study that is related to Human–Computer Interaction (HCI) tasks, aviation, and other areas where cognitive components of tasks dominate. Studying such tasks raises a number of issues that need to be addressed. First of all, there is a need to develop standardized units of analysis of activity and a language for its description; to select stages and levels of analysis; to develop methods for constructing models of activity; to analyze the relationship of the levels of analysis; to identify the relationship between qualitative, formalized, and quantitative research methods, and so on. The conceptual apparatus in AAT does not just lack unified methods of study but also differs significantly from those that are utilized in the West. In this book, the SSAT studies in aviation, computer science, and other areas are described from unitary theoretical perspectives. This perspective is especially important for the specialists who are not familiar with diverse AAT methods of study.

From the SSAT perspective, activity is a goal-directed system, where cognition, behavior, and motivation are integrated and organized by a mechanism of self-regulation toward achieving a conscious goal. There are two types or aspects of activity: object oriented and subject oriented. Object-oriented activity is performed using tools on a material object, where a subject of activity is an individual or a group of individuals engaged in that activity. Subject-oriented activity, also known as social-interaction (Russian word - *obshenie*), involves two or more subjects and is constituted through information exchange, personal interaction, and mutual understanding. During task performance, the object-oriented and subject-oriented aspects of activity continuously transform into one another.

SSAT views activity as a structurally organized self-regulated system rather than as an aggregation of responses to multiple stimuli, or a linear sequence of information-processing stages as described in behavioral or cognitive psychology. Furthermore, it views activity as a goal-directed rather than as a homeostatic self-regulative system. A system is considered goal-directed and self-regulated if it continues to pursue the same goal under changed environmental conditions. The system can reformulate the goal while functioning. Activity as a self-regulated system integrates cognitive, behavioral, and motivational components. A simplified schema of activity is presented on Figure 2.1. This figure shows that cognition, behavior, and motivation influence each other. It also demonstrates that it is not only cognition that regulates behavior but that behavior through feedback also forms and regulates mental processes. The structure of activity also depends on mechanisms of self-regulation.

Comparing the structure of activity with the configuration of equipment or with a user interface is the basic principle which should be used when designing man-machine and human-computer interaction systems. Changes in the material components of work in a probabilistic manner change the strategies of task performance. The description of holistic activity structure during task performance is called morphological analysis. All these aspects of the study of human performance are the basis for its qualitative, formalized, and quantitative assessment. In this book, we also present advanced formalized quantitative methods such as task complexity and reliably assessment. These methods, developed in the framework of SSAT, are considered in Chapters 7 through 9.

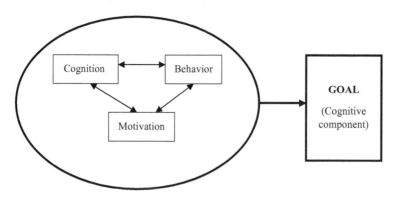

FIGURE 2.1
Simplified schema of activity as a system.

Traditionally, the systemic approach is utilized in work psychology to describe a man-machine system. Here, we talk about human activity that is viewed as a system consisting of independent elements (units of activity) that are organized and mobilized around a specific goal. Hence, here we utilize a systemic-structural analysis of activity. This analysis not only differentiates elements of activity in terms of their functionality but also describes their systemic interrelations and organization. In the absence of the appropriate units of analysis, the whole systemic-structural analysis of activity was impossible. SSAT, which has standardized units of analysis with their precise description, makes systemic-structural analysis possible. For example, SSAT offers a method of precise descriptions of cognitive and behavioral actions and of their extraction from holistic activity. There are also other units of analysis, such as cognitive and mental operations, functional macro-blocks and micro-blocks, and members of a human algorithm, which consists of one or several interdependent actions integrated by a high-order goal. Members of an algorithm have a logical organization and describe the logic of activity performance. The cognitive approach currently dominating in the West treats cognition and behavior only as a process that makes it difficult to study activity and behavior from the systemic-structural perspective. The notion of the process does not allow us to describe activity as a structure. Introducing standardized and unified units of analysis helps to describe activity as a structure that unfolds over time. SSAT does not reject the cognitive approach but uses this approach as a stage of activity analysis.

SSAT invites and empowers the study of the same activity from different points of view and thereby legitimizing the use of multiple approaches to the description of a single object. This implies that in applied research, an adequate description of the same object of study requires multiple interrelated and supplemented models and languages of description.

Therefore, the activity system cannot be described by one best method and calls for numerous stages and levels for description of the same activity. The simplest method is qualitative analysis.

Typically, such methods are complemented by experimental studies.

In SSAT, qualitative analysis also includes the study of activity self-regulation (Bedny and Karwowski, 2004). According to principles of analysis in SSAT, we also distinguish between macro-structural and micro-structural analyses that determine the levels of decomposition. At the micro-structural stage of analysis, cognitive and behavioral actions are described as a system of psychological operations and functional micro-blocks (Bedny and Karwowski, 2007; Bedny, 2015). For example, motor actions can be described as a system of motions. Decomposition of activity into a logically organized system of actions constitutes morphological analysis of activity. Similarly, cognitive actions can be subdivided into mental operations. The starting point of an action is the initiation of a conscious goal of an action (goal acceptance or goal formation), and an action is completed when the actual result of the action is evaluated.

Morphological analysis views cognition not just as a process but also as a logical organization of cognitive actions and operations. Therefore, cognition is described not only as a process but also as a structure. Identification of motor and cognitive actions can be complex. For example, during analysis of computer-based tasks, it may be necessary to use eye movement registration to extract and classify cognitive actions (Bedny and Sengupta, 2005; Bedny, Karwowski and Sengupta, 2006, 2008). Depending on the strategies of activity performance, the same task can contain different actions and their logical organization. The concepts of tool and object depend on a goal of actions and a goal of a task. When transformation of the initial screen into the final screen according to the goal of the task takes place, the initial situation can be considered as an object of activity. However, during the analysis of separate actions, different objects can be extracted. For example, when a subject performs the action "reach and grasp the mouse with the right hand", he/she performs a simple motor action within an approximate distance of 30 centimeters. This action includes two motions (reach "R25A" and grasp "G1A"). Individual motion description can be found in the MTM-1 system. However, a combination of these motions into motor actions is offered only by SSAT (Bedny and Karwowski, 2011; Bedny, 2015). The goal of such action is the mental representation of the desired result (grasp the mouse). In this action, the object of the action is the mouse. When the subject performs the second action "move the cursor to the required position and depress the left mouse button with the index finger for activation of a particular icon" (move an object to the exact location "M7C" and apply pressure "AP1"), the mouse is now not an object of action but is its tool. The pointer is an element associated with the mouse and also a tool of an action. At the same time, an icon on the screen is an object of an action.

There is another important aspect of SSAT actions' classification. It is important to differentiate between activity elements (psychological units) and task elements (technological units) of analysis. Psychological units of analysis represent standardized elements of activity. Anyone who is familiar with these units of analysis can clearly understand what a subject does when he/she performs these actions. For example, in the MTM-1 system, R25A means reach an object in a well-known location within the 25 centimeters distance. Note that the MTM-1 system does not distinguish between psychological and technological units of analysis. This classification has been introduced in SSAT, and a similar classification is applied for cognitive actions. When conducting morphological analysis and designing the time structure of activity and evaluating a task complexity, we need to convert technological units of analysis into psychological ones. In the process of task analysis at the first two stages, a specialist usually utilizes technological units and then transfers them into psychological units. This aspect of task analysis will be described in later chapters in more detail.

The vector motive → goal is a critically important concept in SSAT. Goal is associated with motives and create this vector. The goal is a mental representation or an image of a desired future result and should be differentiated from motives. The goal is a cognitive and motive is an energetic component of activity. Motive, or more generally, motivation is an intentional or inducing component of activity, which is closely connected with human needs. Needs become motives if they are connected with the goal of activity. Motives as an inducing force catalyze a subject's desire to achieve the goal of activity. One of the major difference of SSAT and the theories that consider the concept of goal outside of SSAT is that these theories do not distinguish motivation from goal. For example, Lock and Latham (1990) postulate that goal has both cognitive and affective features. Pervin (1989) stated that a goal can be weak or intensive. These authors insist that the more intense the goal is, the more one strives to reach it.

A difference in the understanding of the goal's position in the structure of activity is seen in the work of Kleinback and Schmidt (1990). They described the volitional process and considered the goal as a source of inducing behavior. The direction of volitional process goes from the goal to behavior. The scheme of the volitional process, according to Kleinback and Schmidt, in a simplified manner is presented on Figure 2.2.

Figure 2.2 demonstrates that a goal pushes activity through intentions. Intentions in this schema should be considered as any desire, plan, or aim that orients a subject to some end state or to acting in a specific manner. Intentions usually include conscience. According to SSAT, intentions, like various needs, only represent a possibility to act in a certain manner. Such needs and intentions can have connections with *potential goals* and create units of information that can be stored in memory (Bedny, 2015). These units of memory can be presented as: *need → potential goal, intention → potential goal*. When a specific *intention → potential goal* vector is activated in memory, it can be transformed into a vector *motive → goal*. In other words, *intention* becomes a *motive*, and *potential goal* becomes a *real goal*. As a result, a subject starts to act according to this vector. Intentions and desires can be not only activated and extracted from memory but can also be developed and formed during activity. Needs and intentions can be integrated into a hierarchically organized needs-intentions system. When needs and intentions are connected with a specific goal they are transformed into motives. As we already described above in AT, motives push activity to reach the goal (see Figure 2.3).

From this figure one can see that motives are not included in a goal. A goal is only a cognitive component that is a mental representation of a desired

FIGURE 2.2
Relationship between the goal and behavior in cognitive psychology.

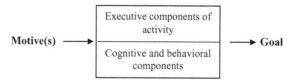

FIGURE 2.3
Relationship between motives, activity and goal in activity theory.

future result. At the same time, there can be a number of different motives. The direction of executive components of activity are determined by the vector motive (s) → goal.

From this figure one can also see that there can be several motives but only one goal of activity during task performance. Awareness of a goal is critically important. There are other aspects of activity, such as motives, that can be conscious or unconscious. A goal of a task or sub-goals of a task are higher-order hierarchical goals in comparison to goals of individual actions. They are often developed consciously. The goals of individual actions are often formed unconsciously or automatically. Such goals could be quickly forgotten after their achievement.

Activity differs from the Purposive Behaviorism of Tolman (1932), where goal-directedness was not distinguished from purposeful behavior. Goal-directed human activity always includes conscious understanding of the intended goal. Purposiveness of animals' behavior does not suggest conscious understanding of a future desired result.

Theorists in psychology and ergonomics do not discuss objectively the given requirements of a task and the process of goal acceptance, or the formulation of the goal. Objective requirements of a task should be conscious and are interpreted by a subject. Only after that can they be transformed into a goal of a task. We also cannot agree with the Leont'ev's (1977, 1978) statement that in certain situations motives can be moved into a goal when a vector motive → goal is transformed into a point and the goal-directness of activity disappears.

Activity is a combination of normative or standardized and variable components. Therefore, it can be described with some approximation. Normative and variable components depend on the rules and prescribed procedures of a task performance. They depend on utilized technology or software, safety requirements, and prescribed methods of task performance. Periodically, these normative requirements can change. However, task performance practices cannot be changed voluntarily according to individual wishes of an operator; a written standardized practice should be maintained and preserved. New methods of task performance can be introduced only after careful analysis of existing methods and enhancements by professionals who are responsible for the work process. At the same time, normative methods have some flexibility, that is, an operator can utilize different individual strategies of a task performance that are restricted by existing technological constraints. Individual strategies can be considered acceptable if they do not violate existing technological constraints. Such strategies in AT are known as individual styles of activity (Bedny and Karwowski, 2007; Bedny and Seglin, 1999a,b). Any activity includes variable components that cannot be totally eliminated. Variability is a feature of not only human activity but also of many complex systems. For example, in mass production and manufacturing, each manufactured part is unique in its size and shape. However, if the variation of sizes and shapes is within an established range of tolerance, all such parts are considered acceptable and identical. Similarly, if the strategies of activity performance fall within an established range of tolerance, these strategies are considered to be the same. Such notions as strategies, constraints, individual styles of performance, variability of activity, tolerance, and so on help us to understand the relationship between normative (standardized) and variable components of activity in the process of activity design. That is why we do not agree with the constraint-based approach of design suggested by Rasmussen and Pejtersen (1995) and Vicente (1999). Their approach ignores the relationship between normative (prescribed) and variable components of activity. An understanding of relationships between normative or standardized and variable elements of activity eliminates contradictions with ideal task models and the models that represent real tasks often discussed in the human-computer interaction field. If variations in the task execution do not exceed the range of tolerance, then such execution of the task can be considered the expected standardized method of task performance.

SSAT considers activity as a goal-directed self-regulative system, and the purpose of functional analysis of activity is to describe it as such (Bedny, Karwowski and Bedny, 2015; Bedny and Karwowski, 2006a,b). These critically important aspects of activity analysis will be discussed further in various chapter of this book. In this chapter, we considered in an abbreviated manner only some aspects of SSAT analysis. The relationship of cognition and behavior is one of them.

Experimental methods of study, observation, survey, measurements of the spatial characteristics of the workplace, and so on can be utilized when the equipment and software already exist. However, when equipment and/or software do not exist yet, there is a need to present design documentation such as drawings, calculations, and so on. Only after that can the designed objects or recommended innovations be worked on. This means that analytical methods play a leading role in the design process.

As we have mentioned, SSAT views activity as a multidimensional structurally organized system, and therefore multiple methods of activity analysis should be utilized. SSAT offers the following method of activity analysis:

The *parametric method* concentrates on the study of different parameters of activity that are treated as relatively independent. The cognitive approach is an example of parametric analysis of activity. The systemic approach also includes morphological and functional analysis of activity. *Morphological analysis* involves the description of the structure of activity. This method uses cognitive and behavioral actions and operations (cognitive acts and motions) as main units of analysis. Based on morphological analysis, activity structure can be described in terms of logical and temporal-spatial organization of actions. Morphological criteria entail representing activity as activity-action-operation. In SSAT, morphological analysis includes algorithmic analysis of activity and analysis of its time structure.

Functional analysis includes the description of activity as a self-regulative system. This method allows utilizing units of analysis such as the function block *or* a mechanism of self-regulation. Describing the specificity of functioning of each block and its interaction with other function blocks facilitates understanding possible strategies of activity. Vector motive → goal is a critical factor that determines strategies of activity performance.

The *qualitative method* involves the verbal description of activity. This method can be used for the traditional objectively logical analysis which includes the verbal description of work, analysis of work space organization, description of work conditions, and so on. The other method studies the sociocultural aspects of activity (Vygotsky, 1978). Culture is regarded as a mediator between the user and technology. It includes beliefs, attitudes, values, social norms and standards, and so on. There is another qualitative method for studying human activity; it has to do with the individual style of performance. This method considers individual characteristics of personality in relation to the objective requirements of task performance (Bedny and Seglin, 1999a; Klimov, 1969).

Quantitative methods are the methods where mathematical procedures are used for activity description. In SSAT, basic quantitative methods include evaluation of task complexity and reliability assessment. Some other quantitative methods of task analysis, such as application of queuing theory in human error analysis (Bedny, 2015) and analysis of abundant actions (Bedny, Karwowski and Bednt, 2015), have been developed within the SSAT framework.

Genetic analysis is used to describe the main strategies of activity at the different stages of activity acquisition. We need to describe the activity structure while subjects are acquiring the method of task performance. If we know how the structure of activity changes during a task acquisition and what the final structure of activity is, then we can improve the efficiency of evaluating the usability of equipment. This method is useful because it is difficult to discover the differences in design solutions when the task performance occurs on a more or less automatic level. The more complex an activity is, the more intermediate strategies are utilized during its acquisition.

SSAT offers the micro-genetic method of studying computer-based and computerized tasks (Sengupta, Bedny and Bedny, 2011; Sengupta, Bedny and Karwowski, 2008; Bedny, Karwowski and Voskoboynikov, 2011). When an activity is already acquired, the users are less sensitive to changes in the equipment or software design. The qualitative differences can be discovered later in unusual circumstances or stressful conditions. A complex task that requires a long and complicated process of skill acquisition is less reliable. Therefore, analysis of the acquisition process or micro-genesis of activity structure development can be used for predicting reliability and efficiency of task performance in adverse work conditions.

All methods of studying work activity are organized into four stages of analysis. These stages are: (1) *qualitative analysis*, which can be parametric and functional. The first one describes different parameters of activity; (2) *functional analysis* describes activity as a self-regulative system (describes the most preferable activity strategies); (3) *morphological analysis* includes algorithmic analysis and time structure analysis. Algorithmic analysis describes activity as a logically organized system of actions. Time structure analysis describes the duration of elements of activity and their unfolding in time; (4) *quantitative method* of activity evaluation focuses mainly on the complexity and reliability of performance. The quantitative method of evaluation of task complexity is a systemic method of analysis because it evaluates activity structure (system of activity) quantitatively. Quantitative measures of complexity outline the critical points of a task that have to be modified to improve the efficiency of its performance. Such measures are also used for evaluation of efficiency of design of equipment or of software.

Stages of analysis can be broken into levels of analysis. All stages and levels of analysis have a loop-structure organization, implying that the result of analysis of one level or stage may require reconsideration of preceding levels or stages of analysis. The logical and hierarchical organization of activity elements determines the structure of activity during the task performance. This structure depends on strategies of activity. If, due to self-regulation, the strategies are changed, the structure of activity also changes. The structure of activity unfolds in time as a process. Such understanding of activity makes it much easier to apply the concept of morphological analysis to the study of human performance.

So SSAT demonstrates that such concepts as *system* and *structure* can be applied not only to the study of man-machine systems but also to the study of work activity in general. Man-machine systems or human-computer interaction systems can be optimized based on the qualitative, morphological, functional, and quantitative analyses of activity structure. To carry out the required studies within the framework of SSAT, standardized terminology has been developed and was first introduced in the Bedny and Meister (1997) book. This terminology differs significantly from the AT terminology because it derives from the SSAT analysis of activity as a structurally organized system.

2.2 Analysis of the Basic Components of Tasks

Professionals such as engineers with various backgrounds, human-computer interaction specialists, economists, and so on recognize that efficient design of performance methods, equipment, tools, and user interfaces is a critical factor in increasing productivity, reliability, and improving usability. Applying psychology to this multidisciplinary field is of distinct importance. All of this raises certain demands on the development of psychological terminology that should be clearly defined and in some cases even standardized. However, existing terminology in psychology is not sufficiently adapted for the study of human performance. As we will further show, professionals with different backgrounds understand psychological terminology in different ways, often try to interpret this terminology in their own way, and suggest elimination of some terms and even seek to introduce their own terminology as a substitute for existing terms. Development of correct terminology is especially important when translating psychology publications into English. When we considered this issue in regards to Activity Theory, it became obvious that many of the Russian-English translations fail to capture the original meaning of activity theory's basic terminology (Diaper and Lindgaard, 2008; Bedny and Harris, 2008; Bedny and Chebykin, 2013).

Psychological studies of human performance are important not only for ergonomists, engineers, or HCI specialists. This area of research plays a critical role in the area of economics that is also involved in the study of human performance. This area of economics includes a wide range of issues and particularly work analysis and design, job evaluation, time study, compensation, and so on. Considering that psychologists collaborate with engineers, HCI specialists, economists, and other professionals, the theoretically based psychological terminology should be clearly defined and scientifically justified. However, concepts such as activity, design, job, task, mental and physical demands of work, task complexity, and so on are not clearly defined outside of SSAT. It should be noted that development of scientific terminology requires a substantial theoretical foundation.

In this chapter, we consider such basic concepts as production or operationally monitoring processes, task, goal, human performance, design, and so on. Usually, work activity can be presented as a production process that can be broken down into separate tasks. In manufacturing, the purpose of a production process is to transform raw materials into finished products. Such production processes include three basic elements: *work activity, or work process; the means of work and associated with them technological process; and the product.* They usually begin with putting the raw materials and proceed through various steps until these materials are transformed into a finished product, or into an intermediate product that fits the purpose of the production process.

Each such production process can be divided into technological and work process. The technological process involves the impact of equipment on an object of work that should be transformed into a product. The work process is an activity of a worker on the object of work and his/her interaction with an equipment during a work process. Thus, production processes include two basic components: work activity and means of work or equipment and tools utilized by a worker. The task as the basic element of the production process has the same components. The structure of production process and task is presented in Figure 2.4.

The work process should always be performed according to technological requirements. As can be seen on Figure 2.4. The technological process includes instructions, means of work, raw materials, and a finished product. The means of work include tools, equipment, and instruments utilized by a worker. Instructions present rules according to which a worker perform various tasks. A work process involves the following components: past experience or knowledge and skills, abilities as individual features of worker, vector "motive → goal" that reflects a worker's purpose, and a method of activity that includes cognitive and behavioral actions necessary for achievement a goal of a task. There are various types of production processes performed by white-collar workers, blue-collar workers, operators (technologists, controllers, inspectors, dispatchers, etc.), and operators involved in performance of computer-based tasks.

There are also production processes performed utilizing automated and semiautomatic systems. In SSAT, these types of production processes are called operational-monitoring processes. The main purpose of this production process is not just transformation of physical material objects but also transformation of information. Although such processes are performed primarily based on specific rules, they can also include creative components.

An operational-monitoring process is defined as a combination of duties essential to accomplish some automatic or semiautomatic system functions. Instead of the term technological process, the term "control process" is used, and instead of raw materials and product, terms such as input and output are utilized. The structure of the operational-monitoring process, with a task as its basic component, is presented in Figure 2.5.

The structure of an operational-monitoring process is similar to the structure of a production process. They both include technological or control processes from the engineering perspective and human activity from the psychological perspective. Although both types of task include physical and cognitive components, in production processes physical components

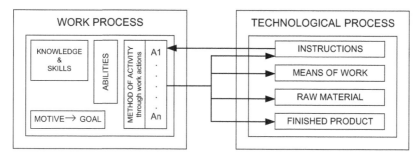

FIGURE 2.4
The structure of a production process and tasks that are its basic components.

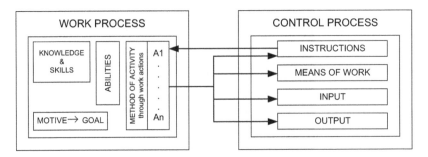

FIGURE 2.5
The structure of operational-monitoring process and tasks that are its basic components.

dominate. In the operational-monitoring process, cognitive components play a leading role. A task as a basic component of production and the operational-monitoring process includes work activity and a technological or control process. In other words, the structure of tasks that are components of a production process is similar to the structure of a production process, and the structure of operational-monitoring tasks is similar to the structure of a operation-monitoring process.

From the SSAT perspective, any task can be presented as a logically organized system of cognitive and behavioral actions that should be performed to achieve a goal of a task. For example, a sequence of different tasks performed by different workers on a production line can be considered as a production process. If a worker performs a strictly determined sequence of different tasks, it is also an example of a production process. Some contemporary computerized tasks emerge in different order and can appear in a probabilistic manner. On the other hand, if a financial consultant performs a number of different tasks that do not have a rigorous order, it is not a production process but a job. A job usually includes a combination of duties essential to accomplish system functions or certain work in a business organization. Thus, the concept of task is a critically important element of a production process or a job. Therefore, we need to consider it in a detailed manner.

The task is the basic component of activity. Task analysis is one of the key steps in developing man-machine or HCI systems. Changing any equipment or software characteristics influences a method of task performance. Figures 2.4 and 2.5 demonstrate that task analysis should be performed in unity with tool, equipment, and software analyses because design of equipment, tools, and interfaces determines the nature of work activity during task performance. If, for example, human activity becomes too complex, this means that utilized equipment, tools, or software are not adequate for task performance. There also is another aspect of task analysis. A subject can use various methods when performing the same task. Therefore, analysis of human activity during task performance can help one to discover the most efficient method of task performance. Task analysis is vitally important for the training process.

The task can be considered as a situation requiring achievement a task's goal in specific conditions (Leont'ev, 1977). The task-problem can be defined as a situation in which a subject needs to find an unknown solution when a solution is not given to a subject (Brushlinsky, 1979, 1999). In SSAT, a task is considered as a logically organized system of cognitive and behavioral actions that is directed to achieve the conscious goal of the task.

Goal, task, performance, model, design, and so on are critically important concepts for different professionals who study human work. As we already mentioned before, some specialists with an engineering background or a background in HCI reject psychological principles of interpretation of these concepts as they are utilized in cognitive psychology and in psychology in general. It is necessary to compare our critical analysis of some of these concepts from SSAT perspectives with the point of view of some professionals with an engineering background. Without such comparative analysis, it is impossible to efficiently conduct task analysis.

The first step of task performance involves presenting to a subject conditions and rules she/he has to follow. There are also requirements for what should be accomplished by performing a task. This information should be understood and accepted by a subject, because a subject perceives and interprets information about task conditions and requirements and such interpretation can be inadequate. A subject might intentionally modify information about requirements to a certain extent. Thus, considering subjects' understanding and interpretation of requirements and their acceptance as a goal is an important step of task analysis.

Specialists with engineering backgrounds misinterpret the concept of goal because cognitive psychology integrates a cognitive mechanism such as a goal with motives. Similarly, insufficiently clear definition of other psychological terms produces their misinterpretation and criticism by such specialists. Let us consider how the concept of goal is criticized by Diaper (2004). His critical analysis of such basic concepts as goal, motives, task, design model, and so on as they are used in cognitive psychology and the study of motivation is interesting from the SSAT standpoint. Let us consider some of his comments. Diaper wrote (Diaper, 2004, p.16):

> *The concept of goals is closely tied to that of tasks, and goals are nearly always described as being psychological, in that it is people who have them. In contrast, in the systemic version of Dowell and Long it is the work system that has goals, which specify the intended changes in the application domain. The advantage of this view of goals is that it recognizes that nonhuman things can have goals.*

And further on the same page he wrote:

> *It is probable that one of the main confusions that has aris[en][...] in task analysis over the years has been the assignment of the different types of goals to individual people rather than to a work system.*

Diaper utilized an example in which he attempted to prove his theoretical position that a human goal is a redundant concept based on analysis of such terms as "related goals" and "unrelated goals". Further, he has stated (Diaper, 2004, p.17):

> *A hierarchy of goals, as used in HTA, consist of multiple related goals, but a person can also perform an action on the basis of unrelated goals. For example, a chemical plant operator's unrelated goals for closing a valve might be (1) to stop the vat temperature rising, (2) to earn a salary, and (3) to avoid criticism from the plant manager. The first might concern the safety of a large number of people, the second is socio-psychological and might concern the operator's family responsibilities, and the third is personal and might concern an operator's self-esteem. These three goals correspond to different analysis perspectives, the sociological, the socio-psychological, and the personal psychological; and there are other possible perspectives as well.*

Let us consider the last example, which demonstrate that scientists often confuse a goal with motives. Diaper and Stanton (2004, p. 611) wrote:

> *The basic idea is that there is some sort of psychological energy that can flow, be blocked, deviated, and so forth. Goals as motivators of behavior would seem to be a part of this type of psychological hydraulics. Given that there is no empirical evidence of any physical substrate that could function in such a hydraulic fashion, perhaps we do not need [the] concept of goal as a behavior motivator.*

Analysis of the selected critical comments about the concepts of goal and motivation presented by Diaper and his colleague cannot be considered as some misunderstanding of these basic concepts and their relationship. Diaper and Stanton are correct when they criticizes these concepts as they are used in cognitive psychology and in the field concerned with motivation. However, in general AT and in SSAT, activity is always viewed as a goal-directed system. In SSAT, goal is considered only as a cognitive mechanism that includes imaginative and verbal-logical components and some of its components should be *conscious*. Furthermore, a goal does not include motivational components. Therefore, there can be *only one goal of a task* and *multiple motives* for its performance. There are also goals of separate actions that are usually not retained in working memory for a long time and can be quickly forgotten after their performance. It is possible to have different relationships between the same goal and different motives. In these cases, goal-directed activity becomes ambivalent. If a goal appears to a subject to be undesirable, she/he is motivated to avoid the goal. Awareness by an individual of possible negative and socially unacceptable consequences of sustaining the same goal-directed activity may cause reformulation of a goal and suppression of the inappropriate motives.

In his critical comments, Diaper wrote about unrelated goals. However, there is no such concept as unrelated goal in AT. Returning to his comments about the chemical plan operator's unrelated goals, we have to point out that there is only one goal there, which is *"closing a valve"*. There are possible motives that drive an operator to close the valve: (1) to stop the vat temperature rising, (2) to earn a salary, and (3) to avoid criticism from the plant manager. These listed wishes or desires (which are described verbally by Diaper) in connection with the goal become motives and together can be presented as the above-described vector. To better understand this interpretation, we can consider in addition one possible motive in this example. Suppose, if an operator does not close the valve it can cause an explosion. Then, it is obvious that an operator would be motivated to close the valve as quickly as possible. The goal in this situation will be a verbally logical and imaginative representation of a future desired result *"closing the valve"*. The main motive will be "to avoid an explosion". Comments presented by Diaper (2004, p. 611) and their analysis are very helpful for the understanding of the concepts of goal and motives in SSAT.

In their last critical comments, Diaper and Stanton consider goal and motivation in the broader context of the information and energy relationship. They do not only reject the concept of goal, and associated with it motivational factor, but they also reject the concept of energy in psychological studies in general. In the previous section we discussed the relationship between informational and energetic aspects of human activity.

Diaper rejects the existence of goal, motivational mechanisms, and "the psychological hydraulics system", stating that there is no evidence that such a system exists. Emotionally motivational components of activity represent energetic components of activity. Motives and goal do not function in a mechanistic manner as a hydraulics system. They are complex psychological subsystems of activity that give activity its goal-directedness.

Here is another example of the energetic aspects of human activity. Klochko (1978) studied contradictions between task elements (conditions or givens) and emotional tension during task performance. Subjects did not know about existing contradictions in task conditions when they read the text of the task and could not report them verbally. It has been discovered that galvanic skin response (GSR) increased when subjects read the text that contained some contradictions. According to our interpretation, this experiment demonstrated the relationship between energetic and informational aspects of activity regulation (Bedny, Karwowski, I. Bedny, 2015). The contradictions in the text made it difficult for subjects to interpret its meaning. This, in turn, required more resources for activity regulation. Thus, SSAT eliminates all the above-listed critical comments by describing the relationship between informational and energetic aspects of activity regulation and between goal and motivation in a totally different manner and by proving them theoretically.

2.3 Cognitive and Behavioral Actions as Basic Elements of Task

Behavioral and cognitive actions can be understood as elements of activity, its main building block and unit of analysis. An action is a discrete element of activity that is directed to achieve a conscious goal of that action. Achievement of its goal and assessment of its result are the end point of an action that separate that action from the following action. From an AT perspective, cognition is not simply a process but also a system. It has a complex structure that consists of cognitive actions and psychological operations as their elements. A standardized description of actions is necessary for depicting cognition and activity in general as a complex structure. Motor and cognitive actions are interdependent, and the first ones include cognitive components. Both cognitive and behavioral actions have beginning and end points and should be considered as self-regulated elements of activity. Motor actions transform material objects, and cognitive actions transform information.

Choosing an action as a unit of activity analysis has the fundamental theoretical meaning. Human knowledge such as images, concepts and propositions are the result of cognitive actions.

Declarative knowledge should be regarded as an object of an action. A subject operates with these objects when performing cognitive actions. At the same time, such manipulation with declarative knowledge can produce new ones that are the result of actions. For example, perception and formation of new images are the result of *perceptual actions*, when a subject manipulates some perceptual data. It is clear that when a subject develops a complex image of an object, formation of such a perceptual image takes more time. Similarly, a concept is a result of a *thinking* action. Actualizing some data in memory is a result of *mnemonic actions*, which sometimes have very short duration.

In order to perform actions, a subject has to acquire knowledge about how to perform them. In cognitive psychology, a concept of procedural knowledge is used to describe this process (Anderson, 1993a, b). However, according to AT, procedural knowledge is also the result of actions. A subject performs thinking, mnemonic, and other actions in order to acquire such knowledge. Moreover, a subject can possess such knowledge but cannot use it because it is not transformed into skills. One can see this in gymnastics, where the gymnast knows the technique, but cannot perform an element without extensive practice. Thus, declarative and procedural knowledge are results of human cognitive and behavioral actions. All of the above are the reasons why cognitive and behavioral actions are chosen as the basic units of activity analysis.

The term action has totally different meanings in cognitive psychology and in ergonomics. For example, in cognitive psychology, action and activity are sometimes used interchangeably. Action has approximately the same meaning in German action theory as activity in AT (Frese and Zapf (1994)).

We would like to start our discussion with analysis of the concept of motor action as it is understood in engineering. Analysis of the effectiveness of production operations (tasks) in industry is accompanied by disassembling them into constituting components. In manufacturing, specialists disassemble operations into their elements or units of analysis. The most important units of such analysis are actions that are described by using common language or technological terms. This allows determining how efficiently a worker performs manufacturing operations. The degree of subdivision depends on the specifics of the operation. A detailed subdivision is used in mass production. Based on such analysis, specialists determine if there are unnecessary actions and how efficiently a manufacturing operation is carried out. Under motor actions, specialists understand elements of production operations such as "take a part", "insert a cylindrical part into a lathe center", "press a button". These are not precise descriptions of actions. Depending on conditions, a worker can utilize different motor activity elements to perform the same above-described motor actions. For example, the distance to the object and its location and the shape of a part and its weight are unknown when the description just states "take a part." Such description assumes further observation, measurement procedures, and so on. Description of actions in such terms in the documentation leads to a lack of clear understanding of what really is involved in their performance. Such shortcomings in the description of the various tasks are especially undesirable when applying an analytical method of study and developing the necessary documentation.

SSAT differentiates between two types of units of analysis. One is called "technological units of analysis" or "typical elements of task." The other units of analysis are called "psychological units of analysis" or "typical elements of activity." Technological units of analysis are commonly utilized at the first step of analysis. At the second step, they are transformed into psychological units of analysis. Such actions as "take a part", "insert a cylindrical part in a lathe center", "press a button" are examples of motor actions description by using technological units of analysis. To avoid inadequate description of actions and to describe them more precisely, SSAT uses psychological units of analysis specifying distance to the object, its shape, position, and so on. Suppose a subject performs the action "take a part" under careful conscious control, when a part is located 40 cm from a subject, the part has an approximately cylindrical shape with a diameter of 15 mm. In SSAT, this motor action is depicted using a standardized description. A motor action consists of motions that are integrated by a conscious goal of this action. It is known that the MTM-1 system offers a very detailed and precise description of individual motions, but it does not use concepts such as motor action and a goal of an action. Thus, we have combined SSAT theoretical data about motor actions with the MTM-1 system data about standardized motions. The above-considered action "take a part" can be described as "R40C + C1C1", which in the MTM-1 system is a standardized description of "reach under careful control at a 40cm distance (R40C), and grasp an approximately cylindrical object with diameter over 12 mm." Having the knowledge of the MTM-1 system, one can precisely understand what's involved in this motor action. As can be seen, the considered motor action is comprised of two motor motions or psychological operations, according to the SSAT terminology. These operations are integrated into a holistic basic unit by the goal of the action. The goal of the action as the desired future result is "to grasp an object." Psychological operations or motions according MTM-1 can be unconscious. For example, when a subject moves her/his arm, he/she does not know the exact trajectory of the movement, or how the fingers move. As it can be seen, a motor action at the first step is described using technological units of analysis (grasp object) and then using psychological units of analysis. Technological units of analysis, psychological units of analysis, the concept of goal, and the concept of action as a psychological unit of analysis are unknown not only in the MTM-1 system but in ergonomics and engineering as well. Utilizing the MTM-1 system in the manner described above allows application of it more efficiently in contemporary task analysis. It is specifically important when creating analytical methods of study for developing models of task performance not only in manufacturing but also for man-machine systems and HCI. For example, a motor action "reach and grasp a computer mouse with the right hand" can be describe in similar terms. A subject can combine some actions into a more complex unit. As will be shown later, the same type of actions can be combined into more complex units by a high-order goal. The quantity of such actions is restricted by the capacity of working memory. In the algorithmic task description, from 1 to 5 actions can be integrated into one member of the algorithm. A member of an algorithm can be described not only verbally but also symbolically.

Similar principles are used in SSAT for the description of cognitive actions. In our later discussion, the standardized description of holistic cognitive actions is utilized because it is not easy to extract mental operations as contents of cognitive actions. Cognitive actions are involved in transforming information based on the goal of the task and the goals of separate actions. For this purpose, externally presented information or information that is extracted from memory is divided into operative units of information (OUI) and can be considered as idealized objects for actions that are internal mental tools of cognitive actions. A subject can integrate, modify, and structurally reorganize operative units of cognitive actions. The thinking process and memory play a leading role in extracting and handling such units of information. Images and verbal-logical expressions also can be used for formation of OUI. As has been mentioned, behavioral or cognitive actions can be described by using technological terms that depict cognitive task elements associated with the considered cognitive action. For example, "taking a reading from a scale indicator with a moving pointer" is a description of a perceptual action using technological units of analysis. Conditions of reading can change. If, in addition, it is described as a *simultaneous perceptual action* with duration of 0.3 s, it gives a clear understanding of what action is performed by a subject.

AAT distinguishes direct connection actions from transformational cognitive actions (Zarokovsky and Pavlov, 1987). Direct connection mental actions are performed with some level of atomization without distinctly differentiated steps and require less attention than transformational cognitive actions. These actions are less consciously registered and often experienced by a subject as instantaneous. Recognition of a familiar object is an example of such actions. Transformational mental actions involve more deliberate examination and analysis of stimulus as, for example, perception of an unfamiliar object in a poorly lit environment. Classification of cognitive actions should always be complemented by analyzing their duration. Duration of cognitive actions can be obtained in psychology or ergonomics handbooks or through special experimental studies. Principles of extracting cognitive actions are described in details in Bedny, Karwowski and Bedny (2015). Verbal actions are another category of actions that perform not only communicative but also regulative functions in human activity.

We discuss these actions in Section 2.3 where verbal-logical thinking is considered.

Mental actions are classified based on the dominant cognitive process and on their ultimate purpose as discussed in the following paragraphs (Bedny et al. 2015; Bedny 2015).

I. Successive perceptual actions

Sensory actions - detecting noise or making a decision about a signal at a threshold level; obtaining information about distinct features of objects such as color, shape, sound, and so on.

Simultaneous perceptual actions - identification of clearly distinguished stimuli well known to an operator that only requires immediate recognition and perception of qualities of objects or events (recognition of a familiar picture).

Direct connection mnemonic (memory) actions - memorization of units of information, recollection of names and events, and so on. Direct connection mnemonic actions include involuntary memorization without significant mental effort.

Imaginative actions - manipulation of images based on perceptual processes and simple memory operations (mentally rotating a visual image of an object from one position to another according to a specific goal).

Decision-making actions at a sensory-perceptual level - operating with sensory-perceptual data like decision-making that requires selecting one out of at least two alternatives (detecting a signal and deciding to which category it belongs out of several possible categories).

II. Mental transformational actions - deliberate examination and analysis of stimulus (perception of an unfamiliar object in a poorly lit environment), exploration of situation based on thinking mechanisms, and so on. They can be classified as:

Successive perceptual actions are defined as recognition of unfamiliar stimuli and creation of a perceptual image of an object that requires deliberate examination and analysis of a stimuli.

Explorative-thinking actions are based on sensory-perceptual information, and are involved in deliberate examination of various elements of tasks, discovering specificity of their interaction, extraction of subjectively significant elements of a situation, interpretation of information obtained, and creation of mental pictures of a situation.

Thinking actions of categorization are based on analysis of features of signals or a situation followed by the logical analysis of their relationship and grouping them into two or more categories or classes (binary or multi-alternative categorization); can be performed based on various strategies of categorization that might change during training or self-learning.

Logical thinking actions are based on manipulation of concepts as well as major and minor premises (deductive actions, syllogisms, reasoning, etc.) using various strategies that can change during the skill acquisition process.

Decision-making actions at verbal thinking level are based on an algorithmic and heuristic level of regulation of thinking processes (after receiving information, an operator has to determine which steps out of several possible steps to take based on logical analysis of the situation).

Decision-making actions that involve emotional and volitional components are performed in combination with verbal thinking components; they include a conflict of motives and volitional process (in a dangerous situation a subject decides "I have to act" or "I don't have to act").

Recoding actions are understood as transformation of one kind of information into another (for example, transformation of meaningful verbal expression from one language to another).

III. High-order transformational actions include a complex combination of thinking and mnemonic actions or creative actions. They can be further distinguished as follows:

Creative-imaginative actions involve a combination of logical and intuitive operation with images.

Combined explorative-thinking and mnemonic actions involve complex manipulation of information in working memory based on mechanisms of thinking, extracting information from long-term memory, storing essential information, and maintaining information in working memory.

Creative actions are operations that generate new knowledge either logically or intuitively; they involve divergent thinking versus reproductive actions that involve convergent thinking.

When specialists study cognitive actions and utilize, for instance, eye movement registration, it is important to know the difference between successive perceptual and explorative-thinking actions. The main difference between them is that the purpose of successive perceptual actions is developing a perceptual image of an object or percept (for example, categorization of objects based on their shape, color, size, etc.) while the purpose of explorative-thinking actions is to discover a functional relationship between elements of a situation based on presented sensory-perceptual information.

Cognitive psychology collected data that demonstrates the unity of motor and cognitive actions. For example, Kosslyn (1973) and Cooper and Shepard (1973) measured the time for manipulating mental images. They found that the time for mental rotation of an image of an object was similar to the time for actual external rotation of that object. According to SSAT, if an individual intentionally turns a mental image of an object to a position according to a required goal, it is *an imaginative action* or an *object-mental action*. Thinking actions are involved in discovering functional property of objects and can only be uncovered through analysis of the relationship between various elements of a situation. This is an important criterion for distinguishing thinking actions from other types of actions.

The concept of actions is also important for *distinguishing knowledge from skills*. A subject can have a wide repertoire of images, concepts, and propositions (*declarative knowledge*) in his/her mind, and a limited repertoire of mental actions required for utilizing this knowledge. Therefore, it is important to distinguish between procedural knowledge and skills.

Procedural knowledge is associated with mental operations that transform such information as images, concepts, and propositions in our memory. Such knowledge is also utilized when a subject performs physical actions. In order to be able to perform assigned procedures, such *knowledge has to be transformed into a skill through practice by using various actions*. For example, knowledge about complex movements in sports is not sufficient for their performance. For this purpose, athletes need extensive training. When students acquire knowledge about a computer language, it does not mean they are ready to write software. They need to practice to develop a certain level of skill in order to be able to do it. Practice transforms knowledge and unskilled behavior into a skill. Skills are understood as cognitive and motor actions that are consciously automated via learning and training. This represents the *first level of skill*. AT recognizes also the *second level* of skill. It represents a high order of skills that reflects an ability of a subject to apply the first level of skill and knowledge in practice. This second level of skill consists of an individual's ability to organize knowledge and the first level of skill into a system and efficiently use it to perform a particular class of tasks or to solve a class of problems.

In such a system, thinking provides flexibility of activity strategies that provide efficiency of performing a particular class of tasks. Combination of the first and second levels of skill and professional knowledge are the foundation of the formation of professional abilities.

Cognitive actions share a number of features with motor action. They are goal directed, have a beginning and an end, function according to the principle of self-regulation, and so on. Motor actions presuppose existence of material objects with which a subject interacts. Cognitive actions transform information rather than material objects.

The material presented in the first and second chapters clearly shows that the use of cognitive and motor actions as units of analysis does not reject the engineering work measurement method (see, for example, Karger and Bayha, 1977) but rather adapts such data for its use in the contemporary task analysis.

2.4 Task Classification and Design

Any work process can be presented as a number of different tasks that have a specific organization. Task classification is an important factor in studying productivity and efficiency of human performance. According to AT, there is a threefold level of activity regulation:

1. Level of stereotypy or automaticity of performance
2. Level of consciousness of activity regulation in terms of acquired rules and familiar strategies
3. Level of regulation of activity based on general knowledge, principles, and heuristic strategies

According to SSAT, these levels of activity regulation are the theoretical basis for task classification. The level first listed above is equivalent to the *skill-based task* performance. The second level of activity regulation covers all tasks that can be *classified as algorithmic or rule-based tasks*. The third level of activity regulation corresponds to *non-algorithmic tasks* and includes problem-solving aspects. This terminology can be compared with Rasmussen's (1983) notions of knowledge-based behavior, rule-based behavior, and skill-based behavior.

According to the threefold level of activity regulation, skill-based tasks, algorithmic or rule-based tasks, and non-algorithmic or problem-solving tasks should be identified. This theoretical analysis helps us to present tasks' taxonomy as it is depicted in Figure 2.6.

This figure shows three main groups of tasks: skill-based tasks; algorithmic- or rule-based tasks; and non-algorithmic or problem-solving tasks. Heuristic tasks are the most complex ones. Skill-based tasks are performed in a rapid

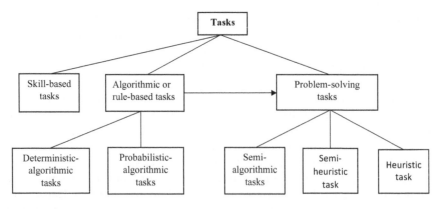

FIGURE 2.6
Tasks' taxonomy in SSAT.

way with minimal concentration. They consist of automatically performed cognitive and behavioral actions, can be performed easily, quickly, and economically with a high level of efficiency and a low level of awareness and concentration of attention (Bedny and Meister, 1999). However, if something happens, the awareness of performed actions as well as the level of attention can be sharply increased. The most representative type of skill-based tasks are simple production operations performed on a production line.

Algorithmic or rule-based tasks are performed according to some logic and rules. They can be divided into two types. The first is deterministic-algorithmic tasks and the second one is probabilistic-algorithmic tasks. The first category of tasks involves making simple "if-then" decisions (logical conditions in algorithmic description of task) based on familiar perceptual signals and relatively simple memorized rules. Each decision usually has two possible outputs. For example, if an even number is displayed, press the red button; if an uneven number is displayed, press the green button. Here, a subject perceives a stimulus and based on the stored-in-memory rules, selects the corresponding button. Decisions in most cases are performed based on perceptual data that is presented at a particular stage of task performance.

The probabilistic-algorithmic tasks are more complex and involve decisions (logical conditions) with three or more outputs, each of which has a different probability of occurrence. Such decisions usually are made based on the analysis of the situation and include thinking operations. A subject knows how to act in particular situations or, in other words, she/he possesses the knowledge that is required to make decisions. The complexity of such decisions also varies depending on the quantity of possible outcomes. All of this can make rule-based tasks extremely complex because they are probabilistic-algorithmic tasks. If algorithmic tasks are correctly described, they have clearly defined rules and a logic of actions to be performed. This should guarantee successful performance if subjects have the necessary knowledge and skills and follow the given instructions. Probabilistic outputs significantly increase an operator's memory workload and the complexity of tasks. An inability to remember some rules that are required for performance of probabilistic-algorithmic tasks and insufficient knowledge about probabilistic characteristics of such tasks can lead to such tasks falling into the category of non-algorithmic tasks (see dashed line in Figure 2.6).

If a task requires conscious consideration of how to accomplish its goal, and a subject does not know how to complete the task, it is a problem-solving task. In the work environment, such tasks include some unfamiliar components of situations. In science, such unfamiliarity is significantly increased. According to Landa (1976), this class can be further divided into three subgroups: semi-algorithmic, semi-heuristic, and heuristic. We recommend the following main criteria for classification of these tasks (a) indeterminacy of initial data; (b) indeterminacy of a goal of a task; (c) redundant and unnecessary data for task performance to consider; (d) difficulty of task performance derived from contradictions between task conditions and task complexity;

(e) time restrictions in task performance; (f) specifics of instructions and their ability to describe adequate performance and restrictions; (g) adequacy of subject's past experience for task requirements.

The tasks that contain some uncertainty and, therefore, a performer has to utilize not just the existing rules but also must create his/her own cognitive and motor actions in order to achieve a goal, are called semi-algorithmic tasks. If uncertainty is even greater and a task requires coming up with some independent solutions, without precise criteria, such tasks are semi-heuristic. This type of task often asks for the explorative components of activity. In order to reduce the degree of subjective and objective uncertainty during performance of problem-solving tasks the efficient performance methods of such tasks should be designed. Purely creative tasks belong to science, art, or situations when engineers need to find an original solution for technical problems. The main criteria for such tasks are undefined field of solution, indeterminacy of initial data, and indeterminacy of the goal of a task.

In our classification, the last two types of tasks are non-algorithmic tasks. Algorithmic tasks are the tasks that can be described as a logically organized system of cognitive and behavioral actions. They can be presented as a combination of standardized verbal and symbolic descriptions in the human algorithm. The method of description of such algorithms has been developed in the framework of SSAT (Bedny, 2015).

With development of the computer industry, games became important, even for adults. Therefore, not just learning and work but also play are essential types of activity. The purpose of playing is not in achieving some useful result of activity but rather the activity process itself. The study of tasks for the computer game industry became an important area of task analysis. According to some scientists, this HCI field shifts its focus from a production environment with clearly defined structure of tasks and their goals to a nonproductive field, where the main purpose is communication, engaging, educating, or playing a game, and the need for such concepts as a task, goal of a task, and so on is eliminated. According to Karrat, et al. (2004, p. 588), these fields don't involve task-oriented activity. Moreover, these scientists wrote that "the science of enjoyment is not capable of defining a goal-directed approach". We would first like to mention that there is no such science as "science of enjoyment". The concepts of task and goal are still important in the considered field. Cognitive and behavioral actions remain the main units of analysis of such tasks (Bedny and Karwowski, 2011; Bedny, Karwowski and Bedny, 2011). Basic principles of the study of tasks in a nonproduction environment are the same as in the traditional fields.

The relevance of a subject's past experience to the task requirements demonstrates that the distinction between all the described types of tasks is relative, not absolute. Let us consider as an example a skill-based task. When a mathematics teacher presents a task to her/his students, for them it a task-problem. However, for a teacher, this is a rule-based task. The rule-based task can sometimes become semi-heuristic or even heuristic for a subject if he/she does not know all the rules for performance of the task and does not possess the required skills to perform it. Here we present the advanced description and classification of tasks according to SSAT.

It is very important to know the relationship between rule-based or algorithmic tasks and problem-solving tasks in the work environment. As we already discussed, there are production- and operational-monitoring processes. In most cases, tasks that should be performed by a subject are rule based or algorithmic tasks. All of them are performed according to existing instructions and prescriptions. However, unexpected deviation from normal conditions can emerge during their performance. In such cases, subjects usually recognize this situation, interpret it, and formulate the problem-solving task. After resolving the problem, the subject returns to performance of the rule-based tasks. This, of course, is a simplified representation of rule-based and problem-solving tasks. For example, a pilot performs various rule-based or algorithmic tasks during flight that can be of various levels of complexity. However, when a pilot detects deviation of actual parameters of the flight from the intended ones, he/she often has to solve a task-problem which can vary in complexity. Usually these are semi-algorithmic or semi-heuristic tasks. Purely heuristic tasks are more specific for science, development of new technologies, and so on.

According to Wickens, Gordon and Liu (1998), cognitive components that are critical for decision-making are selective attention, activities performed within working memory, and information retrieval from long-term memory. It is obvious that these factors influence the complexity of the decision-making. However, the concept of thinking is not considered when analyzing such tasks. These authors describe a number of steps that are involved in performing heuristic tasks, but the thinking process is not discussed. For example, the first step is called "cue reception and integration." This step is defined as a number of cues or pieces of information that are received from the environment and then transferred into working memory. However, in our opinion, even this first step includes the thinking process that plays a leading role in selecting pieces of information and their structuring and organization in memory. We discuss this aspect of thinking when considering the concept of "gnostic dynamic". Similarly, thinking plays a critical role in hypothesis generation, their evaluation and selection, and other steps involved in performing heuristic tasks. It is strange that such leading specialists do not mention thinking in their textbooks *Engineering psychology and human performance* and *Human factors engineering* (Wickens and Holland, 2000; Wickens, Gordon and Liu, 1998). Exaggerating the importance of memory and ignoring the thinking processes are characteristic of the studies in ergonomics and engineering psychology. Thinking should not be reduced to decision-making. Thinking includes decision-making, but a problem-solving situation often requires its analysis and formation of decision-making only as the next step. A goal of a problem-solving task is often formulated by a subject independently. When a problem-solving task is formulated or accepted by a subject, a number of other steps for solving the problem are required. At the first stage, a subject selects the elements most adequate for the solution, performs their structuring, transforms the situation in various ways, and makes the situation more adequate for resolution. At each of these stages, it is necessary to identify the functional relationship between elements of the situation. Only at the final stage does a subject make a decision. At this stage, formation of alternatives and selection of an adequate one is performed by a subject. Decision-making is only one stage of a problem-solving task. Of course, the stage of decision-making also involves the thinking process.

Classification of tasks is needed for task analysis and design. Usually task analysis techniques are divided into two main groups: observation and experiment. The task observation can be carried out on a real system or on a physical model created for this purpose. The direct observation of a real system does not involve interference into system functioning. Experiments, on the other hand, interfere with real systems, but there is some restriction in doing so in the work environment. The direct observation does not use any analytical models of human behavior. Such models are critically important for psychological aspects of task analysis in economics or human factors.

The experiments strive to find out all existing circumstances in a specific task and then measure all the most important characteristics of activity taking place under those circumstances. They often involve the use of physical models that are created for the experiment. There are many circumstances one might like to investigate. There may be an infinite number of circumstances, and most of them are unknown when a new system or new efficient method of performance is at the design stage. Therefore, the type of activity that should be an object of study in an experiment might not be clearly specified. In order to overcome the drawbacks of this main method, one should use *analytical models of activity*. This is a totally new type of model whose purpose is to describe an activity structure. Activity models during performance of various tasks help us to eliminate drawbacks of traditional methods such as observation and experiment, and make it possible to enhance the design process. However, this approach does not reject observation and experiments which should be used as supplementary methods in the design process. Moreover, SSAT offers new methods of qualitative and quantitative analysis and some new techniques of utilizing experimental procedures, combining them with analytical methods (Bedny, 2015).

Task classification is needed primarily for understanding what to study in a particular situation. At the next step, we consider studying the efficiency of human performance. Here, it is necessary to understand the concept of design. The term design derives from engineering and presently there is no precise understanding of what design is outside of this field. The term "design" is used in many different ways and often incorrectly. *Design* should be considered as a process of creation of models of non-existing objects, in accordance with previously set properties and characteristics, with the ultimate goal of materializing these objects (Neumin, 1984). Design is defined similarly by Suh (1990). An example of such models are drawings. At the first step, a qualitative analysis is conducted that follows formalized methods. At the second stage, not only drawings, but also various mathematical, algorithmic, and other models of design can be utilized. Usually creation of one object requires development of a number of different interdependent models. After models of a designed object are created, a prototype of an object is developed and experimental evaluation of a prototype is performed. Thus, design can be presented as the following stages:

1. *preliminary qualitative analysis that includes some intuitive methods of analysis;*
2. *analytical methods that involve creation of designing models and formalized procedures;*
3. *creation of a prototype;*
4. *experimental evaluation of a prototype and introduction of required corrections in drawing or other models of a designed object.*

Only after a new prototype is developed can a cycle of design be repeated. A design is completed when the final design models are created and accepted, and a designed object is going to be created based on the accepted prototype. Then the designed object would be introduced into the production process. In ergonomics, analytical stages of design are eliminated and the design process is reduced to experimental procedures, and such procedures are not theoretically verified. Experimental procedures without analytical design models supporting them can often be inadequate for the purpose of creation of a given object. It is also important to understand that every designed object that is introduced into the production process has special documentation including various analytical models of this object. For example, in manufacturing, the design of an object is always performed according to existing requirements and documented through drawings, mathematical models, algorithmic descriptions, technological descriptions, and so on. Any innovations are conveyed by changes in the documentation. Moreover, any changes in production are preceded by changes in the documentation. This means that the proposed changes should be evaluated analytically beforehand. Software design is also documented by using requirements, flow charts, screen designs, and process simulations. All the changes can be found in existing documentation.

These principles of design are not used outside of SSAT. However, evaluation of productivity, efficiency of production processes, and of human performance in general can be performed using analytical models. Only then, not just ergonomists but also economists, can predict efficiency of a new method of task performance. Therefore, economists, the same as other professionals, can use such models of human performance for their purposes. Analytical models are especially important when studying productivity and efficiency of new task performance. Models of designed objects should be distinguished from models of cognitive processes. The latter helps to understand how cognition works. Design models describe designed objects in a standardized manner. Finally, we want to emphasize that models of activity when performing non-algorithmic problem-solving tasks can be created only if they are transformed into algorithmic tasks or mathematical models of human activity. Although a great deal of effort has been spent by various scientists trying to create such models, in most cases they were not applicable.

Symbolic (verbal, graphical, mathematical) models play a leading role in the design process. The mathematical models include various equations and an algorithmic description of objects. The types of design models and principles of their development should be standardized, and such models should be described using standardized language of description because the design process involves various specialists that have to interpret them adequately. For example, in manufacturing,ˋ models are created by designers (for example, drawings) and this documentation is used by production engineers and workers in the production process. The workers do not always have direct contact with designers (Bedny and Meister, 1997; Bedny, 2015). It is not accidental that ergonomists and psychologists often use only experimentation working with physical models in order to redesigned the objects. However, design is first of all an analytical process, and experiments are usually utilized at the final stage of design for evaluation of a prototype which was developed based on design models.

Task analysis is the focal step in any design, including design of equipment, software, efficient methods of job performance, and so on. Job analysis includes identification of all essential tasks within an existing system, or in a considered production process, accompanied by short and informal descriptions of all tasks, and further creation of activity models for each task listed in the job description. These models depict the structure of activity during performance of the considered task. We are going to demonstrate that one of the basic principles of design is comparison of the activity structure with the structure of a material system. Therefore, description of activity structure is the main purpose of creating human activity models. The technical components of the system affect the structure of activity. Hence, SSAT's basic principle of design is the analysis of activity structure and comparison of the obtained data with the structure of technical components of the system. Changes in the configuration of equipment or interfaces change the structure of activity in a probabilistic manner. Therefore, analysis of the human activity structure is an important tool for the design of means of work or equipment.

Task analysis starts with *qualitative analysis of task performance*. It includes *objectively logical analysis, sociocultural analysis, individually psychological analysis, and functional analysis* (Bedny and Karwowski, 2007). The first method includes a traditional verbal description of the task performance. This stage usually involves analysis of prototypes or any other data that can be useful in the design process. The sociocultural analysis takes its roots in the sociocultural theory developed by Vygotsky (1978). This method focuses on analysis of such components of a task as a subject, an object, internal or mental tools, and the relationship of those components with external tools of activity along with an analysis of the sociocultural context in which a task is performed. The individual-psychological method of study focuses on analysis of the individual style of task performance (Bedny and Seglin, 1999a; Bedny, 2015). The main purpose of such study is analysis of preferable strategies of task performance by individuals with different features of personality and comparison of such strategies with standard requirements of task performance. This data is also instrumental for developing training methods and professional selection. Analysis of activity self-regulation is the basic principle of studying strategies of task performance. SSAT offers models of activity self-regulation (Bedny, Karwowski and Bedny, 2015). Advanced models of self-regulation of activity will be described in this book.

Following the qualitative analysis, the formalized methods of task analysis are allied. The original psychological approach to the morphological analysis of activity is suggested in SSAT for this purpose. The main purpose of this stage of analysis is description of activity structure. *Morphological analysis includes* algorithmic description of activity structure with subsequent development of activity time structure. *Algorithmic analysis of activity* involves description of logical organization of cognitive and behavioral actions. Such algorithms are called human algorithms. This analysis also takes into account the probabilistic features of activity structure and should be considered as the logical sequence of activity elements, with their duration, and a possibility of its elements being performed simultaneously or sequentially. Activity is a structure that unfold in time as a process. Hence, only after developing the time structure of an activity is it possible to utilize the quantitative method of task complexity evaluation which is the main quantitative method of task evaluation in SSAT. All stages of task analysis can be presented in the following way: *Qualitative descriptive analysis → algorithmic analysis → analysis of time structure → quantitative analysis.* Each following stage might require reconsideration of the preliminary stages. When analyzing simple tasks, some stages of task analysis can be omitted. Each stage includes different levels of activity analysis or different level of detail. At each stage of analysis, standardized language of description is used. Thus, an approach developed in SSAT is based on the systemic principles of analysis and design.

Development of various models of activity for a specific task is an analytical approach that allows conducting task analysis even when a real task does not exist yet. Creative problem-solving tasks cannot be designed if they cannot be transferred into rule-based tasks. This type of task assumes that a subject is faced with a new, unknown situation, when there is a need for an immediate solution to an ongoing problem. After making a decision, the subject returns to the normal state of affairs and implementation of a standard mode of a task's performance. Naturally, some components of a task-problem can be performed utilizing standardized and prescribed rules. For example, a pilot performs a set of operationally related tasks during the flight. Sometimes an unexpected event can arise during the flight. A pilot does not have the ready-prepared rules to resolve the task-problem that suddenly appears. For him/her, this is an unknown situation. If this is an experienced pilot, then he/she can solve the problem based on her/his background and intuition, and then return to the performance of the ongoing rule-based tasks. The experience obtained through this process of solving the problem can be taken into account in the future. Special rules for solving the problem in such situations should be documented. Existing methods of analysis of problem-solving or decision-making tasks can be used to transfer the intuitive process of task performance into the normatively prescribed one. This means that models of activity during such task performance can be developed as a follow-up. Then, a problem-solving task is transformed into an algorithmic task. However, when a pilot is faced with such a situation for the first time, and there are no required rules, this situation should be considered as a non-algorithmic problem-solving task, or a decision-making task with problem-solving components.

The task concept can be described only as a very confusing one in the studies of HCI in the nonproduction environment. This is a relatively new, wide area of ergonomics that includes recreational and nonproduction design. The objects of such design are games, recreation activity, education, communication, and so on. Emotionally motivational aspects of task analysis are especially relevant for such areas of activity. Here, such concepts as task, goal, and emotionally motivational aspects of activity regulation are critical factors in task analysis and design. Some scientists use the term "affective design" where the pleasure-based principle of task analysis is the key for criterion of successful design (Helander, 2001). Application of emotionally motivational aspects of affective design requires clear understanding of such concepts as human goal, task, relationship between goal and motives, cognitive and behavioral actions, and so on. Moreover, there are object-oriented and subject-oriented activities. The latter implies that a subject not only interacts with means of activity but also interacts with other subjects (Bedny and Meister, 1997; Bedny and Karwowski, 2007). There are also such types of activity classification as play, learning, and work. Play and games are types of activity that are central in the study of HCI in the nonproduction environment. The purpose of play cannot be reduced to achieving a useful result but rather concentrates on the activity process itself.

At the same time, this factor does not eliminate goal acceptance or goal formation as a stage of activity and its motivational aspects. For example, when a little girl plays with a doll she performs conscious goal-directed actions and finishes feeding a doll when she may decide that feeding is completed according to some rules established by herself. These rules can be formulated in an imaginative form and function as a goal of a task. A child at play operates not only with various images that have the specific meaning for her/him but also uses a verbally logical meaning.

It is no coincidence that children when they play alone can talk to themselves. A child can turn the chair around and place legs on either side of the chair, use a cover of a pot as a car steering wheel and imagine that he/she is driving a car. All these actions are meaningful for a child. Play is associated with pleasure that induces motivational components of activity. For the game, the same as at work, a task can be formulated in advance by others or independently. With the development of the computer industry for a non-production environment, games became important even for adults. We have developed the self-regulation concept of motivation (Bedny and Karwowski, 2006a) that is imperative for analysis of tasks that involve games. According to this concept, there are five stages of motivation: (1) preconscious; (2) goal related; (3) task evaluative; (4) executive or process related, and (5) result related. These stages are the result of the activity self-regulative process, and depending on the specificity of the task, some stages might be more important than others. The process-related stage of motivation is particularly important for development of game and other recreational computer-based tasks. For example, when an operator performs monotonous work, process-related motivation (stage 4) is very low, and it contains negative emotionally motivational components. In such a situation, commitment to the goal (stage 2) and result-related motivation (stage 5) should be positive in order to overcome the negative motivation components. The process-related stage (stage 4) is a critical factor in computer-based games and should be associated with positive motivation. A simple game without the risk of losing reduces positive aspects of the process-related stage of motivation. This is especially relevant for risk takers and gamblers. In a game, a goal of a task is often not clearly defined and is presented in a rather general form. Only at the final stage of the game does the goal become clear and specific. This means that a goal cannot be understood as a clearly defined end state of the system. In contrast, Karat et al. (2004) insists that only in a production environment does a task have a clearly intended purpose or goal. According to these authors, one cannot say that people try to reach a pleasurable state that can be considered as a goal of the game task, meaning that a particular task does not have a goal. Here, we can see that the authors mix the goal of the task with the motives. However, the goal is to reach the desired future result of the game. It is critically important to create sub-goals when designing games (Bedny and Karwowski, 2007; Pushkin, 1978; Newell and Simon, 1972). According to SSAT, a sub-goal is a cognitive and conscious entity. There is also a need for a general motivational state that creates an inducing force to produce these sub-goals. Comparison of the future hypothetical end state with an existing state is provided by the feedback that is accomplished in the mental plane. This demonstrates that thinking is a self-regulative process. A subject promotes hypotheses, formulates hypothetical goals, and evaluates them mentally or practically. Karat et al. (2004, p. 588) also wrote: "[T]he science of enjoyment is not capable to define a goal-directed approach". However, there is no evidence of such science existing. Inadequate theoretical data in the psychology of motivation leads to such erroneous analyses of tasks in nonproduction environment by some HCI professionals.

In conclusion, we can say that goals should not be ignored in game-related or in any other type of task. The goal of a task is a cognitive component of activity, and motives are the energetic component. The vector *"motives → goal"* induces all elements of a task and organizes them around its goal. A goal in its psychological meaning is not an end state of the system. It is a mental representation of the desired future result that can be specified and clarified during task performance. There are also goals of separate actions that can be often quickly forgotten by a subject during task performance. The attempts to eliminate such concepts as task and goal, and to ignore the relationship between energetic and informational components of activity are not productive. This phenomenon can be explained by the fact that in cognitive psychology and theories of motivation, these concepts are not clearly defined.

2.5 Study of Production Process in the Building Construction Industry Using Technological Units of Analysis

As has been mentioned above, two types of units of analysis of activity are distinguished in SSAT. Technological and psychological units of analysis and their relationship are important when analyzing the structure of work activity. In engineering and ergonomics, elements of human work usually are described using a common language or technical terminology. "Turn two-positioning switch up", "take reading from a digital indicator" are examples of such descriptions, and in SSAT these are examples of describing human actions using technological units of analysis. Such descriptions of activity elements do not have clearly defined psychological characteristics, and for more precise descriptions of human actions, technological units of analysis should be transferred into psychological units of analysis. For example, such action as *"taking a reading from a digital indicator"* (technological units of analysis) can be also described as *"a simultaneous perceptual action"* that takes 0.3 seconds to perform, which gives a better understanding of what is involved in completing this action.

A combination of such units of analysis allows description of the activity of any level of complexity with high precision, often using purely analytical methods, which is especially important when actual performance cannot be observed. Moreover, developed documentation can be transferred to other professionals who implement the considered task in production without observing real task performance. Using both the technological and psychological units of analysis is quite complicated, and in some cases it is sufficient to use only technological units of analysis for the description of task performance. The technological units of analysis can be larger than the psychological units of analysis. Utilizing these units of analysis is more effective when performance of the production process can be observed and when this process does not include complex cognitive activity.

In this chapter, we demonstrate using only technological units of analysis for the description of human performance. As an object of study, the difficult and dangerous physical work associated with the reconstruction of an industrial building has been chosen. Specifically, in this chapter we present the original method of study of the production process in the building construction industry. In our study of the existing systems, observation methods that are tightly connected with the time study are utilized.

One of the reasons for selecting the technological units of analysis in our studies is that it is often impossible to use detailed methods of analysis in the building construction industry. Thus, the observation method should be specific and facilitate the purpose of observing the specific production process. Specifically, a dangerous and physically demanding job with operationally related tasks of reconstructing industrial buildings has been selected for our analysis. Operationally related tasks are involved in the stage of the production process essential for accomplishing a specific technological purpose. Hence, the entire production process is divided into stages that have technological completeness. The proposed method is used to improve existing methods of performance. In this case, direct observation, chronometric analysis, questionnaire, and an enlarged technological units of analysis should be utilized. The physical components are the dominant ones for this kind of work. However, if it is necessary, analysis of the cognitive components of work can be also studied by utilizing the above-described methods. The specific feature of the analyzed operationally related tasks is the lack of stable workplaces, the variety of tasks performed in dangerous conditions, and the limited repeatability of those tasks. One of the maim hazards of these tasks is that they are performed over 26 feet above ground level. The work has been well compensated, which provoked workers to ignore safety procedures to increase productivity and receive more money. Team performance methods, when workers coordinate their performance with others, is especially important for the construction industry. Our study revealed that chronometrical methods of study are not well adapted for analysis of task performance in the building construction industry. These methods overload the attention of the observers and make it difficult to record data when it is required to observe several workers' performance at the same time.

Gathering data for individual workers and determining the possibility of coordination of their performance is not clearly defined by the existing methods of study. So, a new graphical data recording method has been proposed for chronometrical analysis in the construction industry. This method allows overcoming the disadvantages of chronometrical analysis and gives an opportunity to utilize the quantitative method of analysis for evaluation of the efficiency of performance. The proposed method is as follows.

Special paper with a millimeter grid and the horizontal lines has been used as an observation form (see Figure 2.7).

Each row is used to register the performance time of various elements of a task by an individual worker. The performance time of separate elements of work is market along the X-axis. The Y-axis is used for denoting individual workers. As can be seen, there are five workers that are involved in this study (see Figure 2.7). Each line has been divided into 10-minute intervals. The length of each interval was 20 mm. This means that two grids along the X-axis designate one minute. The precision of registration using this method of recording of chronometrical data is 0.5 minutes. This allowed taking the precise chronometrical measurement in the building construction environment.

The performance time of the considered element of work with its corresponding number is depicted for each worker in Figure 2.7. The description of the work elements is presented under the figure. Several specialists had to work together to conduct the chronometrical study. Some of them measured performance time; the others registered the measurements utilizing the tables that have been especially prepared for this study. The fragment of the chronometrical observation within the three-hour interval (the first and the third hours of the shift are shown as an example) is presented in Figure 2.7 Personal interviews of the workers has been conducted at the end of the shift. The qualitative analysis of the production process concentrated on the causes

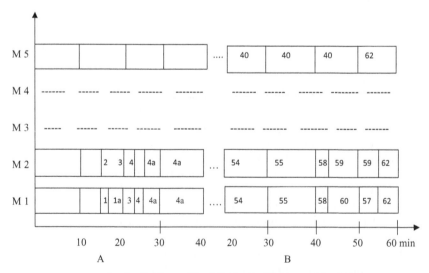

A – the first hour of the shift; B – the third hour of the shift

FIGURE 2.7
Allocation of various elements of work in time.

of ideal time and safety issues. During the interview, special attention has been paid to the critical points of the work process. At the following stage of analysis, the data obtained in the chronometrical study has been compared with the data collected by interviewing the workers. Representatives from the engineering department that were responsible for development of the considered work process were also involved in the observation. They conducted their own observation, and their opinion has been taken into account in the further analysis of the considered fragment of the work process.

Figure 2.7 shows that some workers were not ready to start working at the beginning of the shift. Specifically, worker number 5 was absent for the first 40 minutes of the shift. Subsequently, a standardized classification of all types of work with the following determination of duration of these types of work has been conducted. Chronometry measurements at the first stage and classification of types of work at the following stage allow collection of data during the industrial building construction work process.

We have also developed a special classification of various work types that can be used for qualitative and quantitative analysis of the obtained data. The following work types have been identified: (1) the main work process when a worker is directly involved in the task performance; (2) ancillary work – preparation stage for performing the main work process; (3) nonproductive work – a worker performs unnecessary elements of work that can be excluded from the task performance, because he/she is not sufficiently skilled or violates technological restriction.; (4) unnecessary waiting periods or downtime that can be avoided; (5) performance of tasks accompanied by violation of safety requirements; (6) additional main work that depends on the specificity of conditions in which reconstruction is performed in comparison to the conventional forms of construction; (7) additional supplementary work that depends on the specificity of conditions in which reconstruction is performed in comparison to conventional forms of construction.

When a group of workers performs operationally related tasks together, chronometrical data is usually collected for each worker separately. At the next step, the performance time of specific types of work is determined for all workers involved in the team performance individually and for the whole team. This helps to determine the efficiency of team performance and the efficiency of individual workers. Reconstruction of industrial buildings is very dangerous and difficult work. It often involves ancillary work types unlike the similar types of work that involve construction of a new industrial building. Such types of work should be designed in such a way that methods of work in reconstruction conditions are similar to the processes utilized during new building construction. This is the reason for selecting the above-described work types numbers 6 and 7. The production process can be considered as an optimal one when ancillary work, unnecessary waiting periods, and nonproductive types of work are reduced to a minimum, and violations of safety requirements are eliminated.

After determining the time spent on each type of work, it is possible to convert qualitative data into quantitative data by utilizing a set of specially designed coefficients. It is important to notice that for this purpose we use only technological units of analysis. Psychological units of analysis are not utilized here. In this study we utilized, classified according to psychological principles, members of algorithm such as cognitive and behavioral actions and their elements. The following coefficients are offered for a quantitative evaluation of the production process (operationally related tasks).

The fraction of time spent performing the main work processed in the total work time is determined by the following coefficient:

$$\Delta T_{mw} = T_{mw} / T_g, \tag{2.1}$$

where:
T_{mw} is time spent performing the main types of work
T_g is duration of the shift.

The fraction of the additional main work during the reconstruction in comparison to the conventional construction is defined according to the formula:

$$\Delta T_{mw}^{ad} = T_{ad\,mw} / T_{mw}, \tag{2.2}$$

where:
$T_{ad\,mw}$ is time of additional main work in reconstruction conditions
T_{mw} is time of main work in conventional conditions.

The fraction of the ancillary work time in the total work time is:

$$\Delta T_{anc} = T_{anc} / T_g, \tag{2.3}$$

where:
T_{anc} time of ancillary work
T_g is general work time.

The fraction of the ancillary work in reconstruction conditions in comparison to ancillary work in conventional construction conditions is defined according to the formula:

$$\Delta T_{anc}^{ad} = T_{anc}^{ad} / T_{anc}, \tag{2.4}$$

where:
T_{anc}^{ad} time of additional ancillary work in during the reconstruction
T_{anc} time of ancillary work in conventional conditions.

The fraction of nonproductive work in the total amount of work is

$$\Delta T_{n-pr} = T_{n-pr} / T_g, \tag{2.5}$$

where:
T_{n-pr} performance time of nonproduction work
T_g is the duration of the shift.

The fraction of time involved in the violation of safety requirements in the total time of work is defined as:

$$\Delta T_{sv-r} = T_{sv-r} / T_g, \tag{2.6}$$

where T_{sv-r} – time, during which violations of safe requirements has occurred.
The measure of the safety of work is defined as:

$$M_{s-r} = 1 - \Delta T_{sv-r} \tag{2.7}$$

The fraction of unnecessary waiting periods or downtime that can be avoided in efficient work conditions can be determined as:

$$\Delta T_{un-wait} = T_{un-wait} / T_g \tag{2.8}$$

The principle of development of the above-discussed coefficients is based on the classification of various types of work according to the above-considered criteria. Such analysis helps to determine how efficiently the work process is organized in the construction industry. The use of these coefficients is especially effective when there is a large amount of data. The coefficients can be calculated for each worker individually, or for the whole team.

Analysis of the obtained data and their comparison allows getting detailed information about the labor organization. For example, a low value of ΔT_{anc} (ancillary work) is a positive factor. However, in this situation large values of coefficients as ΔT_{n-pr} (nonproductive work) and ΔT_{sv-r} (violation of safety requirements) can be used as evidence that reduction of ancillary work is achieved through violation of technological and safety requirements. In one case of our studies in the reconstruction conditions, it has been discovered that ΔT_{mw}^{ad} (time of additional main work in reconstruction conditions) and ΔT_{anc}^{ad} (time of additional ancillary work in the reconstruction conditions) are equal to zero. However, comparison of these coefficients with ΔT_{sv-r}, T_{s-r} and also ΔT_{n-pr} demonstrated that this led to violation of safety and technological requirements.

Analysis of these coefficients for individual workers helped to evaluate the effectiveness of each of them. Below we present as an example quantitative measures of assessment of the efficiency of the production process that evaluated team performance:

$\Delta T_{mw} = 0.32; \Delta T_{mw}^{ad} = 0; \Delta T_{anc} = 0.26; \Delta T_{anc}^{ad} = 0; \Delta T_{n-pr} = 0.37; \Delta T_{un\text{-}wait} = 0.05;$

$\Delta T_{sv-r} = 0.51; M_{s-r} = 0.49.$

The data obtained show that the fraction of T_{mw} (main work) is only 0.32, which is very low for the fraction of the main productive work. Workers also spent little time on ancillary work (ΔT_{anc}), which led to an increase in nonproductive work (ΔT_{n-pr}). The causes of such phenomena are easy to understand. Ancillary work should be at the optimal level according to the specificity of performed tasks when all necessary preparations for the main work should be completed.

As a note, $\Delta T_{mw} + \Delta T_{mw}^{ad} + \Delta T_{anc} + \Delta T_{anc}^{ad} + \Delta T_{n-pr} + \Delta T_{un\text{-}wait} = 1$ representing all the components of the considered production process. Safety violations occurred during the performance of all above-listed components of this process.

Both ΔT_{mw}^{ad} and ΔT_{anc}^{ad} equal zero, and it could be considered as a positive factor. However, comparison of this data with safety requirements demonstrates that $\Delta T_{sv-r} = 51$ and $M_{s-r} = 0.49$. This means that 51% of the shift time workers violated the safety requirements. It can be explained by the fact that scaffoldings, platforms, and other safety equipment were practically absent. Their use in the production process is time-consuming, especially when workers perform their tasks at high altitudes. Productivity and compensation are in conflict with safety requirements, and the workers are motivated to ignore them. All team members that were involved in the reconstruction received very high salaries. ΔT_{mw}^{ad} and ΔT_{anc}^{ad} required preparation time in order to create safe work conditions. Despite the fact that all work tasks had to be performed at over 26 feet above the ground, the workers made little use of the safety equipment.

The material presented demonstrates that in order to evaluate the efficiency and safety of the work process, all coefficients obtained should be considered and compared.

The above-suggested new method of analysis of work processes in the building construction industry is based on a graphical method of recording chronometrical data and their classification according the developed criteria. The graphical data recording method does not overload the attention of specialists, and in some cases, it allows them to conduct observation of several workers at the same time. This method can be used for analysis of the efficiency of individual workers and of the team performance. In this study, technological units of analysis are used. When it is necessary, psychological units of analysis can be utilized for the more detailed analysis of task elements.

For example, in this study we had to select the best safety belt for the workers who worked at over 26 feet above the ground. For this purpose, workers have to put on different safety belt, and close and open the lock. All their actions have been filmed, and at the next stage motor actions have been described using the MTM-1 system as described above. Based on such study, the best safety belt has been recommended.

References

Anderson, J. R. (1993a). Problem solving and learning. *American Psychologist*, 48(1), 35–44.

Anderson, J. R. (1993b). *Rules of the Mind*. Hillsdale, NJ: Lawrence Erlbaum Associates, Publishers.

Bedny, G. Z. (1981). *The Psychological Aspects of a Timed Study during Vocational Training*. Moscow, Russia: Higher Education Publishers.

Bedny, G. Z. (1987). *The Psychological Foundations of Analyzing and Designing Work Processes*. Kiev, Ukraine: Higher Education Publishers.

Bedny, I. S. (2004). General characteristics of human reliability in system of human and computer. *Science and Education*, Odessa, Ukraine, (8–9), 58–61.

Bedny, G. Z. (2006). Activity theory. In Karwowski, W. (Ed.). *International Encyclopedia of Ergonomics and Human Factor*. Volume 1 (pp. 571–576). Boca Raton, FL and London, UK: CRC and Taylor & Francis.

Bedny, G. Z. (2015). *Application of Systemic-Structural Activity Theory to Design and Training*. Boca Raton, FL and London, UK: CRC Press and Taylor and Frances.

Bedny, G. Z. and Chebykin, O. Ya. (2013). Application the basic terminology in activity theory. *IIE Transactions on Occupational Ergonomics and Human Factors*, 1(1), 82–92.

Bedny, G. Z. and Harris, S. (2005). The systemic-structural activity theory: Application to the study of human work. *Mind, Culture, and Activity: An International Journal*, 12(2), 128–147.

Bedny, G.Z, and Harris, S. R. (2008). "Working sphere/engagement" and the concept of task in activity theory. *Interacting with Computers. The Interdisciplinary Journal of HCI*, 20(2), 251–255.

Bedny, G. Z. and Karwowski, W. (2004). Activity theory as a basis for the study of work. *Ergonomics*, 47(2), 134–153.

Bedny, G. Z., Karwowski, W. (2006a). The self-regulation concept of motivation at work, *Theoretical Issues in Ergonomics Science*, Vol. 4, 413–436.

Bedny, G., Karwowski, W. (2006b). General and systemic-structural activity theory. In W. Karwowski (Ed.) *International Encyclopedia of Ergonomics and Human Factors*. London, UK, Taylor and Francis, Vol 3, pp. 3159–3167.

Bedny, G. Z. and Karwowski, W. (2007). *A Systemic-Structural Theory of Activity. Application to Human Performance and Work Design*. Boca Raton, FL and London, UK: CRC and Taylor & Francis.

Bedny, G. Z. and Karwowski, W. (2008a). Application of systemic-structural theory of activity to design and management of work systems. In Gasparski, W. W. and Airaksinen, T. (Eds.). *Praxiology and the Philosophy of Technology. The International Annual of Practical Philosophy and Methodology*. Volume 15 (pp. 97–144). New Brunswick, NJ, and London, UK: Transaction Publishers.

Bedny, G. Z. and Karwowski, W. (2008b). Activity theory: Comparative analysis of Eastern and Western Approaches. In Chebykin, O. Y., Karwowski, G. Z. Bedny and Karwowski, W. (Eds) *Ergonomics and Psychology. Development in Theory and Practice* (pp. 221–246). London, UK, and New York, NY: Taylor & Francis.

Bedny, G. Z. and Karwowski, W. (2011). Introduction to applied and systemic-structural activity theory. In Bedny, G. Z. and Karwowski, W. (Eds.) *Human-Computer Interaction and Operators' Performance. Optimization of Work Design with Activity Theory* (pp. 3–30). Boca Raton, FL and London, UK: CRC Press and Taylor & Francis.

Bedny, G. Z., Karwowski, W., Voskoboynikov, F. (2011). The relationship between external and internal aspects in activity theory and its importance in the study of human work. In G. Z. Bedny and W. Karwowski (Eds.) *Human-Computer Interaction and Operators' Performance. Optimization of Work Design with Activity Theory.* (pp. 31-62). Boca Raton, FL: CRC Press.

Bedny, G., Meister, D. (1997). *The Russian Theory of Activity: Current Application to Design and Learning.* Mahwah, NJ: Lawrence Erlbaum Associates.

Bedny, G., Meister, D. (1999) Theory of activity and situation awareness. *International Journal of Cognitive Ergonomics.* Vol 3(1), pp. 63–72.

Bedny G. and Seglin M. (1999a). Individual style of activity and adaptation to standard performance requirement. *Human Performance*, 12(1), 59–78.

Bedny, G. and Seglin, M. (1999b). Individual features of personality in former Soviet Union. *Journal of Research in Personality*, 33(4), 546–563.

Bedny, I. S. and Sengupta, T. (2005). The study of computer based tasks. *Science and Education*, Odessa, Ukraine, 1–2(7–8), 82–84.

Bedny, G. Z., Karwowski, W. and Bedny, I. S. (2011). The concept of task for non-production human-computer interaction environment. In Kaber, D. B. and Boy, G. (Eds.). *Advances in Cognitive Ergonomics* (pp. 663–672). Boca Raton, FL and London, UK: CRC Press and Taylor & Francis.

Bedny, G. Z., Karwowski, W. and Bedny, I. (2015). *Applying Systemic-Structural Activity Theory to Design of Human-Computer Interaction Systems.* Boca Raton, FL and London, UK: CRC Press and Taylor and Frances.

Bedny, G., Karwowski, W. and Sengupta, T. (2006). Application of systemic-structural activity theory to design of human-computer interaction tasks. In Karwowski, W. (Ed.). *International Encyclopedia of Ergonomics and Human Factor.* Volume 1 (pp. 1272–1286-576). Boca Raton, FL and London, UK: CRC and Taylor & Francis.

Bedny, G. Z., Karwowski, W. and Sengupta, T. (2008). Application of systemic-structural theory of activity in the development of predictive models of user performance. *International Journal of Human–Computer Interaction*, 24(3), 239–274.

Brushlinsky, A. V. (1979). *Thinking and Forecasting.* Moscow, Russia: Thinking Press.

Brushlinsky, A. V. (1999). Subject-oriented concept of activity and theory of functional system, *Questions of Psychology*, (5), 110–121.

Chebykin, O. Y., Bedny, G. Z. and Karwowski, W. (Eds.). (2008). *Ergonomics and Psychology. Developments in Theory and Practice.* Boca Raton, FL and London, UK: CRC Press and Taylor & Francis.

Cooper, C. S. and Shepard, R. N. (1973). Chronometric studies of the rotation of mental images. In Chase, W. G. (Ed.), *Visual Information Processing.* New York, NY: Academy Press.

Diaper, D. (2004). Understanding task analysis for human-computer interaction. In Diaper, D. and Stanton, N. (Eds.). *The Handbook of Task Analysis for Human-Computer Interaction* (pp. 5–47). Mahwah, NJ: Lawrence Erlbaum Associates, Publishers.

Diaper, D. and Lindgaard, G., (2008). West meets East: Adapting activity theory for HCI & CSCW applications? *Interacting with Computers. Interdisciplinary Journal of Human-Computer Interaction*, 20(2), 240–286.

Diaper, D. and Stanton, N. A. (2004). Wishing on a sTAr: The future of the task analysis. In Diaper, D. and Stanton, N. (Eds.) *The Handbook of Task Analysis for Human-Computer Interaction* (pp. 585–602). Mahwah, NJ: Lawrence Erlbaum Associates Publishers.

Frese, M. and Zapf, D. (1994). Action as a core of work psychology: A German approach. In Triadis, H. C., Dunnette, M. D. and Hough, L. M. (Eds.). *Handbook of Industrial and Organizational Psychology* (pp. 271–340). Palo Alto, CA: Consulting Psychologists Press.

Helander, M. G. (2001). Theories and methods in affective human factors design. In Smith, M. J., Salvendy, G., Harris, D. and Koubeck, R. J. (Eds.). *Usability Evaluation and Interface Design. Proceedings of HCI 2001*, Volume 1 (pp. 357–361). Mahwah, NJ: Lawrence Erlbaum Associates.

Karat, J., Karat, C. M. and Vergo J. (2004). Experiences people value: The new frontier for task analysis. In Diaper, D. and Stanton, N. (Eds.), *The Handbook of Task Analysis for Human-Computer Interaction*. Mahwah, NJ: Lawrence Erlbaum Associates.

Karger, D. W. and Bayha, F. H. (1977). *Engineering Work Measurement* (3rd ed.). New York, NY: Industrial Press.

Kleinback, U., and Schmidt, K. H. (1990). The translation of work motivation into performance. In Kleinback, V., Quast, H.-H., Thierry, H. and Hacker, H. (Eds.), *Work Motivation* (pp. 27–40). Hillsdale, NJ: Lawrence Erlbaum Associates, Inc.

Klimov, E. A. (1969). *Individual Style of Activity*. Kazan: Kazahnsky State University Press.

Klochko, V. E. (1978). Goal formation and dynamic of evaluation of problem solving tasks. Ph.D. dissertation, Moscow, Russia: Moscow University.

Konopkin, O. A. (1980). *Psychological Mechanisms of Regulation of Activity*. Moscow, Russia: Science.

Kosslyn, S. M. (1973). Scanning visual images: some structural implications. *Perception and Psychophysics*, 14, pp. 90–94.

Kotik, M. A. (1974). *Self-Regulation and Reliability of Operator*. Tallinn, Estonia: Valgus.

Kotik, M. A. (1978). *Textbook of Engineering Psychology*. Tallinn, Estonia: Valgus.

Landa. L. M. (1976). *Instructional Regulation and Control: Cybernetics, Algorithmization and Heuristic in Education*. Englewood Cliffs, NJ: Educational Technology Publication. (English translation).

Landa, L. M. (1984). Algo-heuristic theory of performance, learning and instruction: Subject, problems, principles. *Contemporary Educational Psychology*, 9, 235–245.

Leont'ev, A. N. (1978). *Activity, Consciousness and Personality*. Englewood Glifts, NJ: Prentice Hall.

Leont'ev, A. N. (1977). *Activity, Consciousness, Personality*. Moscow, Russia: Political Publishers.

Locke, E. A. and Latham, G. P. (1990). Work motivation: The high performance cycle. In Kleinbeck, V., et al. (Eds.), *Work Motivation* (pp. 3–26). Hillsdale, NJ: Lawrence Erlbaum Associates.

Neumin, Y. G. (1984). *Models in Science and Technic.* Leningrad: Science Publishers.

Newell, A. and Simon, H. A. (1972). *Human Problem Solving.* Englewood Cliffs, NJ: Prentice-Hall.

Pervin, L. A. (1989). Goal concepts, themes, issues and questions. In Pervin, L. A. (Ed.), *Goal Concepts in Personality and Social Psychology* (pp. 173–180). Hillsdale, NJ: Lawrence-Elbaum Associates.

Platonov, K. K., (1970). *Problems of Work Psychology.* Moscow, Russia: Medicine Publishers.

Ponomarenko, V. A. and Zavalova, N. D. 1994. *Applied Psychology. Problems of Safety in Aviation.* Moscow, Russia: Science Publishers.

Pushkin, V. V. (1978). Construction of situational concepts in activity structure. In Smirnov, A. A. (Ed.). *Problem of General and Educational Psychology* (pp.106–120). Moscow, Russia: Pedagogy.

Pushkin, V. N. and Nersesyan, L. S. (1972). *Psychology of the Railroad.* Moscow, Russia: Transportation.

Rassmussen, J. (1983). Skills, rules, and knowledge; signals, signs. And symbols, and other distinctions in human performance models. *IEE Transactions on System, Man and Cybernetics, SMC*-15, 234–243.

Rasmussen, J. and Pejtersen, A. (1995). Virtual ecology of work. In Flach, J., Hancock, P., Caird, J., and Vicente, K. J. (Eds.), *Global Perspectives on the Ecology of Human-Mashine Systems* (pp.121–156). Hillsdale, NJ: Lawrence Erlbaum Associates.

Rubinshtein's (1922/1986). Principles of creative independent activity. (Reprinted from the first publication). *Questions of Psychology*, 4, 101–107.

Rubinshtein, S. L. (1957). *Existence and Consciousness.* Moscow, Russia: Academy of Science.

Rubinshtein, S. L. (1958). *About Thinking and Methods of Its Development.* Moscow, Russia: Academic Science.

Schedrovitsky, G. P. (1995), *Selective Works.* Moscow, Russia: Cultural Publisher.

Sengupta, T., Bedny, I. S., and Karwowski, W. (2008). Study of computer based tasks during kill acquisition process, *2nd International Conference on Applied Ergonomics Jointly with 11th International Conference on Human Aspects of Advanced Manufacturing* (July14–17, 2008), Las Vegas, NV.

Sengupta, T., Bedny, I., and Bedny, G. (2011). Microgenetic principles in study of computer-basewd tasks. In G. Z. Bedny, W. Karwowski (Eds.) *Human-Computer Interaction and Operators' Performance. Optimizing Work Design with Activity Theory* (pp. 117–148). Boca Raton, FL and London, UK: CRC Press and Taylor & Francis.

Strelkov, Y. K. (2007). Operationally- semantic structure of professional experience. In Bodrov, V. A., *Psychological Foundations of Professional Activity* (pp. 261–268). Moscow, Russia: Logos Publisher.

Suh, N. P. (1990). *The Principles of Design.* New York, NY, Oxford, UK: Oxford University Press.

Tolman, E. C. (1932). *Purposive Behavior in Animals and Men.* New York, NY: Century.

Vicente, K. J. (1999). *Cognitive Work Analysis: Toward Safe, Productive, and Healthy Computer-Based Work.* Mahwah, NJ: Lawrence Erlbaum Associates, Publishers.

Vygotsky, L. S. (1971). *The Psychology of Arts.* Cambridge, MA: MIT Press.

Vygotsky, L. S. (1978). *Mind in Society. The Development of Higher Psychological Processes.* Cambridge, MA: Harvard University Press.

Wickens, C. D. and Hollands, J. G. (2000). *Engineering Psychology and Human Performance* (3rd Ed.). New York, NY: Harper-Collins.

Wickens, C. D., Gordon, S. E. and Liu, Y. (1998). *An Introduction to Human Factors Engineering.* Authors, Christopher D. *Wickens*, Sallie E. *Gordon*, Yili *Liu.*, Upper Saddle River, NJ: Longman Publisher.

Zarakovsky, G. M. (1966). *Psychophysiological Analysis of Work Activity. Logical-Probability Approach.* Moscow, Russia: Science.

Zarakovsky, G. M., and Pavlov, V. V. (1987). *Laws of Functioning Man-Machine Systems.* Moscow, Russia: Soviet Radio.

Zarakovsky, G. M., Korolev, B. A., Medvedev, V. I., and Shlaen, P. Y. (1974). *Introduction to Ergonomics.* Moscow, Russia: Soviet Radio.

Zinchenko, V. P. (1978). Functional structure of executive perceptual-motor actions. In Zinchenko, V. P. and Munipov, V. M. (Eds.), *Study of Cognitive and Executive Actions.* Volume 16 (pp. 3–40). VNITE.

3

Concept of Activity as Self-Regulative System and Its Role in Studying the Pilot's Performance

3.1 Concepts of Orienting Activity and Situation Awareness: Comparative Analysis

In systemic-structural activity theory (SSAT), cognition is considered as a process (cognitive approach) and as a system of cognitive actions and operations (activity approach). Cognition also can be considered from the functional perspective as a goal-directed, self-regulative system, where various stages of self-regulation are presented as functional mechanisms or function blocks (Bedny, Karwowski and Bedny, 2015). This system integrates cognitive, behavioral, and emotionally motivated components. Studying activity self-regulation is also called functional analysis of activity. This analysis is not concentrated on separate cognitive processes. Instead, it considers cognitive processes in their unity at various stages of activity regulation, because cognitive processes are interrelated and studying them in isolation during task analysis is to some extent artificial. Therefore, the cognitive analytical approach should be combined with integrative functional analysis when cognitive processes are considered in their interrelation with other cognitive processes. When studying activity as a self-regulative system, function mechanisms or function blocks are utilized as main units of analysis. The model of self-regulation consists of a number of different function blocks that interact with each other through feed-forward and feedback connections. Each function block integrates various cognitive processes in a particular way, depending on the specificity of activity regulation at the considered stage of performance. A function block can also be described in terms of mental or behavioral actions. Thus, we can define a function block as a coordinated system of sub-functions that has a particular purpose in activity regulation. For example, there is a function block that is responsible for creation of a goal of activity during task performance. The function block *subjectively relevant task conditions* is involved in creation of a mental model of the situation. Each block in the model has a rigorous scientific justification. The self-regulation model can be viewed as an information processing model that has a loop-structured organization. When analyzing a particular task, a specialist should select the most relevant function blocks in order to discover and describe the most preferable strategies of task performance.

A model of self-regulation can be interpreted as an interdependent system of windows (function blocks) from which a specialist can observe task performance strategies at various viewpoints. For example, a specialist can open the window called "Goal" and describe such stages of activity as goal interpretation, goal acceptance, goal formation, and so on based on existing theoretical data in SSAT about the goal analysis and description. Then, another function block, for example, "*subjectively relevant task conditions*" can be opened. At this stage of analysis, a specialist would study such aspects of activity as formation of "*operative image of situation*" and "*situation awareness*" and their interrelation based on theoretical data offered by SSAT. Similarly, some other function blocks can be selectively considered. At the following stage of analysis, the data obtained should be analyzed, engaging the feed-forward and feedback interconnection between the considered blocks. Therefore, the model of activity self-regulation developed in SSAT is a powerful tool for systemic analysis of strategies of task performance. Thus, each function block determines the range of issues that are connected with this block and should be examined in activity regulation during analysis of the considered task performance.

Stages of activity in a simplified manner can be presented as follows:

> 1) *reflection of a situation* → 2) *final decision about a situation* → 3) *decision about performance* → 4) *performance.*

This schema demonstrates that there are two types of activity. The first type of activity precedes real execution and includes only the first two stages. The purpose of this activity is to receive information about the situation and interpret this information. This is *orienting activity*. In general, activity includes all four stages. So there are two models of activity self-regulation in SSAT (Bedny, Karwowski and Bedny, 2015). One of them is called "*model of self-regulation of orienting activity*" and the other is "*general model of activity self-regulation.*" The last model also includes a decision about performance and performance. Thus, orienting activity is the stage of activity that precedes decision-making and associated with it executive cognitive and behavioral actions. Orienting activity can also be an independent type of activity that does not include the real executive stage of task performance. This type of activity, based on its purpose, is to some extent similar to the stage of task performance that is defined as situation awareness (Endsley and Jones, 2012). It is often the case that the orientation in a situation is more important than the task execution. In cognitive psychology, the orientation stage of activity is known as situation awareness (SA) (Endsley and Jones, 2012).

SA is recognized by cognitive psychology as an important theoretical concept that describes human activity before the actual execution of an action. SA is the same as the orienting activity that precedes the executive stage of activity, and therefore they both have relevance to the study of human performance. Let us consider the relationship between the concept of SA in cognitive psychology and the concept of self-regulation of orienting activity in SSAT from a theoretical perspective. The concept of SA has been introduced by Endsley (1995, p. 36), who defined it as "the perception of the elements in the environment within a span of time and space, the comprehension of their meaning and the projection of their status in the near future". In a simpler way, the author defined SA as knowing what is going on around you. There are some other definitions of SA, but these are the most recognized ones as they are given by Endsley. This concept entails the perception of critical factors in the environment (Level I); understanding those phenomena, or how people combine, interpret, store, and retain information (Level II), and understanding what can happen with a system in a given situation in the near future (Level III). It is obvious that this concept is especially important for the analysis of a complex task when an operator has to evaluate a lot of information in dangerous and time-restricted conditions.

We will now consider SA from the SSAT perspective, which will facilitate a better understanding of this concept and make it more useful for practical application. SA is treated as the operator's internal mental model of the state of the environment in a considered situation.

Unfortunately, the model of SA described by Endsley is not logically consistent. For example, her model depicts SA as a separate box which includes three overlapping boxes. These boxes interact with other boxes outside of SA, such as "decision," "performance of actions," "information processing mechanisms," "long-term memory," "automaticity," "ability, training," and some others. However, the selection of these boxes and their relationship are not scientifically justified. All of these boxes or stages suggest involvement of various psychic processes without which none of these stages can function. Depending on the specifics of the task being performed, the content of the psychic processes that are involved in each box changes. Therefore, the box labeled *Information Processing Mechanism* in this model cannot be presented as a separate mechanism. Moreover, we cannot find in this model such processes as thinking, working memory, emotionally motivational mechanisms, and so on.

Feedback in this model is introduced only after the stage of execution. However, the internal mental feedback during mental exploration of the situation is critically important in SA. Also, the goal, actions, tasks, and other basic psychological concepts are not clearly defined, and some of them are even explained incorrectly.

We disagree with the statement that a dynamic mental model of a situation is constructed based only on perception. Various cognitive processes that interact with each other are involved in creation of such model.

One can speak about the dominance of one cognitive process over the other. Operative thinking and working memory play a special role in the creation of a mental model. Imagination is also a critical factor in the creation of such a model. An analysis of interaction between elements of the situation cannot be reduced to considering just the perceptual mechanisms. This is, first of all, the result of the operative thinking that involves imaginative, verbal-logical mechanisms, interaction with working memory, and emotionally motivational processes. All cognitive processes are tightly interconnected during every task performance, and SA is, first of all, the result of integration of these processes. Cognitive processes perform reflective, regulative, and evaluative functions. Reflection can be considered as a mental representation of reality. The human ability to consciously and unconsciously reflect a state of the environment has been widely studied in psychology (Konopkin, 1980; Kotik, 1978; Kahneman and Tversky, 1984, and others). Thus, the *conscious reflection of the situation* has exactly the same meaning in AT as SA has in cognitive psychology. SA ignores unconscious components of reflection of the situation.

The concept of goal is important for both SA analysis and for the orienting activity, but in AT this concept is understood in a completely different way. Endsley and Jones (2012, pp. 68–69), in their section "Goals versus tasks" wrote:

> The GDTA goals seek to document cognitive demands rather than physical tasks. Physical tasks are things the operator must physically accomplish such as filling out a report or calling a coworker. Cognitive demands are the activities that require expenditure of higher-order cognitive resources, such as predicting the enemy's course of action (COA) or determining the effects of an enemy's (COA) on battle outcome. Performance of a task is not a goal, because tasks are technology dependent. A particular goal may be accomplished by means of different tasks depending on the system involved.

According to SSAT, a goal is not a cognitive demand that is specific only for cognitive tasks. A goal also exists during the performance of tasks when physical or motor actions are dominating components. An achievement of the goal of a task can be less or more demanding.

A goal of a task is not a cognitive demand but a desired future result of a subject's own activity that is connected with motives. It includes cognitive, imaginative, and conscious components. There is a goal of a task, a goal of actions, a goal of a subtask, and so on. A goal is dynamic and can have various personal interpretations. It can be reformulated by a subject independently in a more or less precise manner. A goal cannot be considered as a cognitive demand, because only a task that is either complex or simple presents various cognitive demands for its execution. A goal can be presented by instructions, can be formulated and reformulated, and can be accepted and rejected by a subject.

From the SSAT perspective, a task analysis always includes consideration of a task's goal and associated with it other goals that are formed during a task's performance. A goal of a task can be presented to a subject in a ready form as an objective requirement. The key point is that the goal can be interpreted correctly or can be misinterpreted, and it can be accepted or rejected by a subject. It is well known that a person can reformulate the goal based on her/his subjective preferences. It can also be formulated by a subject independently. The question is how adequate such a goal is to the specific situation. All of these factors also depend on the *emotionally evaluative* and *motivational components* of activity. A goal does not exist without motives. It is a cognitive component, whereas motivation is an energetic component of activity. The goal of a task manifests itself as a desired future result that should be achieved by performing the task. Goal and motive create a vector "motive → goal" which gives direction to activity. As has been mentioned in Chapter 2, it's a psychological vector that has certain properties that allow comparing it with vectors in physics.

There are requirements and conditions for every task. Anything that is presented to an operator or known by her/him about a task constitutes the task conditions. Requirements are what should be achieved by the task performance. When requirements are accepted by a subject, they become the subject's personal goal. There are goals of a task and also goals of cognitive and behavioral actions. A goal is a conscious component of activity that is connected with motives. Motive is an energetic, whereas goal is a cognitive, component of activity. Physical or cognitive human activity is goal directed. Therefore, we cannot agree with Endsley and Jones' assertion that a physical task does not have cognitive demands and goals. These authors have a section in their book titled "Goals versus tasks." However, this opposition is unacceptable from the SSAT perspective. The goal of a task is a requirement that is accepted, interpreted in a specific way, or formulated independently and should be achieved during a task performance. The same situation can produce a totally different SA depending on acceptance and/or formulation/reformulation of a goal by a subject. In the field of study known as SA, there is no clear understanding of the concept of action. We can only infer that the term action is used for explaining the stage of performance.

SA drives decision-making and performance (Endsley and Jones, 2012, p. 11).

SA is described as a system of operations in perception and memory that enables a person to formulate a hypothesis about a current and future state of a situation. After that, he/she makes a decision and specifies a required course of actions. Thus, SA does not include the decision-making and performance. We will show below that decision-making, as well as cognitive actions and explorative motor actions, are important components of orienting activity that precedes the real executive stage of activity. Therefore, cognitive and behavioral actions are basic elements of activity that are involved in formation of SA (Bedny, 2015; Bedny, Karwowski and Bedny, 2015;

Bedny, Karwowski and Jeng, 2004). Cognitive actions are involved in mental manipulation of information according to the goal of a task to be performed. Operators do not choose a ready-made response from memory, and a response cannot be viewed as a reaction to stimuli. An operator actively forms a program of motor actions, adapts it to the changing situation, and, based on this program, performs required motor actions. Motor and cognitive actions are performed based on principles of self-regulation. Feedback can be used at various stages of the action execution.

According to the SA model, a motor action can be evaluated only after it is completed, and therefore an error can be corrected only after it has occurred. Anticipation of errors and prevention of them in the course of actions is virtually impossible here. Immediate feedback during performance of the motor actions is not considered, either. The SA model ignores the ability of a person to perform cognitive actions and evaluate the result by using mental feedback. Hence, mental feedback is important after performance of not only motor actions but the cognitive actions as well. Mental feedback is especially important at the stage of formation of SA. Explorative activity, which can be not just mental but also external or motor, plays an important role in SA.

Endsley (2000, p. 149) wrote: "[C]ompletely different tasks may be carried out to perform the same goal in two systems." A goal is a conscious future result of activity and it cannot be "performed." It can or cannot be achieved. If the goal is the same, but conditions of a task are different, then these are different tasks that are performed by a subject. Although the task can be the same (the same goal and conditions of the task), the strategies of a task performance can differ. Endsley and Jones (2012) utilize the term "mental model" in an unusual manner. According to them, a mental model helps to develop a situation model or SA. They describe an example of a mental model of a flight as follows: "[A] pilot develops not only a mental model of how the aircraft operates, including its many subsystems and its aerodynamic performance in the physical environment, but also a mental model of a flight operations including air traffic control (ATC) procedures and expected behavior associated with interacting with ATC and with other pilots." However, this is not a mental model. It is a *conceptual* model of a flight. For example, when a pilot is about to fly from New York to London, he/she develops the conceptual model of the flight. Such a model includes a general understanding of possible tasks that should be performed during the flight, including understanding of environmental conditions, the specifics of the functioning of technical components of the aircraft, and so on. A mental model is task-specific, and it includes SA (Ponomarenko and Bedny, 2011).

Endsley and Jones (2012) studied the SA requirements for various types of unmanned vehicle operations. As an example, they presented special tables for the SA requirements for the task that involved operating the unmanned vehicles. These tables include approximately 180 factors about the equipment and environment (Endsley and Jones, 2012, p. 224–225). However, these are not the SA requirements yet, because such a list of factors tells us very little about

SA for a particular task performance. The SA analysis cannot be reduced to the collection of such unrelated requirements, which have meaning only in the context of specific tasks. According to SSAT, SA is associated with certain stages of a task performance. First, tasks which are to be performed by an operator should be identified, and then factors that can influence SA when performing a particular task should be listed. Some factors can be unimportant because the awareness of them during the task performance does not create any doubts. However, other factors related to the equipment and to the environment can be important for SA; they should be taken into account in the context of specific tasks.

In this schema, *Situation awareness → Decision-making → Performance*, the decision-making stage is outside of SA. According to AT, the individual assesses the situation by making a series of explorative cognitive and motor actions. This is a gnostic or orienting activity and various stages of its performance should be evaluated, and an appropriate decision about the situation should be made (Bedny and Meister, 1999).

This kind of decision is not included in the performance or execution that transforms the externally presented situation. Thus, decision-making can be included into the stage of orientation. So, with some approximation, activity should be broken down into three stages:

Orientating or Gnostic Activity → Executive Activity → Performance Evaluation

In this schema, the executive stage of activity means performance that is involved in the transformation of an external situation according to the goal of activity. Based on orienting activity, people develop a subjective model of reality, which should be adequate to the goal of activity. If a goal of orienting activity changes, then a mental model or mental representation of the situation also changes. In the above-presented schema, orienting activity is independent from executive activity. The distortion of the orientating components results in misunderstanding of the situation and causes a ripple effect on all other stages of activity. In some situations, the first stage of orientating or gnostic activity can be the main component of operators' performance. For example, an operator serves primarily as a monitor, which requires vigilance. An orienting or gnostic stage of activity is dynamic and can change in an operator's mind over time in spite of the constancy of the external situation. Continual change of the situation in the subject's mind in light of its external constancy is called *gnostic dynamic* (Pushkin, 1978). This means that the mental model of a situation is dynamic.

The material presented above brings us to the following conclusions.

Currently, the concept of SA receives increasing attention in psychology and ergonomics. There is also interesting data in the studies of distributed SA (Salmon et al., 2008). Despite the popularity of this concept, analysis of the literature on the topic reveals that this construct is not clearly defined. There are different understandings of this construct, and some authors even question its existence (Dekker et al., 2010). We consider SA as one of a number of other functional mechanisms of activity regulation that integrate cognitive processes in various manners that are important for a particular stage of activity regulation. In AT there is a concept of orienting activity, a component of activity, that is involved in reflection of a situation in accordance with a goal of activity. The concept of orienting activity existed long before SA was introduced in ergonomics. Orienting activity is an important psychological concept that can be applied in the studies that are aimed at improving efficiency of performance, and SA just strives to explain one possible mechanism in activity regulation.

3.2 Self-Regulation Model of Orienting Activity in Systemic-Structural Activity Theory

In this chapter, we present the latest version of the model of the self-regulation of orienting activity (activity that precedes the execution stage). As can be seen in Figure 3.1, the model consists of various functional blocks that are dedicated to specific functions in the activity self-regulation.

Comparison of each block with a window through which we look at the same situation allows us to understand how we use the presented model. The model includes conscious and unconscious channels of activity self-regulation. Let us consider the unconscious channel of activity self-regulation (channel 2). The incoming information activates the orienting reflex (block 4). This functional mechanism produces such responses as moving one's eyes or head to the stimulus, altering sensitivity of different sensory organs, a change in the galvanic skin response, a change in blood pressure, and so on. The orienting reflex plays an important role in the functional mechanism of involuntary attention. It produces some electrophysiological changes in brain activity. Even a brief analysis of the isolated functions of the orienting reflex demonstrates the range of issues that can be considered when utilizing this block in the study of task performance.

However, the considered function block does not work in isolation. It interacts with some other function blocks. This raises a further range of issues related to the consideration of this function block. For example, the orienting reflex provides an automatic tuning to the external influences and effects general activation and motivation of a subject, which is depicted by the connection between blocks 4 and 6. Function block 4, "mechanism of orienting reflex", also interacts with block 5 "afferent synthesis". The horizontal arrow between blocks 4 and 5 reflects the main information that influences block 5. There is also some additional (situational) information that also influences block 5. Hence, block 5 can also receive irrelevant stimuli (diagonal arrows). For example, in aviation it can be a non-instrumental information,

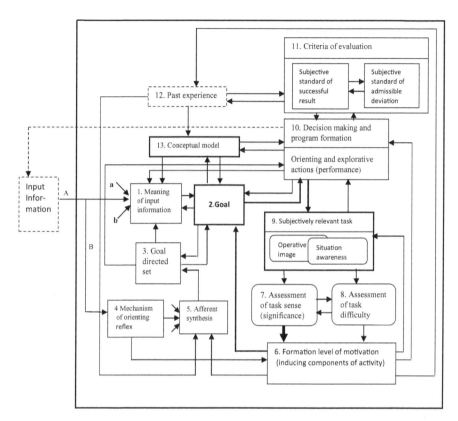

FIGURE 3.1
Model of self-regulation of orienting activity.

or irrelevant environmental stimuli such as noise, vibration, and so on. Thus, this model takes in consideration the fact that the main stimulus or the instrumental information that determines the orienting reflex never exists in isolation. An analysis of the interaction of blocks 4 and 5 demonstrates that instrumental information and non-instrumental information can be combined, and this factor influences functioning of the *afferent synthesis* block. Block 5 (afferent synthesis) is also affected by blocks 6 (motivational block) and block 12 (past experience). Block 5 performs integrative functions combining temporal needs and motivation. It also involves mechanisms of memory that represent the relevant past experience, irrelevant (non-instrumental) information, and the most significant (instrumental) information.

Therefore, a person does not react to some isolated stimulus as it is stated by classical conditioning, and he/she does not perceive information from isolated stimulus that are considered as input from the environment in cognitive psychology. Therefore, thanks to the afferent synthesis, the unconscious channel determines with some accuracy and selects the main stimulus from an infinite variety of stimuli based on a person's temporal need, motivation, and past experience. Selected and integrated in a specific way, information becomes available for the next function block. Afferent synthesis promotes formation of block 3 (goal-directed set). There are two types of sets: general stable system of sets that a person forms as a result of his/her life experience, which is an important personal characteristic of a subject (Uznadze, 1967).

A goal-directed set is more specific. It is formed by instructions and by the specificity of a particular situation. This type of *set* is an internal state of a person that is close to the concept of a goal. However, it is not sufficiently conscious or completely unconscious. Such a set determines the constant and goal-directed character of an unconscious component of activity. A person is not well aware of it in a specific situation. A goal-directed set is an unconscious regulator of activity that preserves the goal-directed tendency in a constantly changing situation without a subject's awareness of a conscious goal. It plays an important role in preserving a dynamic tendency to complete interrupted goal-directed activity at the unconscious level of its regulation. This type of set influences the conscious meaningful interpretation of information (block 1) and goal (block 2). This is how the unconscious channel effects the conscious channel of information processing.

The unconscious set also influences block 10 (decision-making and program formation). If, for example, a goal-directed set is inadequate, interpretation of a situation can also be incorrect. As a result, associated with function block 10 explorative actions can occur without a purpose or a goal and the actions become chaotic. This situation has been revealed in an emergency and dangerous conditions when an operator performed undesirably and not fully consciously recognized explorative cognitive and behavioral actions. As a result, an operator might lose time in the time-restricted conditions.

It is important to point out that a set (block 3) can affect block 1 (meaning) when block 2 (image-goal) is not activated. In such conditions, meaning in block 1 is primarily non-verbalized. Pushkin (1978) called such meaning *situational concept of thinking*. According to the functional model of activity regulation (Figure 3.1), it can be formed under the influence of a set and associated unconscious operations, and not through conscious explorative cognitive and behavioral actions. Not sufficiently conscious or unconscious meaning is an important mechanism of creating non-verbalized hypothesis in emergency situations.

The unconscious channel of self-regulation can function in parallel with the conscious channel and can modify its goal-directed set depending on the specificity of functioning of the conscious channel of self-regulation. Therefore, a certain amount of information processing related to the specifics of the task performance can go through the unconscious channel. This channel has supplementary function in the task performance. Its importance depends on the specificity of task and the past experience of a performer. This channel enables noticing when it is required to modify some aspects of information processing in the main conscious channel. For example, an operator might notice some dangerous aspects of the task performance that at the first stage emerged as some ambiguous feelings. Thus, unconscious processing can be partly with some modification made available to the conscious information processing.

A goal includes verbalized components and should be conscious. A goal has a high hierarchical position in orienting activity. A subject does not receive a goal in a ready form because the requirements that are given to a subject though instruction are interpreted by this subject and accepted as a goal or rejected. This can produce variations in interpretation of a goal and its acceptance. Different individuals may have an entirely different understanding of the goal when identical requirements or instructions are given. A subject receives requirements based on instruction, interprets them, accepts them in his/her own subjective way, or rejects them. All these aspects of activity should be taken into consideration when function block 2 (goal) is the object of analysis in the goal formation process. Such understanding of a goal is totally different from its consideration in cognitive psychology. In SSAT, a goal is always associated with some stage of activity that includes interoperating requirements, accepting them, formulating a goal, and so on. Analysis of these stages of activity is associated with consideration of function block 2, which is totally ignored in cognitive psychology. For example, when an experimenter gives instructions to the subjects in experimental conditions, he/she expects that all subjects receive these instructions and strictly follow them. However, even at the stage of goal acceptance, subjects can interpret the goal in different ways and modify it according to such interpretation. A goal also influences other function blocks of activity regulation. For example, it interacts with "conceptual model" (block 13) and "decision-making and program formation" (block 10). A goal is a boss in the process of self-regulation. All other blocks are subordinates of the goal and in various ways coordinate their functions in order to achieve the goal of self-regulated process. This is another aspect of analysis of the goal (block 2).

It is important to pay special attention to the interaction of a goal (block 2) with the goal-directed set (block 3). This interaction facilitates the fast involuntary goal formation process. Not just the goal-directed set can be transformed into the conscious goal but also the conscious goal can be transformed into the unconscious set (see interaction between block 2 and 3). For example, a subject performs task 1 and suddenly a new situation emerges and a subject formulates new goal of task 2 that is adequate to the new situation. The goal of task 1 does not disappear but is transformed into not sufficiently conscious set. When the second task is completed, the goal-directed set is quickly transformed into the conscious goal (see feed-forward and feedback interaction between block 2 and block 3). Block 2 (goal) cannot exist without interaction with block 6 (formation level of motivation). Interaction of these blocks creates vector "motive → goal". This mechanism gives the whole activity its goal-directed character during task performance. During the unconscious information processing (channel 2), goal (block 2) is not yet activated. Block 1 is involved in actualization of mostly non-verbalized information.

At this stage of information processing, various elements of the situation are not yet integrated into a holistic system or a dynamic mental model.

Formation of a dynamic mental model is a critical stage of functioning of the orienting activity. This becomes possible only after formation of the conscious goal (block 2). Block 9 (subjectively relevant task conditions) is involved in formation of *dynamic mental model of the situation* during task performance. The thinking process plays a critical role in creating and exploring a mental model of a task performed by a subject. According to Bruner (1957), a mental model enables the thinker to go beyond the perceptually available information when solving a particular problem. Creation of an adequate mental model helps with selecting the most efficient course of action for reaching a desired goal without overt trial and error.

Block 9 includes two sub-blocks: verbal-logical block or *situation awareness* and imaginative sub-block or *operative image*. Therefore, not only verbally logical but also imaginative components play a critical role in formation of the mental model of the situation. The imaginative mechanisms of reflection of the situation can be largely unconscious and quickly forgotten. Imaginative and conceptual sub-blocks partly overlap. Only overlapping part of an imaginative subsystem can be conscious. The non-overlapping part of the sub-block "operative image" provides the unconscious dynamic reflection of a situation. Conscious and unconscious components of dynamic reflection of a situation can to some degree convert into each other (Bedny and Meister, 1999). A non-overlapped part of the imaginative reflection of a situation can be viewed as the preconscious reflection. By shifting attention from one sub-block to another within block 9, this type of reflection can be transformed into a conscious reflection. In contrast to dynamic reflection of the situation, *conceptual model* (block 13) is a stable and slowly changing model. It reflects various scenarios of possible situations that are relevant to the particular type of tasks. In contrast to *past experience* (function block 12), the conceptual model is more specific and relevant to a possible task that can be encountered by an operator during performance of their job or duty.

Conscious and unconscious processing of information involves not only cognitive but also emotionally motivational mechanisms. When a main conscious channel is activated, information goes through channel 1 (conscious channel of self-regulation) to block 1 and block 2. Goal influences blocks 13 and 10 and activates conscious decisions. Under the influence of blocks 10 and 6, function block 9 (subjectively relevant task conditions) is activated. As a result, the dynamic mental model of the situation is developed in contrast to cognitive psychology where the mental model is a result of purely cognitive functions. According to the model of the orienting activity, the motivational block 6 plays an important role in development of the mental model and in formation of some other function blocks. Thus, the mental model is a result of interaction of blocks 10 and 9, and motivation block 6. The goal of activity (block 2) plays a critical role in formation of the dynamic mental model (block 9) or the subjectively relevant task conditions. It determines the specificity of the program formation and associated with it explorative cognitive and behavioral actions that are important in development of the dynamic mental model (block 9). The dynamic mental model can be developed differently depending on the accepted or formulated goal of activity. Motivation block 6 also influences the afferent synthesis and therefore the unconscious level of activity regulation.

Function block 7 (assessment of sense of task or its significance) and block 8 (assessment of task's difficulty) play an important role in the formation of motivational block 6. Block *sense* influences the process of selection of information and its meaningful interpretation. Therefore, it can influence the dynamic mental model (block 9) through the motivation block and feedback from it. The more significant a situation is for a subject, the more he/she is motivated to reach the goal of a task (see vector motive → goal). Function block 8 (difficulty) is also a critical factor in activity regulation. From the functional point of view, there are *objective complexity of task* and *subjective assessment of its difficulty*. These is a probabilistic relationship between these two characteristics of a task. The more complex the task is, the higher is the probability that it can be evaluated by a subject as a difficult one. Cognitive task demands during task performance depend on a task's complexity. A subject might underestimate or overestimate the objective complexity of a task. This is important for the analysis of activity regulation. For example, a subject can overestimate a task difficulty and reject it as a result, in spite of the fact that objectively this task is not difficult for this subject. Moreover, if a task is accepted but still perceived as very difficult, a subject can reduce the quality of the task performance. This can be traced when analyzing the interaction of blocks 8, 6, 10, and 11.

Block 11 determines *criteria of evaluation*. This block includes two subblocks: 1) *subjective standard of successful result* and *subjective standard of admissible deviation*. A subjective standard of successful result can significantly deviate from the standard presented through the instructions. The last can be modified by a subject during performance. Such modification can be performed under influence of block 10, which includes decision mechanisms that can correct the functioning of two above-mentioned subblocks. Moreover, a subject can modify the goal (block 2) through interaction of blocks 11, 10, and 2. A subjectively accepted goal can be used as a subjective standard of success. However, a goal often does not match a subjective standard of successful result. For example, a worker can have the goal to produce 100 parts during two hours. If tired, the subject can sacrifice the quality to achieve this goal of activity. The *subjective standard of admissible deviation* is also an important evaluative criterion. The criteria of evaluation of the performance result may vary within the certain limits. This variation is determined by a subjective standard of admissible deviation and can be seen as a range of tolerance for the subjective standard of the successful result.

The functional analysis considers activity as a self-regulative system whose main units are function blocks. Each functional block includes various cognitive processes. This means that each function block can be described in terms of cognitive processes, their interaction, and the importance and specificity of their functioning. These blocks can also be described in terms of cognitive and behavioral actions that are included in the corresponding function blocks. Predictive features of this model can be explained by the fact that at this stage of analysis cognitive processes are considered in the context of the specific mechanisms that are associated with certain stages of activity regulation. This stage of analysis does not exclude studying individual cognitive processes.

In conclusion, we want to emphasize that traditionally the term *conscious verbal reflection of the situation* is utilized in AT instead of the term *situation awareness*. *Imaginative reflection of the situation* is also an important concept in AT. In our model, instead of the term *conscious verbal reflection of the situation*, we used the term *situation awareness*. The last concept plays a critical role in cognitive psychology (Endsley, 1995). However, the presented model of orienting activity demonstrates that not only verbalized awareness of the situation but also imaginative mechanisms play a significant role in reflection of the situation and creation of its dynamic mental model. Moreover, an adequate dynamic reflection of the situation, including current and near future events, cannot be sustained by an isolated mechanism that is called situation awareness. In SSAT, orienting activity that is viewed as a self-regulative system performs this function.

We described briefly, as an example, self-regulation of orienting activity. The general model of self-regulation has also been developed in SSAT. It includes a number of function blocks that describe regulation of the executive component of activity (Bedny, Karwowski and Bedny, 2015).

The model of self-regulation presented in SSAT helps not only to conduct a qualitative stage of task analysis from systemic perspectives but also to critically analyze some basic laws that are widely accepted in cognitive psychology.

The concept of self-regulation in SSAT is totally different from its understanding outside of this theory. For example, Bandura (1977) describes the process of self-regulation as a consequence of such stages as observing, judging, rewarding, and regulating oneself. This is not a correct psychological description of the self-regulation process. There is no clear description of mechanisms of self-regulation and the principles of their interaction. The concept of self-regulation becomes meaningful only when scientifically proved models of self-regulation are developed.

3.3 Examples of Studies of the Pilot's Activity in Emergency Situations from the Functional Analysis Perspective*

In this chapter, the main purpose of study is the complexity of a task and its influence on pilots' performance. The model of orienting activity is utilized for studying these tasks. The participants of the study were 20 test pilots who were 30–40 years old and had over 15 years of experience on the job.

Participants were asked to perform some test flights of varying complexity (Ponomarenko and Karwowski, 2011). All tasks were divided into four groups of complexity. Group 1 was the simplest, and group 4 included the most complex tasks. The tasks were explained to the pilots so they could independently evaluate the upcoming flights beforehand. Such subjective evaluation can be viewed as transformation of objective complexity of the tasks into their subjectively evaluated difficulty.

Before performing each task, pilots' physiological functions were measured. The results of the measurements are presented in Table 3.1. Thus, the purpose of the experiment was to evaluate the impact of the factor of anticipation of the future flight of different complexity on the pilots' physiological state. According to the model of self-regulation, there are function blocks "assessment of task's sense/significance" (block 7), and "assessment of task's difficulty" (block 8). These two blocks influence other functional blocks and especially influence each other. For example, the more difficult the tasks were evaluated by pilots, the higher was the chance that such tasks would become significant for pilots. The significance should be understood as the emotionally evaluative mechanism of task performance. This mechanism can also influence motivation of performance. The more significant the task is for a subject, the more he/she is motivated to perform it. However, in some cases it is possible to observe a situation when the task difficulty is evaluated

TABLE 3.1

The Effect of the Predicted Difficulty and Significance of the Upcoming Flight on the Pilots' Physiological State

	Physiological Indexes					
	Hart Rate (bpm)		Breath Frequency (cycle/min)		Concentration of Sugar in Blood (mg %)	
Difficulty Level	Base Line	In Cabin before the Flight	Baseline	In Cabin before the Flight	Baseline	In Cabin before the Flight
1	64 ± 6	75 ± 5	12 ± 2	12 ± 4	65 ± 3	58 ± 2
2	64 ± 6	84 ± 2	12 ± 2	12 ± 6	65 ± 3	76 ± 4
3	64 ± 6	84 ± 5	12 ± 2	12 ± 4	65 ± 3	92 ± 6
4	64 ± 6	92 ± 4	12 ± 2	14 ± 5	65 ± 3	120 ± 12

so high that pilots are not motivated to perform it or can even reject the idea of performing the task at all.

As it can be seen in Table 3.1, in the considered experiment, all test pilots accepted the tasks with different levels of difficulty. This brief explanation should help readers to understand how the functional model of activity self-regulation can be utilized. The effect of the predicted difficulty and the significance of the upcoming flight on the pilots' physiological state is presented in Table 3.1.

The results of these measurements demonstrate that performing the task of the highest complexity (level 4) manifests itself in the most observable changes in the physiological state of the pilots. This data was statistically significant.

It is important to note that pilots did not verbally express their awareness of the complexity of the situation. Physiological data obtained before the actual flights demonstrated physiological reactions at the stage of orienting activity. Pilots were under the assumption that the measurements would be taken only during the flight.

The study demonstrates that the orienting reflection of the situation includes conscious and unconscious components. Experienced pilots could clearly identify the objective complexity of the future flight and evaluated its subjective difficulty that, in turn, impacted the personal significance of the task. They accepted all levels of flights' complexity and were motivated to conduct them according to the requirements. At the same time, inexperienced pilots' subjective reflection of the flights' complexity might either overestimate or underestimate it. It is also possible for pilots to consciously or unconsciously reflect this kind of situation in an inadequate manner.

Preparedness for a possible emergency situation depends primarily on the psychological mechanisms of activity regulation, which in turn affect the physiological mechanisms of self-regulation. This predisposes pilots' bodies for an unexpected situation. It is also a component of unconscious self-regulation of activity that can help to correctly respond in a given situation. The special training can make such adaptive reactions more useful for acting in possible emergency situations. Therefore, such functional blocks as "past experience" and "conceptual model" are also very important in such situations. The conceptual model of flights is a relatively stable model that changes slowly in time. It, in contrast to past experience, is more specific and helps pilots to adapt to the possible tasks that pilots can encounter during the flight. The task-specific mental model is facilitated mainly by such blocks as "goal" and "subjectively relevant task conditions". The last block includes two sub-blocks. One of them is called "operative image" and the other is "situation awareness". The latter is responsible for the conscious reflection of the situation. The sub-block "operative image" provides conscious and unconscious reflection of the situation. Only the part that is overlapped by "situation awareness" can be conscious (Bedny, Karwowski and Bedny, 2015; Ponomarenko and Bedny, 2011; Makarov and Voskoboynikov, 2011).

The function block "goal" is critical in creation of the dynamic mental model of the task. Thus, an evaluation of the situation is the main purpose of orienting activity. Pilots' orienting activity involves performing various explorative actions, evaluating these actions, making decisions based on such evaluations, and so on. Mental feedback plays a leading role in such activity. However, in more complex situations, pilots can perform even external explorative motor actions for a better understanding of the situation.

All data contained in SA sub-block can be *verbalized*. Data in the sub-block "operative image" can be either *verbalized partially* or *not verbalized* at all because some of its aspects are associated with unconscious processing of information (there are no verbal equivalents), and others can be quickly forgotten. Therefore, a dynamic reflection of a situation cannot be reduced to an interview, questionnaire, or other verbal methods of study.

A mental model is not just a function of memory, because it provides an adequate interaction with the outside world through cognitive actions and operations. Thanks to this, activity becomes object-oriented. Cognitive and behavioral actions connect a subject with the external environment. Actions can be perceptual, thinking, imaginative, mnemonic, and so on (Bedny, 2015), which means that not only mechanisms of memory, and in particular their activation processes, but also various cognitive processes are involved in formation of mental models. Thanks to cognitive and behavioral actions and operations, people actively manipulate operative units of information extracted from memory and units of information allocated in the external environment.

Here, we would like to explain in an abbreviated manner how orienting activity and its mechanism, *situation awareness*, can be studied. The questionnaires and verbal responses should not be the main methods of study of *situation awareness* and orienting activity. An experiment that has a theoretical basis plays a leading role in such studies.

The mental model of the flight is always task-specific and is adequate to the goal of the task. Operative thinking is the most important mechanism in the formation of such a model. Several functions of pilots' mental models that are connected with operative thinking can be distinguished at the orienting activity stage. These functions are to extract the subjectively relevant elements and their structural organization from the situation, and, finally, to diagnose the situation in general. The mental picture of the situation obtained can be modified with time. Therefore, such a model is dynamic, not static, while the goal of the task remains the same. Due to this, pilots can extract diverse relationships between the elements of the same objectively presented situation.

The ability of the pilots to extract the most adequate elements of the situation in conditions of uncertainty in an emergency situation and to develop a holistic mental picture that is adequate to the goal of the task is the first stage of orienting activity. At the next stage the pilot, based on the data obtained, has to forecast the dynamic of future events and attempt to predict how such a dynamic can influence goal attainment. These stages contribute to reducing uncertainty in an emergency situation and to selecting adequate actions at the following stage of activity execution. The factor of uncertainty in an emergency situation is vitally important in orienting activity. This is an especially relevant characteristic of an emergency situation that is accompanied by non-instrumental information. The data presented shows that the methods of pilots' training should include special procedures that promote development of operative thinking and of maintaining its effectiveness in stressful conditions. The development of mental skills should be carried out based on an analysis of possible emergency situations.

As an example, let's consider an incident that occurred on January 15, 2009, and involved US Flight 1549. At 3:27:33 p.m. EST, Captain Chesley Sullenberger called New York Terminal Radar Approach Control and informed the operator that as a result of "hit birds", they lost thrust in both engines. He reported that they were "unable" to turn back to LaGuardia or land in New Jersey's Teterboro Airport. About 90 seconds later, at 3:31 p.m., Capt. Sullenberger made an unpowered water landing in the Hudson River. This very well-known incident was even dramatized in the movie called "Sully." We will consider only some points of this incident investigation that are interesting from our perspective.

The U.S. National Transportation Safety Board (NTSB) used a flight simulator to test the possibility of returning the plane safely to LaGuardia, and only seven of the 13 simulated returns were successful. Furthermore, the NTSB report called this simulation unrealistic. The simulation with an immediate turn made by the pilots did not include the time necessary for an analysis and interpretation of the situation that could have led to a correct decision. In the testimony before the NTSB, Sullenberger maintained that there had been no time to land the plane at any airport. The further simulations that have been conducted, with the pilots' being delayed by 35 seconds, demonstrated that the considered flight would crash.

Investigators utilized an experimental method as an important tool in the investigation. A number of questions about the investigation can be raised that would demonstrate that the simulations conducted at the first stage of the investigation were not realistic:

(1) Why were the pilots fully briefed before they attempted to perform the landing on the simulator? (2) Why did the *experts* not take into account the fact that emergency situations always happen unexpectedly? (3) Why did the *experts* not take into account that the subject had an opportunity to repeat the same simulation multiple times? (4) Why did the *experts* ignore the fact that pilots had to perceive and interpret the situation first, and only after that, could they make the decision about how to act in the situation?

Finally, the NTSB reported that only seven of 13 simulated returns to La Guardia succeeded. However, the board did not provide the information about the sequence of successful and unsuccessful attempts. If this additional information were available, it could explain if the successful attempts were the result of the training, which would negatively affect the true value of this experiment.

Therefore, subjects had to participate in each simulation only once. Information about landing should imitate a real situation, be given to subjects unexpectedly, and the time delay due to the situation interpretation and decision-making should be taken into account in the first few simulations.

Only one *expert* was right, the first pilot, Sullenberger. He is not only an excellent pilot but also holds a master's degree in industrial psychology.

The analyzed situation had to include an emergency situation when both engines unexpectedly failed. All stages of the task performance should be clearly described and the graphical scale of the considered task performance should be developed. Below, in Figure 3.2, we present a simplified version of such a scale, where only two stages are distinguished. The first stage is the orienting activity (T_1), and the second stage depicts the executive activity (T_2).

Experts had information about the time when the birds collided with the airplane. They could determine how much time the pilots had for landing in general. This time, at the first stage of investigation, was accepted as equal to T_2. The NTSB used the flight simulator and discovered that this time was sufficient for the pilots to land the airplane in the airport. However, Figure 3.2 shows that the total time is $T = T_1 + T_2$.

Therefore, at the first stage of the experiment, the NTSB did not take into consideration the fact that only time for executive activity T_2 should be allowed for the simulation. However, the total time T has been given for simulation at the executive stage of landing instead, and the pilots have been fully briefed before they attempted to perform the landing on the simulator, as it is stated in the NTSB report. Only after performing the orienting activity could the crew of the flight 1549 make a decision on how to land. Time T_1 on the above-presented figure shows how much time the pilots spent on receiving information and on its interpretation. This time is necessary for creating the mental model of the situation. Hence, the crew of the flight 1549 had only time T_2 for landing. Hence, it becomes obvious that the airplane could not turn around and fly back to the airport immediately after it hit the birds.

In some cases, it is useful to investigate events that precede an emergency situation. Here, we are talking about studying the pilots' conceptual model of the flight during the specifically defined time period. We would like to repeat that such a mental model is relatively stable and changes slowly in

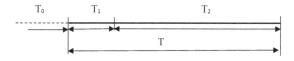

FIGURE 3.2
Time scale of task performance: T_0 – period of time that precedes the incident; T_1 – time for orienting activity; T_2 – time for executive activity.

time. In a pilot's activity, a conceptual model includes some basic data that is stored in a pilot's memory about the flight from one airport to another. In our schema, this stage of activity is presented by a dashed line and designated by T_0. This is only a fragment of the conceptual model of the flight when the plane took off from the airport and before it was struck by birds. Such information can be useful for an investigation. The suggested time scale of the task performance gives only a general idea about the incident investigation from the SSAT viewpoint. Thus, utilizing such concepts as task, mental model of the situation during the task performance, orienting and executive activity, and stages of task performance is helpful for designing the experiments that are utilizing flight simulators and investigating flight incidents in general.

This chapter is prepared together with Ponomarenko, O. A.

3.4 Analysis of Pilot Strategies When Engines Can Fail*

Ponomarenko (2004, 2006) and his colleagues studied pilots' performances in real flights when one out of four motors failed. Together with Ponomarenko, we have considered these experimental data from the concept of activity self-regulation analysis perspectives, when the more preferable and effective strategies of task performance can be discovered.

Information about the state of the engine has been presented by meters and four warning lights. The pilot also received non-instrumental cues, such as aircraft movements, engine sound changes, changes in the engine thrust, and so on. Based on exchanges with the pilot via the radio messages during the flight, it has been discovered that the mental model of the situation has been first formed as the hypothesis that regulated the strategy of researching the situation. Responding to this situation, the pilot observed her/his instruments for somewhere between 0.3–0.8 seconds. This data has been obtained based on the eye movement registration method. However, because of the inherent slowness of the instruments' response, a pilot was unable to detect failure-related information quickly enough. As a result, a pilot often developed an incorrect hypothesis of the situation, and the task and the mental model of the situation have been formulated incorrectly. A pilot started to search for information that would prove his/her incorrect hypothesis of the situation.

Consequently, it has been discovered that the real strategies of searching for information about the failure substantially differed from the strategies that have been given by the formal instructions (see Table 3.2).

This table reveals that in the experimental flights, pilots first pay attention to the non-instrumental signals. Only at the last step do they pay attention to warning-light signals. This significantly reduces the reliability of a pilot's

TABLE 3.2

Comparison of Formal Instructions with Actual Pilot's Performance When One of the Four Engines Has Failed

Stages	Prescribed by Formal Instructions Stages of Task Performance		
	1	2	3
Suggested sequence of gathering information	Increasing turn and deviation of aircraft	Signal light	Reading of engine instruments
Expected stages of task performance	Detect engine failure	Comprehend of the cause of failure	Specify number of the failed engine and perform required actions

Stages	Real stages of task performance			
	1	2	3	4
Actual sequence of gathering information	Non-instrumental signals: feeling of angle of acceleration, changes in effort on controls, etc.	Instrumental readings (during first second after failure)	Analysis not adequate signals	Turning attention to warning lights (30% cases)
Real stages of task performance	Suspect some engine failure	In 70% of the cases the pilot did not receive prove of his/her suspicion and turned attention to other possible causes; in 30% of cases the causes have been confirmed.	Continued search for required information	Received prove and began to act

actions in the accidental conditions. The discrepancy between the real strategy and the strategy formally prescribed by the instructions has been caused by incorrect instructions and the inefficient design of an instrument panel. This results in the incorrect formulation of the task and an inadequate mental model of the situation.

The mental model of the situation developed should be compared with the informational model that is based on reading the instruments and observing the environment.

The pilot develops a hypothesis and a mental model of the situation based on direct sensations. At the next stage, he/she does not search for information in an emergency situation but rather compares the developed mental model of the situation with the received information in order to prove the existing hypothesis and the corresponding mental model. The orienting activity precedes the executive activity.

However, due to the inertia of the instruments, the true state of the engines is not yet displayed and therefore is not presented in the information model. As a result, the mental model of the situation becomes inadequate to the information presented. A pilot can overlook the warning indicator and shut the engine off, because the warning conflicts with his/her mental model of the situation. Obviously, developing hypotheses about what is wrong and adequate executive actions derived from them is of the most importance in emergency situation. When the pilot's initial hypothesis is not confirmed, the pilot starts to use exploratory mental and behavioral actions and analyze their consequences. In stressful situations, these auxiliary exploratory actions are often triggered automatically. Such actions can be especially undesirable in time-restricted conditions. It has been discovered that the main purpose of formal instructions is to describe executive actions that have to be taken to transform the external situation.

Such instructions are not helpful for a pilot at the stage of performance of orienting activity, when he/she has to formulate the task's goal correctly and build the related mental model of activity. The non-instrumental information is also ignored in the instructions. It has been discovered that after receiving non-instrumental information, a pilot turns his/her attention to the instruments. Only at the last stage of information processing, in 30% of all cases, did a pilot look deliberately at the warning-light signals. Hence, during performance of the considered task, a pilot often ignored the warning signals.

Based on the analysis of preferable strategies of task performance, scientists made the conclusion that the instrument panel is not adequate for the real strategies of task performance. After the study of the real strategies of task performance, more efficient instructions on how a pilot should act when an engine fails have been developed. It has also been suggested that the position of the warning lights that indicate engine failure should be changed. A light indicator that displayed the word *danger* has been placed on the instrument panel in front of the first pilot. Another indicator with the words *engine off* and the engine number has been placed above the configuration of engine displays. Later, in a simulator, and then in the actual flight, this design configuration has been tested, and pilots immediately attended to these warnings.

Therefore, studies of activity self-regulation and strategies derived from them of task performance and training are important for developing training procedures and for enhancing equipment design. Goals and mental models of tasks include verbally logical and imaginative components. Hence, such method take into consideration situation awareness and some unconscious or not fully conscious aspects of human performance. An emergency situation is accompanied by stress. The functional block responsible for assessment of task significance reflects self-regulation in a stress situation. The formation of an adequate strategy of task performance is always performed under the influence of this block.

In an emergency situation, a pilot does not start performing executive actions right away but begins with an analysis of the situation at hand. Consequently, the mental model of the problem situation is formed based on the strategies of gnostic or orienting activity. Thinking actions play a leading role in formation of the mental model in the emergency situation. Only after that are executive actions and an adequate strategy of transforming the situation developed. The general model of activity self-regulation should be used for analysis not only of the orienting but also of the executive stage of task performance. For this purpose, a general model of activity self-regulation should be used (Bedny, Karwowsi and Bedny, 2015). As can be seen, we are not talking about a pilot's reactions, responses, and so on but are considering strategies of a pilot's activity in emergency situations.

The self-regulation process of developing various strategies can be observed even when a pilot reads individual aircraft instruments. For example, the strategy of reading a specific instrument depends on the significance of the instrument. It has been discovered that an aircraft's attitude indicator has a rough scale, the distance between scale elements is about 5° and the distance between the numerals is about 15°. This instrument is very significant for pilots, because they independently learn to read the horizon of an aircraft with an accuracy of about +/− 3°, which is much higher than the reading precision of other displays that have a more detailed scale (Kotik, 1978).

Let us consider other examples (Ponomarenko and Bedny, 2011). The main source of information about the situation is a number of displays or instruments on a dashboard in operators' and explicitly in pilots' activity. The information gathered from various instruments about the external situation and about the controlled object is referred to as the *informational model*. The better this model presents essential information to an operator, the more effective an operator is at receiving and interpreting this information. Based on the informational model, a pilot develops a goal and a mental model of the real situation, performs an analysis of the situation, and develops strategies of task performance. Signals about the flight conditions can be *instrumental* and *non-instrumental*. For example, during a flight, the pilot receives information about the aircraft's position based on information she/he observes outside of the cabin. In other cases, non-instrumental information is less definite. To this type of information can include vibration of the aircraft, changes in the background noise, resistance of controls, and so on. Non-instrumental signals are particularly important in emergency situations. The informational model and the external situation include both instrumental and non-instrumental information and *relevant* and *irrelevant* information. Therefore, the strategies of a pilot's activity can be very flexible. The more flexible the human activity is, the more important the concept of activity self-regulation becomes. An ability to synthesize information and create a mental model of a situation that is adequate to a goal of a task formulated by a pilot is an important aspect of pilot's activity. All of the above shows that a pilot is often confronted with *ill-defined problem-solving tasks*. Thinking plays a leading role in solving such tasks. The functional model of activity self-regulation helps to describe a pilot's *dynamic mental model of the flight* more precisely (Ponomarenko and Bedny, 2011). It includes *"sub-model of space position"*, *"feeling of an aircraft"*, and *"instrumental model."* These sub-models interact with *"sensory model"* and *"motor model"*. The structure of the pilot's dynamic mental model is presented in Figure 3.3.

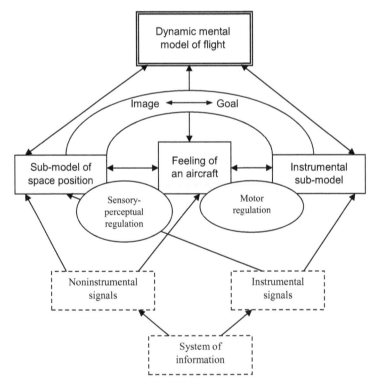

FIGURE 3.3
Structure of the pilot's dynamic mental model of flight.

The dynamic mental model is situation-specific. The function blocks "goal", "conceptual model" (block 13), and "subjectively relevant task conditions" (block 9) are the main mechanisms that are responsible for creation of the dynamic mental models of the flight (see self-regulation model of orienting activity, Figure 3.1). Sometimes the term *image-goal* is utilized when analyzing a pilot's task performance, because not just verbal but also imaginative components are equally important in the pilot's activity. A pilot's interpretation of the situation can be better understood based on the dynamic mental model of the flight. The same as pilots' executive actions or performance program (see Table 3.2 in Bedny et al. (2015), general model of activity self-regulation), identification and interpretation strategies also depend on characteristics of the dynamic mental model of a flight.

Let us consider "image ↔ goal" in this model. There are situations when it is difficult for a pilot to retain the image ↔ goal in her/his mind for a sufficiently long period of time. In these circumstances, it is necessary to develop specific external means that help sustain the image ↔ goal during performance. This is especially relevant when a pilot has to interrupt his/her current task, switch to another one, and then return to complete the interrupted task. To facilitate performance of the interrupted task, the goal of this task can be presented to a pilot as a picture or verbal description or as a voice-induced command.

The sub-model of the "space position" regulates the spatial orientation of a pilot during task performance. It helps him/her to be aware of the aircraft's position in relation to Earth. A contradiction between perception and thinking and between external, visual, and internal receptors' signals (see sub-model "feeling of an aircraft") can produce illusions. In such circumstances, a pilot tries to repress the false sensations by concentrating his/her attention and forcing activation of the logical components of thinking.

Let us consider the sub-model "feeling of an aircraft" in more detail. Its formation is based on the influx of non-instrumental signals. Although these signals may cause errors, they are important for the feeling of the aircraft. For example, the lack of muscular feelings during piloting with the automatic stabilizer of the aircraft's position increases the duration of visual fixation on the displays. This is due to the fact that the use of a stabilizer disrupts the level of resistance of the controls, affecting the pilot's feel of the aircraft. The controlling actions of the pilot change the position of the aircraft, which leads to changes in the instrumental and non-instrumental signals. Such signals give information not only about the position of an aircraft but also about result of the pilot's motor actions. Based on the consequences of the pilot's own actions, she/he can change his/her strategy of task performance and correct the dynamic mental model of the flight.

The "instrumental sub-model" is created by a pilot based on the perception of information from instrumental displays. This sub-model reflects the discrepancy between the required and real flight regimen based on perceived difference between set parameters of the flight and the real position of the indexes. The instrumental sub-model is very pragmatic and regulates motor actions. It facilitates the automaticity of activity, which also supports speed and accuracy of performance. However, such a mental model might lower the reliability of task performance in an unexpected situation.

Sensory-perceptual regulation is achieved primarily on the basis of inter-action of sub-models of space position and feeling of an aircraft. The motor regulation is provided primarily based on interaction of sub-model of feel-ing of an aircraft and the instrumental sub-model.

The presented pilot's dynamic mental model shows that non-instrumental signals impact "sub-model of space position" and "feeling of an aircraft." Instrumental signals have impact on "sub-model of space position" and "instrumental model."

We define five identification and interpretation strategies of the faulty situation.

The *first strategy* is instantaneous identification, when the pilot's dynamic mental model of the situation coincides with the circumstances of the cur-rent situation. Signals are compared with the mental model already devel-oped in the past, and the *situation is identified instantly.*

The *second strategy* is identification and interpretation of the situation after the mental search for alternatives is evaluated. Operative thinking facilitates the development of a mental model that unfolds over time in the internal mental plane without addressing the external stimulus.

The *third strategy* takes place when a pilot needs additional information in order to evaluate the incoming information and develop a mental model of the flight. A pilot can identify information only by examining the equip-ment. The identification and interpretation rely on additional signals and perceptual and thinking actions. The dynamic mental model of the situa-tion that derives from experience is not completely adequate and, therefore, identification and interpretation of information cannot occur in the internal mental plane.

The *fourth strategy* is the identification and interpretation based on not only perceptual and thinking actions but also on motor actions. The pilot cannot develop an adequate dynamic mental model of the flight based on previ-ous experience and mental analysis of situation alone. In addition, he/she needs to use the trial-and-error method, which includes motor actions and analysis of their consequences. Motor actions perform explorative functions in this case.

The *fifth strategy* can be viewed conditionally as identification and interpre-tation based not only on perceptual and thinking actions but also on motor actions. It's close to the fourth strategy, but the difference is that it does not lead to a correct interpretation of the situation.

These strategies have been revealed in experimental conditions and gath-ered utilizing multiple real examples. A pilot can transfer from one strategy to another depending on the specificity of the situation at hand. For instance, a pilot can utilize the mental actions (third strategy) and because it was not successful he/she would start using external explorative motor actions (fourth strategy). This was the trial-and-error method. Such actions do not always help in obtaining the required information. All considered identification and interpretation strategies do not eliminate erroneous actions in the process of transformation of vague uncertain signals into definite certain ones. The strategies described above can be clearly identified and evaluated based on models of activity self-regulation. Explorative strategies are primarily associated with function block 10 (see Figure 3.1). The emotionally evaluative block also plays an important role in the time limit conditions. Analysis of self-regulative strategies demonstrates that internal and external explorative actions can become chaotic, and a pilot can lose time and cannot complete the task correctly.

Another example also involved the study of pilots' actions in an emergency situation in real flight conditions. The aircraft was equipped with a special failure-simulation panel. Simulation of the engine failure was included in the test during the flight. One of the important characteristics of the efficiency of the pilot's actions during the breakdown is the time spent detecting the emergency situation. The diagnosis time includes the detection time and the time for correct interpretation of an engine failure. The study showed that a pilot uses three types of information about engine breakdown. The first type of information is gathered by looking at various instruments that show the parameters of the engines of the aircraft, such as revolution per minute, gas and oil pressure, temperature of the exhaust gases, revolving moment, and so on. The second type of information is obtained from light-bulb indicators. Two bulbs were located on the ceiling of the cockpit. One of them alerted the pilot about the "engine breakdown" and the other one showed that the "pump is working." The third bulb was located on the panel near the left pilot and notified the pilot about the state of the "propeller." The fourth one was on the panel near the right pilot and signaled "ice formation." It has also been discovered that pilots utilize non-instrumental types of information. The pilots obtained this information via their subjective feelings. Such information is associated with turning force, changing of pedals' resistance, changing sounds, deviation of body position from vertical, and so on. As has been mentioned above, because of the inherent slowness of the instrument response, the failure-related information has not been displayed quickly enough. So, the pilot often incorrectly interpreted the meaning of input information because the non-instrumental information is often associated with vague feelings. In such circumstances, the unconscious channel of information processing becomes important (see Figure 3.1). If block 3 (set) affects block 1 (meaning) with insufficient activation of block 2 (goal), meaning reflected by block 1 is mostly non-verbalized. Unconscious mental operations from block 10 (decision-making and program formation) are involved in formation of such meaning that helps in creating non-verbalized hypothesis in emergency situation. If a goal-directed set does not match the situation, it slows down the goal formation process. Because the goal of a task and its mental model are often formed incorrectly at the first stage of information processing, a pilot searches for information that can prove his/her incorrect hypothesis about them. At the next stage, a pilot transfers his/her attention to instrumental signals, which allows the transfer of non-verbalized meaning into the conscious verbalized meaning. This facilitates formation of a conscious goal of a task and, based on block 10 that activates conscious orienting and explorative actions, the mental model of a task is developed (block 9, subjectively relevant task conditions). Thus, a pilot transfers predominantly unconscious processing of information into the conscious one. The data obtained in this study discovered that the real strategies of searching for information about a failure is essentially different from the strategies prescribed by the formal instruction (see Table 3.3).

Table 3.3 shows that in experimental flights that have simulated failure of engines, pilots initially paid attention to non-instrumental signals. Only at the last step did they consider the warning signals (warning lights). This significantly reduced the reliability of a pilot's actions in the event of the aircraft malfunctioning. Therefore, a discrepancy between the real strategies and the formally developed strategies is caused by the inadequate instructions and by an inefficient design of an instrument panel. This results in the incorrect formation of the mental model of the situation and inadequate performance of the diagnostic tasks. The pilot also might process information from his/her instruments without analyzing the context in which it appears. Obviously, in an emergency situation, the most important things are to search for information to develop hypotheses about what is wrong, and then select one hypothesis upon which to act. When the pilot's initial hypothesis cannot be confirmed by the stimuli, the pilot strives to test other hypotheses, and *auxiliary explorative actions* are performed automatically. These actions can be motoric and mental. In time limit conditions, this can lead to a failure.

TABLE 3.3

Prescribed and Real Strategies of Information Gathering during Engine Failure

Sequence of Information Gathering to Standard Procedural Manual	Real Sequence of Information According to Standard Procedural Manual
Turning of the aircraft progressively increases. Progressively increasing pitch	Feeling of acceleration, resistance on control column and pedals, changing position of body, sound changes
Light indicator "ENGINE BREAKDOWN" is turned on	Decreasing RPM
Decrease in the turning momentum	Decrease in the turning moment
Pressure drop in the fuel system	Pressure drop in the fuel system
Decreasing RPM	Decrease exhaust gases temperature
Decreasing exhaust gases temperature	Light of signal bulb (alarm signal) about engine failure
Pressure drop in the oil system	Pressure drop in the oil system

In conclusion, let us consider the pilot's spatial orientation during the flight. The pilot should continuously assess the aircraft's position in space and the progress of the aircraft's movement toward the goal of the flight. Consequently, the image of the flight is an important component of the mental model of the situation. During the flight, the pilot is constantly faced with the requirement to perform two tasks. One of them is related to the flight control, and the other requires an assessment of the aircraft position in space. To improve the accuracy of holding any flight parameters, a pilot should detach him/herself from the information related to the spatial position of the aircraft. There's also a different scenario when a pilot has to estimate the aircraft position in space, and thus to ignore the signals to which he/she needs to respond for precise control of the aircraft during the flight. So piloting means constantly switching from one considered task to another. In the meantime, the human body continuously experiences an unusual physical impact of the flight conditions which are not encountered on Earth. However, such impacts are critical during the flight. Acceleration, changes in altitude, and so on. act on the sensory organs of the body and are perceived by a pilot as gravity. These impacts do not match the usual influences of gravity when orientating in space and is not constant either in direction or magnitude. All this violates the natural human orientation. Contradictions between the visual and interceptive signals, between perception and thinking, subjective feelings, and intellectual evaluation of the pilot's body position emerge. This influences the evaluation of the real position of the aircraft in space.

The position of the aircraft in space based on visual signals from various instruments on the panel does not always match the subjective sensitive feelings of a pilot. If this discrepancy is recognized by a pilot, then he/she would try to suppress false sensory-perceptual feelings. If the pilot does not have the proper intellectual assessment of the aircraft position in space, and he/she follows only sensory-perceptual feelings, he/she can make a serious mistake when controlling the flight. A brief analysis of the above-presented material allows the conclusion that a pilot's activity includes complex strategies that are required for adequate task performance. It is obvious that such concepts as self-regulation, a goal, a dynamic mental model of the flight, conscious and unconscious reflection of the situation, and so on. play an important role in analysis of pilot's task performance.

Dynamic reflection of a situation enables a subject to discover different aspects of the situation, its changes over time, as well as changes in the relationship between conscious and unconscious human activity. The function block "subjectively relevant task conditions" plays an important role in the dynamic reflection of the situation. Let us consider as an example the relationship between conscious and unconscious reflection of the situation in this function block. This block has two sub-blocks. One is called "situation awareness" and the other "operative image". These two sub-blocks partially overlap each other (see Figure 3.4).

The figure allows identification of three areas in the considered block based on the level of consciousness of information processing. The situation awareness (SA) block is involved in conscious reflection of the situation. Sub-block "operative image" includes two areas. The area that is overlapping with block SA is also conscious.

The non-overlapping area is involved in unconscious reflection of the situation. Thus, when studying pilots' activity, one has to pay attention not only to conscious but also to unconscious principles of information processing. The concept of SA ignores this factor. In contrast, the model of orienting activity takes it into consideration. Such a model permits discovering the specificity of the strategies of task performance. Thus, in this chapter some studies that demonstrate application of the concept of activity self-regulation in real flight settings are presented in an abbreviated manner.

In concluding this section, we would like to outline the specifics of the relationship of SA and orienting activity concepts for pilots' performance. The mechanism that is labeled as SA is task-specific. It is important for constructing a mental model of a task, but the methods of SA analysis ignore this factor. Most methods of SA studies are based on comparative evaluation of systems. Such evaluation is achieved through ranking of these systems by expert.

Various questions and pilot responses to them are used as a preliminary method of study. However, analysis of pilots' verbalized responses ignores imaginative and unconscious components of orienting activity.

According to SSAT, the main method of studying orienting activity is experiment and analysis of strategies of task performance under various conditions. Functional analysis utilizes experimental procedures in combination with observation and pilots' subjective judgment. Such analysis allows discovery of not one but several acceptable strategies of task performance in emergency conditions. A strategy implies flexibility and plasticity of the activity during task performance based on the evaluation of the situation, activity outcome, and internal state of the subject. Not only cognitive

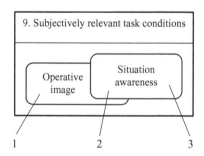

FIGURE 3.4
Function block "subjectively relevant task conditions": 1 – unconscious imaginative reflection of situation; 2 – combination of conscious imaginative and verbal-logical reflection of situation; 3 – conscious verbal-logical reflection of situation (SA).

but also emotionally-motivational aspects of activity regulation should be taken into account when analyzing strategies of orienting activity. The decision-making mechanism is excluded from SA. However, this mechanism is involved in the evaluative stage of SA.

Conscious and unconscious reflection of the situation is facilitated by orienting activity. This activity precedes execution and involves various mechanisms, including SA. Orienting activity that is responsible for the dynamic reflection of various situations is an important mechanism to consider not only when studying pilots' performance but for research involving other professions as well. It provides dynamic orientation in a situation, and an opportunity to reflect not only the present but the past and the future. It also deals with both actual and potential features of situations.

SA ignores the fact that the feedback can be accomplished in the internal mental plane. Orienting activity takes into consideration the existence of external and internal or mental feedback. Situation awareness is just one mechanism out of other multiple mechanisms responsible for the dynamic reflection of the situation that is, in turn, task-specific. It is an important mechanism for constructing a mental model of a task. However, the methods utilized in the SA analysis ignore this fact. Not only cognitive but also emotionally motivational aspects of activity regulation are important for analysis of strategies of orienting activity. The decision-making mechanism should not be excluded from SA, because it is also important for the formation of SA.

The pilot's mental model of the flight is dynamic and changes as the situation changes. It also depends on evaluation of the consequences of the explorative actions of a pilot. This means that objectively the same situation presented can be interpreted in different ways and the mental model of situation can vary. A mental model of the flight is a result of the interaction of various function blocks, such as goal, subjectively relevant task conditions, assessment of the task significance and difficulty, and so on. The impotence of each function block depends on the specificity of a task performed by a subject. In any particular situation, the most relevant function blocks for the study should be selected. There are not only conscious and unconscious levels of self-regulation. Therefore, not all aspects of activity self-regulation can be verbalized. Mental models have operative features, enabling operators to understand a system and predict its future state by mentally manipulating its parameters.

Operators utilize various sources of information during task performance, such as the external environment and various instruments and tools. When an operator can directly operate the object under control, he/she can receive not only instrumental but also non-instrumental information. For example, information can be obtained through aircraft vibration, resistance of controls, changes of engine noise, environment outside of the cabin, and so on. The informational model includes both instrumental and non-instrumental information, and relevant and irrelevant information. All this data is important in creation of the mental model of the flight. Therefore, not just verbally logical and conscious processes but also non-verbalized processes are important in pilots' performance. Analyzing pilots' interactions with the informational model during performance of orienting activity, and discovering utilized strategies of task performance, are powerful tools for introducing a more efficient method of task performance, and also for developing design solutions of various types of equipment utilized by pilot and other professionals.

The functional analysis utilizes experimental procedures along with observation and recording pilots' subjective judgments. Due to such analysis, several acceptable strategies of task performance in emergency conditions can be revealed.

3.5 Examples of Eye Movement Analysis of Pilots' Performance

Depending on the nature of the task, the same instruments or indicators can be used in a different manner during flight. In some situations, they are used for checking or examining parameters of the flight, and in others for controlling and/or managing the flight. For example, attitude display and variometer (vertical speed indicator, VSI) is frequently utilized for control and correction of the flight functions. Other devices are used for these functions less often. Devices that are utilized for checking purposes are usually perceived by the pilot via short fixations of the eyes.

If a pilot evaluates information as a deviation from the normal flight mode, she/he initiates motor actions that are associated with aircraft control and adjustment of the flight parameters. In such situations, the gaze fixations on the instruments associated with the control take 1–2 seconds. The direction of the eye movements around the instruments are determined by the content of the information and its significance in a particular situation. The analysis of eye movements and of the duration of eye fixation is an important method of analyzing any visual task and particularly tasks performed by a pilot, when the visual information is of foremost importance. As an example, Table 3.4 depicts average duration of visual fixation on various devices during a flight depending of the specificity of performed tasks (Ponomarenko, Lapa and Chuntul, 2003).

The table reflects the duration of eye fixations and shifting of attention only. The other frequently used index that is based on eye movements registration is frequency of eye fixations on various instruments. The collected data shows that the pilot at a certain stage of the flight draws attention to such parameters as aircraft horizontal position and course. At the second stage of the flight, pilots pay attention to the aircraft variometer and altimeter.

TABLE 3.4

Temporal Data of Visual Control of Various Devices during the Flight Depending of the Specificity of Performed Tasks

| Flight Section | Indicators of Control | Devices | | | | | |
		Attitude	Variometer	Directional Indicator	Altimeter	Other Indicators
Output from the calculated turn	Percentage of time from total time of flight	49	15	30	4	2
	Duration of visual fixation in sec	0.7	0.5	0.7	0.5	0.6
Decrease from a height of 1000m to 400m.	Percentage of time from total time of flight	14	30	27	12	17
	Duration of visual fixation in sec	0.5	0.9	0.9	0.6	0.6

The strategy for collecting information is performed for further managing the motor actions.

This is useful information for the task analysis. However, this example demonstrates that the method of eye movement interpretation that has been utilized for a long period of time is not sufficiently informative. The existing method of eye movement registration was developed by Yarbus (1965, 1969). Currently, from a technical perspective, the method developed by Yarbus of eye movement registration has been significantly improved, and there are much more advanced devices for eye movement registration. However, there are theoretical problems associated with interpretation of the collected eye movement data. For example, Vertegaal (1999) wrote that eye fixation provides some of the best measures of visual interest and do not provide a measure of cognitive interest. According to this author, eye movement simply determines whether a pilot or a computer user is observing certain visual information and how often he/she does this. This is demonstrated by data presented in Table 3.4. Such data simply show shifting of attention during task performance (Rayner, 1998). Goldberg and Kotoval (1999) stated that usability researchers do not have a strong theory for performing eye movement analysis. Presently, there is a contradiction between the technical improvement of the eye movement registration devices and the underdeveloped old methods of interpretation of the data collected using these devices.

Not just the eye movements and fixations and determining their duration should be of interest. Also critically important are discovering cognitive processes and associated cognitive actions (perceptual, mnemonic, thinking, decision-making) and their logical organization when performing various tasks. This is made possible by the new method of eye movement interpretation (Bedny, Karwowski and Bedny, 2015; Bedny, Karwowski and Sengupta, 2008).

Here, in an abbreviated manner, we describe our new method of eye movement interpretation which can be useful not only for the study of pilots' performance but also for analysis of computer-based tasks and any task where visual information is of vital importance. This method allows extraction of cognitive actions and their logical organization based on eye movement analysis. Not just perceptual but also other mental processes are involved in visual analysis of the information. Hence, considering eye movement and eye fixation time in the respective areas facilitates discovering various cognitive actions and their logical organization. The traditional principles of eye movement analysis are not discussed here.

In contrast to the traditional method of eye movement analysis, we suggest dividing the cumulative scan path of the eye movement into segments that correspond to one or several cognitive actions, which can be visually extracted by the expert.

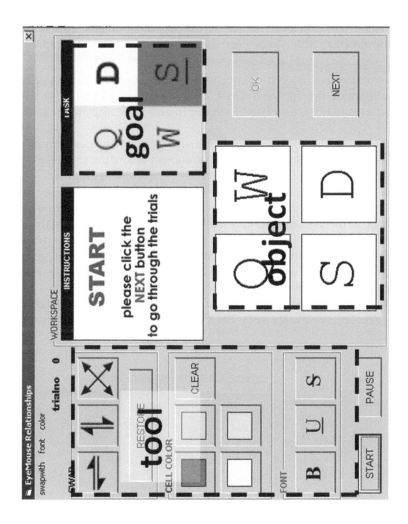

FIGURE 3.5
The interface with areas of interest.

The hypothetical task with interface that is presented in Figure 3.5 has been developed to demonstrate this method. According to functional analysis of activity, the following areas of the screen were selected:

1. *Tool area* displays icons that could be clicked, imparting the desired features to various elements on the screen and changing their arrangements.
2. *The object area* consisted of the elements whose states should be manipulated in order to achieve the final given arrangement in the goal area.
3. *The goal area* is changed during task performance. This area only demonstrates the future desired arrangement of elements in the object area that the user has to achieve in order to complete the task.

Figure 3.6 below presents the cumulative scan path during performance of the completed computer-based task by one subject according to the selected strategy. This data is used for the traditional eye movement analysis.

The basic characteristics of eye movement analysis are saccades (i.e., rapid intermittent eye movements occurring when the eyes are fixed on one point after another in the visual field) and fixations. They include scan paths, frequency of fixation, fixation duration, and transition between areas of interest. Specialists usually pay attention to the general length of scan path, general duration of fixation, average duration of one fixation, frequency of fixations. Such data is useful for analysis of the subject's attention only.

In SSAT, *cumulative scan path is divided into small segments* that determine some fragment of eye movements. Each segment includes a small number

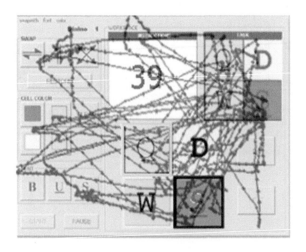

FIGURE 3.6
Cumulative scan path during performance of the computer-based task.

of eye movements, and eye fixations associated with them which can be easily interpreted. These segments reflect logically completed sets of cognitive actions that are performed based on visual information.

Such subdivision provides an effective way of eye movement interpretation and allows uncovering cognitive actions that are involved in performance of the considered segment of a task. For example, in Figure 3.6, the cumulative scan path has been divided into segments between two mouse clicks. Instead of one cumulative scan path, 12 separate fragments have been identified. It made it easy to trace eye movement and eye fixations. Each eye movement and eye fixation usually are considered as separate cognitive actions. Qualitative analysis of the obtained data provided classification of cognitive actions that were performed based on visual information. Table 3.5 presents the first fragment of the eye movement track and associated with it cognitive actions.

Unity of cognition and behavior is an important principle of activity analysis in task performance (Bedny et al., 1996, 2001). Therefore, it is recommended that both eye and hand movements should be used in eye movement interpretation. For example, in our analysis of the computer-based task, eye and mouse movement were registered together because it helped to interpret eye movement data. The rationale for such recommendation was that every motor action involved in the performance of visually based tasks is based on preceding mental processes. Analysis of the following motor actions helps in understanding the purpose of preceding cognitive actions. Each fragment is completed when a subject performs a mouse click.

The following are some rules of eye movements' interpretation:

1. One complete eye movement and one complete fixation that follows this eye movement should be roughly estimated as one cognitive action. The type of the cognitive action is distinguished based on the descriptive analysis and the duration of the action as well as on the analysis of a purpose of an action and its relevance to the considered stage of task performance.

2. Eye fixation or gaze can be associated with the following cognitive actions: (1) receiving information or perceptual actions; (2) interpretation and mental manipulation of information or thinking actions; (3) selection of the following actions or the decision-making actions; (4) extraction of information from long-term memory or keeping information in working memory or mnemonic actions.

3. Eye movement interpretation should be combined with analysis of the preceding and following eye movements and fixations as well as the motor actions associated with these eye movements.

4. The sum of the time spent on eye movement and eye fixation is the total approximate time of the cognitive action.

TABLE 3.5

Action Classification Table (Fragment with One Image of the Scan Path)

1		2	3	4		5	6	7
Eye Move And Final Position		Activity between successive mouse events (clicks) Mental/motor Actions Involved	Mouse events	Time (ms)		Total Action Time (a + b)	Classification of cognitive actions	Scan path generated/duration
From	To			a. Approximate eye movement time to reqd. position (ms)	b. Approximate Dwell time at position (ms)			
Start	G_Q	Goal acceptance and formation, comparison of object and goal areas for creation of subjective model of situation relevant to accepted goal:		150	180	330	Simultaneous perceptual actions	
G_Q	O_q			150	220	370	Simultaneous perceptual actions	
O_Q	T_{CB}	Formation sub-goal (selection of element in object area O_D for subsequent task execution. (includes simultaneous perceptual actions with combination with thinking actions)).		180	150	330	Simultaneous perceptual actions	
T_{CB}	G_S			180	220	400	Thinking action based on visual information	
G_S	G_D			150	190	340	Thinking action based on visual information	
G_D	O_Q			210	220	430	Thinking action based on visual information	
O_Q	O_W	Comparison of object and goal in relation to the program of performance.		150	330	480	Thinking action based on visual information	
O_W	O_D	Formation program of performance according to developed mental model and accepted goal.		150	190	340	Thinking action based on visual information	
O_D	G_D			210	190	400	Thinking action based on visual information	
G_D	O_D	Decision-making about possible motor actions. Execution of action.	Click element D in object area.	210	630	840	Decision-making action at sensory-perceptual level with simultaneous motor action "MB + G5 + AP1".	

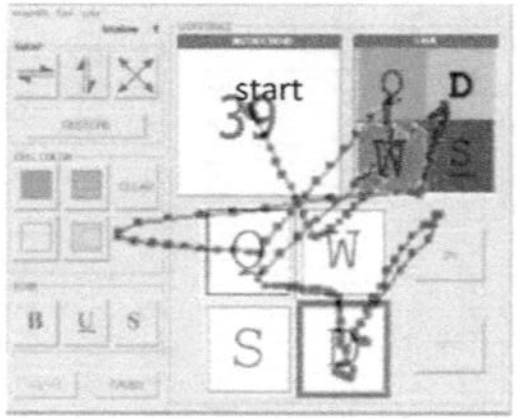

Image 1

5. Sometimes a subject might perform a complex perceptual action that is required for creation of a *perceptual image*. Here, a series of eye movements and eye fixation should be integrated into one complex perceptual action whose goal is creation of a perceptual image.

In Table 3.5, elements of image 1 in the right column have the following areas: *"start area"*; *"tool area"* on the left side of the screen; *"object area"* at the bottom middle position; and *"goal area"* in top right corner. Each area contains multiple elements such as the letters Q, W, S, and D in the object or goal areas and such elements as T_{CR} (*"tool area, color red"*) or T_{CG} (*"tool area, color green"*) in the tool area, and so on. The left column of this table shows that eyes move from the Start position down to G_Q (goal area, element Q) and then from this position to O_Q (object area, element Q) and so on. The purpose of the first three actions is to receive information about the position of the considered elements. Hence, these are perceptual actions. After that, a subject moves his/her eyes to various elements on the screen in order to understand the relationship between considered elements. Discovering the relationship between elements is the thinking function. Thus, all other actions are thinking actions which are performed based on visual information (see column 6 - classification of actions). The last two actions that are combined with motor actions include eye movement from G_D to O_D. They are *"decision-making actions at a sensory-perceptual level* which are combined with simultaneous performed *motor action*. The duration of these actions is significantly longer than the duration of other actions.

The material presented shows that qualitative analysis of each cognitive action involves the following steps. The task is divided into meaningful, logically completed segments of activity; a selected segment of activity is divided into smaller elements such as cognitive and motor actions; the goal of each action is determined; the tools that are utilized in these actions are defined; the transformation of an object of activity through each action and how this transformation or result corresponds to the goal of the action is considered; the purpose of actions that are performed before and after the considered action are analyzed; and the duration of actions is determined. As a final step of this analysis, the goal of the motor action that follows the considered sequence of cognitive actions is defined. All of these steps of analysis allow inferring what kind of cognitive actions are performed by a subject. These steps help to *penetrate the subject's mind during task performance*.

Comparison of Tables 3.4 and 3.5 clearly demonstrates the differences in the data presented. The first one tells us how a pilot's attention shifts from one instrument to another and the duration of the attention concentration. In general, the traditional method that has been used for over 50 years utilizes such measures of eye movement as scan path length (in pixels), cumulative dwell time or average fixation time in the respective areas of interest, and attention shift strategies. This data is useful, but it is not sufficient for analysis of the existing methods of task performance and the development of more efficient methods. In addition, the suggestions by the SSAT method presents information about accompanying cognitive actions. It also allows determination of the logical organization of cognitive actions and development of the human algorithm of task performance at the following stage of task analysis. Based on such data, a quantitative analysis of task performance can be conducted.

The above-presented material brings us to the following conclusion.

1. The traditional method of eye movement interpretation has existed without any change for about 50 years. Improvements in this area are associated with the development of more sophisticated methods of eye movement registration. However, the best method does not guarantee efficient *interpretation* of data. For eye movement data interpretation, specialists utilize the cumulative scan path, and its length, as a measure of search behavior; cumulative dwell time or average fixation time in respective areas as a measure of difficulty of information analysis; average eye movement time; and so on. Such data can be used, but it is not sufficient for the task analysis.

2. SSAT developed a new method of eye movement interpretation. It includes the following basic stages of eye movement analysis.

   a. In order to simplify eye movement interpretation, the cumulative eye movement scan path should be divided into (1) task segments or subtasks that can be easily interpreted, and (2) logically completed sets of actions that can be extracted based on analysis of considered segment.

   b. At the following stage, action classification analysis is conducted. For this purpose, each segment is further used for identification of cognitive and behavioral actions during task performance. The basic criterion for this purpose is a *goal of an action*. Actions are classified based on principles developed in SSAT. The duration of individual actions is determined. Motor actions that are associated with cognitive actions should be considered. Interaction of actions, their structural organization, and interaction are analyzed.

   c. The human algorithm of task performance is developed based on an analysis of the logical organization of cognitive and motor actions. All described stages are aimed at penetrating the performer's mind and uncovering mental actions and their logical organization during task performance.

In general, we presented here the totally new method of eye movement interpretation that can be used for analysis of cognitive actions and their organization in task performance. Such analysis makes eye movement much more effective and useful.

3.6 Qualitative Analysis of Eye and Mouse Movement Relationships

Observation of the video of the subjects' task performance during the experiment revealed various types of mouse-eye coordination. Generally, three types of such coordination have been observed.

1. The eyes preceding the mouse. This was a result of the eyes detecting the target and then the mouse being quickly brought closer to the target.

2. The mouse residing near the target while the eyes look for the next items.

3. The target being activated without any foveating based completely on peripheral vision. This occurred generally in the post-learning stage when the subjects had considerable knowledge of the visual layout of the interface as well as of the task.

Based on these types, seven categories of eye-mouse movement relationship have been identified, and their distribution over the whole experiment has been analyzed.

This correlation has been determined by observing the movements of the eyes and the mouse before, after, or during an action with the mouse. This correlation is used in order to understand the behavior during task performance rather than when only searching, reading, or comprehending the text.

The seven distinct categories of the observed correlations are presented in Table 3.6. The left column shows the name of the movement strategy and the next column contains the relation of the eye and the mouse movement at a particular target location on the time scale. The grey bars indicate the mouse and the black bars indicate the movements of the eyes.

Eye preceding mouse: This type of eye-mouse relationship is the most commonly observed one. In this case, the eye movement is a precursor to the mouse movement. The eyes fixate on the path that the mouse follows. Since most icon-based interfaces utilize the point-and-click method of using the mouse, this type of eye-mouse movement relationship is common for the tasks that involve interfaces with icons or menu options.

Mouse preceding eyes: This category is similar to eye preceding mouse, but the mouse is preceding the eyes. This one is generally observed when the users have considerable training and have memorized the location of the target so that the cursor speed to the desired location is increased. Since a reduction in cognitive overload can be attributed to knowledge of the location of the searched item, it can be suggested that the users' stress level is lower while performing this kind of operation.

TABLE 3.6

Relationship between Eye and Mouse Movement

	Movement Strategy	Graphical Representation along Timeline	Timeline
1	Eye preceding mouse.	Eye entering target area / Mouse entering target area	
2	Mouse preceding eye.	Mouse entering target area / Eye entering target area	T IM E
3	Eye and mouse together (match).	Mouse and eye entering target area	
4	Eye overlap (eyes continue to stay, whereas mouse moves).	Eye entering target area / Mouse entering target area / Mouse leaving target area / Eye leaving target area	
5	Mouse overlaps (mouse stationary with eyes moving).	Mouse entering target area / Eye entering target area / Eye leaving target area / Mouse leaving target area	
6	Peripheral vision (no use of eyes to point mouse).	Eye entering target area / Eye leaving target area / Mouse entering target area / Mouse leaving target area	
7	Back track (eyes moving back and forth between target and cursor or other areas).	Eye entering target area / Mouse entering target area / Eye leaving target area / Eye entering target area / Mouse leaving target area	

Eye and mouse together (match): This is extremely rare, and there is always a lag between the eye and the mouse movement in terms of time as well as space while transiting to a particular target. However, in the ballistic phase of mouse movement, this type of relationship can be observed. The studies of hand movement toward a particular target suggests that there are two phases: one is a ballistic phase, and the other one is an adjustment phase. This type of eye-mouse movement relationship is generally possible in the ballistic phase when the mouse moves with considerable acceleration, and the speed of the eye movement is reduced due to more fixations for reasons varying due to other mental processes or distractions.

Eye overlap (eyes continue to stay, whereas mouse moves): Mostly encountered with reading or interpreting graphical features. This occurs when the user has found the target but still has to process the presented information to decide whether the chosen action is going to produce the desired result. This type is mostly associated with operational icons, in this case swapping or changing the position. In real cases, this can be readily observed in link activations, menus, or ActiveX controls. These controls are based on certain procedures or macros that require the user to imagine the process beforehand. The mouse automatically moves away to ease the reading or comprehension.

Mouse overlaps (mouse stationary with eyes moving): This type of relationship is generally observed when the user wants to both make sure that the chosen operation is going to give the desired results and check the current state once again so as to verify the choice. If feedback is not adequate, then this type of movement might increase.

Peripheral vision (no use of eyes to point mouse): This is the most desired form of mouse movement, as it requires no visual supervision. This can occur when the users are extremely familiar with the spatial features of the interface as well as the task at hand. Mostly, the cursor should just stay in the area of focus, which is enough for the eyes to coordinate the mouse to the particular location.

Backtrack (eye moving back and forth between target and cursor or other areas): This type of movement is generally observed when the eyes are tracing the cursor after it has an overshoot. It can also occur when the eyes backtrack to obtain feedback about the previous fixation region or any distractions in the path.

The eye movement preceding the mouse movement and the mouse preceding the eye movement toward the target had been already reported in studies of eye-mouse coordination (Smith et al., 2000). However, due to the fact that our study presented a comprehensive task-oriented situation, additional relationships have been observed. The distribution of these eye-mouse movement correlations for all the tasks is presented in Figure 3.7.

The bar chart in Figure 3.7 suggests that most of the eye-mouse movement correlations were of the eyes preceding the mouse or the mouse preceding the

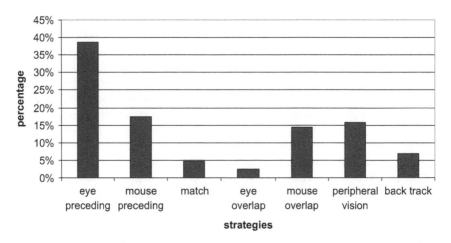

FIGURE 3.7
Percentage distribution of different eye-mouse coordinations over tasks.

eyes. The complete match of the eye and mouse movements were extremely rare and can be considered as negligible.

The mouse overlap type increases in the post-learning stage, so the use of peripheral vision and the backtracking is reduced at this stage (see Figure 3.8). Some of the data collected was the most prominent at the exploratory learning stage, while the remainder was most useful for the analysis of the post-learning stage.

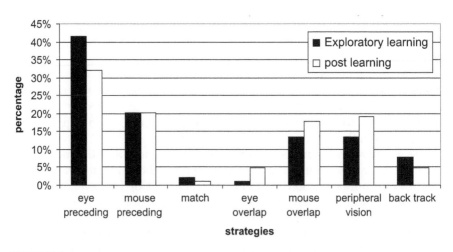

FIGURE 3.8
Percentage distribution of different eye-mouse coordination in the exploratory and post-learning stages.

The chapter presented the analysis of various types of eye and mouse movement relationships via studying tasks with different characteristics of the interface usability. The eye and mouse movements had various qualitative relationships due to differences in the eye-mouse coordination during task performance. The suggested method of eye movement analysis and interpretation is important for comprehensive data collection and especially for analysis of cognitive activity in general.

References

Bandura, A. (1977). *Social Learning Theory*. Englewood Cliffs, NJ: Prentice-Hall.

Bedny, G. Z. (2015). *Application of Systemic-Structural Activity Theory to Design and Training*. Boca Raton, FL and London, UK: CRC Press, Taylor & Francis.

Bedny, G., Polyakov, L. and Stolberg V. (1996). Russian psychological concept of work activity. *National Social Science Journal*, 8(2), 145–158.

Bedny, G., Karwowski, W. and Young-Guk Known. (2001). A methodology for systemic-structural analysis and design of manual-based manufacturing operations. *Human Factors and Ergonomics in Manufacturing*, 11(3), 233– 253.

Bedny, G. Z., Karwowski, W. and Jeng, O-J. (2004). The situation reflection of reality in activity theory and the concept of situation awareness in cognitive psychology. In G. Z. Bedny (invited editor). Special Issue. *Theoretical Issues in Ergonomics Science*, 5(4), 275–296.

Bedny, G. Z., Karwowski, W. and Bedny, I. (2015). *Applying Systemic-Structural Activity Theory to Design of Human-Computer Interaction Systems*. Boca Raton, FL and London, UK: CRC Press, Taylor & Francis.

Bruner, J. S. (1957). On going beyond the information given. In Bruner, J. S. (Ed.). *Contemporary Approaches to Cognition* (pp. 41–69). Cambridge, MA: Harvard University Press.

Dekker, S., Hummerdal, D. and Smith, K., (2010). Situation awareness: some remaining questions. *Theoretical Issues in Ergonomics Science*, 11(1–2), 131–135.

Endsley, M. R. (1995). Toward a theory of situation awareness in dynamic system. *Human Factors*, 37, 32–64.

Endsley, M. R. (2000). Theoretical underpinnings of situation awareness: A critical review. In Endsley, M. R. and Garland, D. J. M. (Eds.). *Situation Awareness Analysis and Measurement* (pp. 3–32). Mahwah, NJ: Lawrence Erlbaum Associates.

Endsley, M. R. and Jones, D. (2012). *Design for Situation Awareness. An Approach to User-Centered Design*. 2nd edition. Boca Raton, FL and London, UK: CRC Press, Taylor & Francis.

Goldberg, J. H. and Kotval, X. P. (1999). Computer interface evaluation using eye movements: Methods and constructs. *International Journal of Industrial Ergonomics*, 24, 631–645.

Kahneman, D. and Tversky, A. (1984). Choices, values and frames. *American Psychologist*, 39, 341–350.

Konopkin, O. A. (1980). *Psychological Mechanisms of Regulation of Activity*. Moscow, Russia: Science.

Kotik, M. A. (1978). *Textbook of Engineering Psychology*. Tallin, Estonia: Valgus.

Makarov, R. and Voskobojnikov, F. (2011). Methodology for teaching flight-specific English to nonnative English-speaking air-traffic controllers. In Bedny, G. and Karwowski, W. (Eds.). *Human-Computer Interaction and Operator Performance. Optimizing Work Design with Activity Theory* (pp. 277–306). Boca Raton, FL and London, UK: CRC Press and Taylor & Francis

Ponomarenko, V. and Bedny, G. (2011). Characteristics of pilots' activity in emergency situation resulting from technical failure. In Bedny, G. Z. and Karwowski, W. (Eds.). *Human-Computer Interaction and Operators' Performance. Optimizing Work Design with Activity Theory* (pp. 223–255). Boca Raton, FL: Taylor & Francis.

Ponomarenko, V. and Karwowski, W. (2011). Functional analysis of pilot activity. A method of investigation of flight safety. In Bedny, G. Z. and Karwowski, W. (Eds.). *Human-Computer Interaction and Operators' Performance. Optimizing Work Design with Activity Theory* (pp. 255–276). Boca Raton, FL; London, UK; and New York, NY: Taylor & Francis, CRC Press.

Ponomarenko, V. A., Lapa, V. V. and Chuntul, A. V. (2003). *Activity of Flight Crews and Flight Safety*. Moscow, Russia: Scientific Research Center in Aviation and Astronautics.

Pushkin, V. V. (1978). Construction of situational concepts in activity structure. In Smirnov, A. A. (Ed.). *Problem of General and Educational Psychology* (pp. 106–120). Moscow, Russia: Pedagogy.

Rayner, K. (1998). Eye movement in reading and information processing: 20 years of research. *Psychological Bulletin*, 124, 372–422.

Salmon, P. M., Stanton, N. A., Walker, G. H., Jenkins, D. Baber, C. and Mcmaster, R. (2008). Representing situation awareness in collaborative systems: A case study in the energy distribution domain. *Ergonomics*, 51(3), 367–384.27. Smith, B. A., Ho, J., Ark, W., Zhai, S. (2000). *Hand Eye Coordination Patterns in Target Selection. Proceedings of the 2000 Symposium on Eye Tracking Research & Applications*, AMC, NY, 117–122.

Uznadze, D. N. (1967). *The Psychology of Set*. New York, NY: Consultants Bureau.

Vertegaal, R. (1999). The GAZE groupware system: mediating joint attention in multiparty communication and collaboration (pp. 294–301). *Proceedings of the ACM CHI' 99 Human Factors in Computing Systems Conference*, Boston, MA: Addison-Wesley/ACM Press.

Yarbus A. L. (1965). *The Role of Eye Movements in the Visual Process*. Moscow, Russia: Science Publishers.

Yarbus, A. L. (1969). *Eye Movement and Vision*. New York: Plenum.

4

Concept of Self-Regulation of Cognitive Processes in Systemic-Structural Activity Theory

4.1 Self-Regulative Model of Perceptual Process

According to SSAT, cognition can be studied as a process, as a system of cognitive actions, and as a goal-directed, self-regulated system. The first approach is the traditional one in cognitive psychology. SSAT does not reject the cognitive approach but uses this approach as one of the stages of activity analysis. The last two approaches have been developed within the SSAT framework. They complement each other and facilitate more efficient study of cognitive processes that are involved in the performance of specific tasks. In morphological analysis, the main units of analysis of cognitive processes are cognitive actions. Functional analysis of cognitive processes considers them as goal-directed, self-regulating systems, and its units of analysis are function blocks that include various cognitive processes. The content of each block is described in terms of cognitive actions or cognitive processes. Sometimes motor actions can also be considered because they are used as a tool for analyzing the result of the transformation of the problem-solving situation. When cognition is viewed as a self-regulated system such approach pays special attention to the goal, emotionally evaluative and motivational components and feed-forward and feedback connections between various function blocks that represent specific cognitive process.

Only the most important function blocks can be considered in any particular case. In this section, for the first time, all cognitive processes are described as self-regulative systems, where major units of analysis are function blocks. Some models are presented for the first time and others enhanced based on the newly obtained data.

Let us first consider the self-regulative model of the perceptual process. It can be utilized for the description of strategies of the perceptual process. For this purpose, we need to present our analysis of sensation and perception that facilitated development of the self-regulative model of the perceptual process. Analysis of sensation and perception in cognitive psychology and SSAT demonstrates they are tightly interconnected and can be divided into a sequence of sub-processes or stages. They can also be presented as functional mechanisms or function blocks where each function block interacts with other blocks based on feed-forward and feedback connections. A function block represents a coordinated system of sub-functions that has a specific purpose in the regulation of activity. In applied activity theory (AAT) perceptual process is divided into four stages: detection, discrimination, identification, and recognition. The demarcations between described stages are not always clearly defined. The scheme of the stages of perceptual process can be presented as follows (Zaporozhets et al., 1967; Zinchenko, Velichkovsky and Vuchetich 1980):

Signal → Detection → Discrimination → Identification → Recognition

However, in this area of research there is no clearly defined terminology. Scientists in AAT do not distinguish between the stages of perceptual process and perceptual actions and operations. In one instance, detection, discrimination, identification, and decoding are stages, and in another one these stages are defined as types of perceptual actions. Each stage includes various actions and operations that are classified into groups according to the names of the corresponding stages. In the above-mentioned publications, the term "action" is considered in general terms. Usually, the discussion is concentrated on the relationship of hand and eye movement or between vision and the sense by touch, or on the role of object-oriented actions in mental development, etc. In these studies, such terms as mental and physical actions are not clearly specified. For example, when talking about perceptual or mnemonic actions, we have to know the beginning and the end point of these actions, and their clear specifications, relationship, and logical organization. The above-mentioned authors utilize such terms as "perception as an action", "memory as an action", and so on. Perception, memory, and other cognitive processes usually include a number of various cognitive actions and operations. Therefore, instead of the terms as "perception as an action" or "memory as an action", we use such terms as "perception as activity" or "memory as activity", and so on. In reality, perception, memory and other mental processes usually include a number of mental actions during performance of a specific task. Terminological inaccuracy is common not just for Zinchenko and his colleagues but also for some other authors working in the areas of general and applied activity theory (AAT).

According to SSAT, all stages of perceptual process are performed via conscious perceptual actions that can be verbalized and unconscious perceptual operations that are components of behavioral and cognitive actions. SSAT allows us to clearly distinguish various actions and operations within the perceptual process. For example, decoding is a system of actions and operations that is involved in the creation of a perceptual image and its interpretation. Identification is a system of actions and operations that facilitates comparison of stimulus with the templates that are already stored in memory. Thus, perception is often not an action of creating a perceptual image as stated by Zinchenko and his colleagues. Perception of a complex stimulus can involve a wide range of perceptual actions and operations that are necessary for creation of a perceptual image.

Moreover, perceptual process is closely related to other mental processes, such as the goal of and motives of activity, and significance of perceived information and isolated consideration of perception is only conditional. Engineering psychologist Rubakhin (1974) conducted an experiment that studied the process of deciphering army reconnaissance codes. Two groups of military experts with an average skill level were chosen to decipher the codes. The first group was told they were determining the position of a "dangerous" enemy, whereas the second group was informed that they were to locate a "weak" enemy. Deciphering information about the weak enemy resulted in a decrease in the correctness of identification, but deciphering information from the dangerous enemy resulted in a sharp increase in false alarms.

According to SSAT, the model of perceptual process has to include a goal of perceptual process, significance of information (emotionally evaluative component), and motivation. This is especially true for complex vigilance tasks, when perception is the main component of human activity. Thus, the perceptual process should be presented as a goal-directed, self-regulated system that includes stages with feed-forward and feedback connections and some additional function blocks, as seen in Figure 4.1.

When utilizing the presented model, the rules listed in Figure 4.1 should be followed. Specificity of each function block depends on the specificity of the considered task. Each block determines the scope of questions to be considered at each stage of analysis. The task under consideration determines what type of function blocks should be involved in its analysis. The first four function blocks are defined as follows:

Detection is an initial stage of the perceptual process. At this stage, a subject can only determine if there is a stimulus in the visual field. The detection stage can be simple and performed in a very short time period. Nevertheless, it is well known in psychophysics that detecting a signal when background noise is present can be challenging. Under normal conditions, this initial stage is quickly transformed into the next stage of perceptual process.

Determining a sensory threshold is one of the key aspects of activity analysis. At this stage of the perceptual process, a specialist strives to determine an absolute or difference thresholds by utilizing the existing methods of psychophysics. The threshold data is affiliated with the effort exerted in detecting a signal in extreme conditions.

The term operative threshold can be applied to the absolute or difference threshold that defines an optimal effort for detecting or discriminating various stimuli during task performance. In task analysis, operative threshold is determined by multiplying the value of absolute or difference

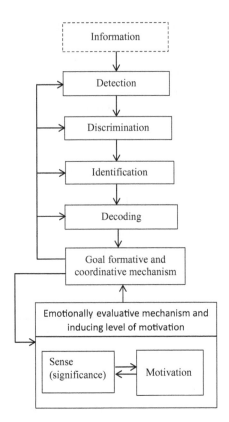

FIGURE 4.1
Self-regulation model of perceptual process.

thresholds by a value from 10 to 15 (Dmitrieva, 1964). Signal detection is determined by the sensitivity of the sense organs and the criteria used by a person. In psychophysics, there is the notion of the "threshold region", which means that the threshold is not a constant. Bardin (1982) demonstrated that the observer, while performing the task of signal detection of noise, can use additional criteria corresponding to various axes of the sensory space. The subjects began to use additional criteria such as perceiving previously unnoticed qualities in the acoustical stimuli, which they used as discrimination criteria for signal differentiation in the noisy environment. A noisy environment can be defined as a situation in which two signals are very difficult to discriminate on the basis of loudness. For example, the subjects reported that the sound seemed to be dimmed, resonant, dull, and so on.

According to SSAT, this means that a subject changes his/her strategy of signal detection based on mechanisms of self-regulation. This brief overview of detection stage demonstrates that a number of questions can be raised at this stage of perceptual analysis. A more detailed analysis of activity strategies during performance of the signal detection tasks is presented in Bedny, Karwowski and Bedny (2015).

Discrimination is the selection of individual features of the object in accordance with the task's goal and the formation of a perceptual image. At this stage, sensory features of a stimulus are selected in accordance with the specifics of a stimulus, or in addition, based on the requirements of a perceptual task. After that selected features are integrated into a structurally organized system. The end result of this stage is the formation of an image of a stimulus or an object. This stage combines active cognitive processes and innate mechanisms of the perceptual system. The aforementioned stage of perceptual processes is particularly important when the subject is involved in visually perceiving an unfamiliar object. A subject then develops a counter of the object and constructs its perceptual image.

Other types of perceptual process are identification and recognition. As an example, we can consider the ability of a subject to relate a perceived object to previously presented objects. A subject is striving to match current stimuli with ones stored in memory. The perceptual process is integrated with the function of memory during recognition. Recognition should be distinguished from identification.

Identification always suggests dividing all presented stimuli into two classes: those that are identical according to all features to the stored in memory templates (positive identification) by all features, and those that are not identical to those templates at least by one feature (negative identification). Sometimes a template can be presented externally, and identification is not accompanied by discovering the inner contents of the stimulus. In this case, the recognition is in play instead.

Decoding or Recognition is the ability of a subject to relate a presented object to previously perceived objects. Subjects match current stimuli with the ones that are stored in memory. In recognition, perceptual processes are integrated intensively with the function of the long-term memory, which might involve thinking operations. Conscious processes are important at this stage.

According to Shekhter (1967), recognition is distinguished from identification because it includes additional categorization (classification) and verbal designation. For example, in the visual perceptual process, recognition involves selection of verbal templates that can correspond to their visual equivalents. The success of perception to some extent depends on the number of verbal templates, compatibility of verbal and visual templates, capacity of working memory, past experience, and training. Taking this information into account is important for the development of training programs for the operators.

The decoding stage involves discovering the relationship of the sign and the object which is denoted by this sign. Decoding is the process of transforming a signal's image into an image of an object. An abstract code makes this process more difficult. The speed, precision, and reliability of such process depends on the relationship between an "alphabet of signals" and an "alphabet of objects." Associating an abstract code of a signal with its meaning takes a long time and requires a lot of mental effort. Requirements for how the information should be presented can change depending on what stage of information reception is critical for the considered perceptual process.

Sometimes it is difficult to draw clear boundaries between detection, discrimination, and identification stages in order to identify the particular stage in the perception of familiar objects. Here, a genetic method helps to clearly define the aforementioned stages. This method considers cognitive processes during their development (Vygotsky, 1978). The study of the acquisition process is an important tool for analyzing cognitive processes in general.

According to SSAT, the consideration of cognitive processes or activity as a goal-directed, self-regulative system is regarded as functional analysis of activity (Bedny, Karwowski and Bedny, 2015; Bedny, 2015) in which informational and energetic components of activity are interdependent. Hence, the presented model of perceptual process includes cognitive and emotionally motivational mechanisms. Additionally, it includes goal formative and coordinative block. The decoding stage, and especially goal formative and the coordinative stage, are tightly connected with the thinking process. The last one is important in the creation of a perceptual process goal (or sometimes an independent perceptual task), and in the coordination of all other blocks' functioning.

The *evaluative and inducing level of the motivation block* consists of two sub-blocks that influence each other. The sub-block "sense" (significance) is responsible for the evaluation of personal significance of the received information and of the perceptual features of a situation. The sub-block "motivation" refers to the inducing components of a perceptual process. The regulative integrator coordinates the functioning of processing blocks. Hence the model of self-regulation of the perceptual process is goal directed and integrates informational and energetic components of activity. The contents of *the goal formative and coordinated block* depends on the specifics of the task at hand.

4.2 Self-Regulative Model of Memory

Before we start to discuss the self-regulative model of memory created in SSAT, it is necessary to give a brief analysis of the memory studies in general and AAT. The cognitive process of storing or remembering information, and later retrieving it, is known as memory. We can remember episodes, general facts, and skills or procedures. First and foremost, memory is a goal-directed process that involves various mnemonic actions and operations. We need to consciously pay attention to information that needs to be memorized.

For example, the experiment conducted by P. Zinchenko (1961) suggested that the classification of information depends on the goal of the activity. This study used pictures and numbers on cards. Subjects were instructed to organize the cards either by pictures or by numbers that have been presented on the cards. The subjects who were instructed to organize cards by the pictures were unable to recall the numbers. In fact, some insisted there were no numbers on the cards, while those who were instructed to organize the cards by numbers could not recall the pictures. This experiment demonstrated that memorization depends on the particular feature of the stimulus and on the way the material is used. In other words, memorization of material is stipulated by motives, goals, and the method of activity performance. We need to consciously pay attention to information that needs to be memorized. Activity theory (AT) distinguishes between voluntary and involuntary memory. Cognitive psychology utilizes such terms as deliberate or intentional memorization and incidental memorization, which often takes place without awareness of the memorization process. This method of memorization is not as productive as intentional memorization.

The other aspect of memory is storing information. In order to be remembered, our experiences must be recorded by our brain. This record is known as memory trace that should be held in stable form for later use. This information can gradually fade away or can stay in memory forever. The final aspect of memory is the retrieval of required information for later use.

Mnemonic actions are the basic elements of memory that transform held in our memory units of information and determine strategies of memorization and memory functioning in general. Mnemonic actions determine strategies of memorization and memory functioning in general. Examples of mnemonic actions are manipulation of information in working memory, extracting information from long-term memory, maintaining information in short-term memory, and so on. Operative units of memory, known in cognitive psychology as "chunks", are used for mnemonic action performance. In contrast, mnemonic operations are performed automatically without awareness. Therefore, information processing in memory is considered not only as a process but also as a system of conscious mnemonic actions or unconscious automatized mnemonic operations. Motor components of activity also help to organize external material. In other words, the memorization of material is stipulated by motives, goals, and the method of activity performance.

Memory is closely related to other mental processes and is used in perception, thinking, language, decision-making, and so on, which creates some difficulties when studying memory as an independent cognitive process. AT identifies various types of memory depending on the characteristics of activity. There are three basic criteria for the classification of memory: (1) based on the nature of mental activity, memory is divided into motor, emotional, imaginative, and verbally logical; (2) based on the goal of activity, it's divided into voluntary and involuntary; (3) after acquiring ideas of cognitive psychology, general AT scientists started distinguishing memory based on the duration of retention of information (sensory, short-term, long-term, and operational or working memory).

Some scientists (Anderson, 1984) suggest that short-term and long-term memory are not separate components but they simply present activation states of single memory. According to this understanding of memory, all information stored in memory can be in an active or inactive state. Information that is currently active includes incoming information and some information previously stored in memory. A subject pays attention to this information.

The goal of *mnemonic tasks* is memorization of some information. Such tasks are important for memorizing instructions and various data for the further performance of various tasks. However, there are *tasks with the increasing memory workload*. For example, if an item of information is presented to an operator, she/he evaluates the importance of the elements of this item, puts the elements in a specific order, and then finally keeps this information in the short-term memory during the task performance. Such tasks complicate the task performance due to the memory involvement. When a person performs purely mnemonic tasks, the term *short-term* memory is used. When an operator is involved in performance of *tasks* in a work situation *with increasing memory workload*, the term *operative or working memory* is utilized. Studies of memory functioning during performance of such tasks are important for reduction of errors and stress, and for improving productivity.

The operative or working memory is utilized to examine, to compare, and/or transform information according to the goal of a task. Operative or working memory also uses information acquired through immediate perceptual processing or information that has been extracted from long-term memory. Thus, thinking is tightly connected with working memory as well as with the other types of memory. Information from operative or working memory can be transferred to long-term memory. Information can be transferred from operative or working memory to the long-term memory. It can be also transferred from sensory memory into short-term memory for the further detailed analysis.. This type of memory uses information in verbally logical and imaginative forms. Studies demonstrate that operative or working memory combined with thinking is responsible for execution of the cognitive, motor actions, and operations that are involved in task performance. Memory also interacts with emotionally motivational processes, and such interaction is a critical factor for the functioning of memory.

All of the above data demonstrate that the memory system includes processes of control and monitoring involved in organizing, transforming, and structuring information. Such a memory system can be depicted by interconnections of separate stages or blocks of memory. Feed-forward and feedback interconnections between various blocks facilitate transformation and distribution of information in memory between different memory blocks. These transformations are associated with such functions as thinking, decision-making, and mechanisms of attention. Therefore, memory can be also described as a goal-directed, self-regulative system.

The self-regulation of mnemonic activity should be taken into account when developing the memory model. The above-presented material leads us the consideration of the proposed model of memory that can be used for analysis of tasks that involve memory functions. This model reflects only some basic aspects of memory functioning during task performance (see Figure 4.2). At the same time, the model helps us to select adequate theoretical data useful in studying memory during performance of a specific task.

The model presented can be interpreted primarily as a multi-store model of memory. However, there are other concepts of memory. According to the levels of the processing model of memory, incoming information is processed by a central processor (Craik and Lockhart, 1972) that has a number of processing levels. The deeper the level of processing is, the better the storage of information in memory is. The deeper the level of analysis in memory is, the more important function blocks 3 and 5 are. We consider our memory blocks to be relevant to levels of information processing in memory. Of course, these blocks are not strictly identical to the described levels of processing. However, it is clear that the storage of memory or levels of processing in this model can be described similarly. The above-presented material clearly demonstrates that memory is involved in the functioning of all cognitive processes. Therefore, considering memory separately from other cognitive processes is artificial to some extent.

Thus, in the model suggested by us some blocks reflect the close connection of memory with such cognitive processes as perception, thinking (intellectual block), and emotionally motivational processes (evaluative and inducing levels of motivation).

Let us consider how this model describes interaction of memory with perception and thinking. The relationship of sensory processes with perception that is demonstrated in our model is well established. Such stages as detection and discrimination are considered as sensory processing stages. In our model, they are associated with the functioning of sensory memory (SM). However, identification, recognition, and decoding are all related to the perceptual process that also involves the functioning of short-term memory. Perception always provides some structural organization to the incoming information. Moreover, organizing incoming information structurally cannot often be prevented and is performed automatically. This means that perception includes automated processing mechanisms and the decoding or interpretation of information during which a subject gives a signal a certain meaning.

For example, when a spot appears on the radar screen of a ship, an operator not only identifies it as a spot but she/he also interprets it as a friendly or enemy ship. Interpretation is always carried out in accordance with a certain system of rules. Therefore, special instructions that are given to an operator are important. Context plays a significant role in the interpretation of incoming information. Two streams of information interact during perception: one coming from the external environment and another one from the depth of

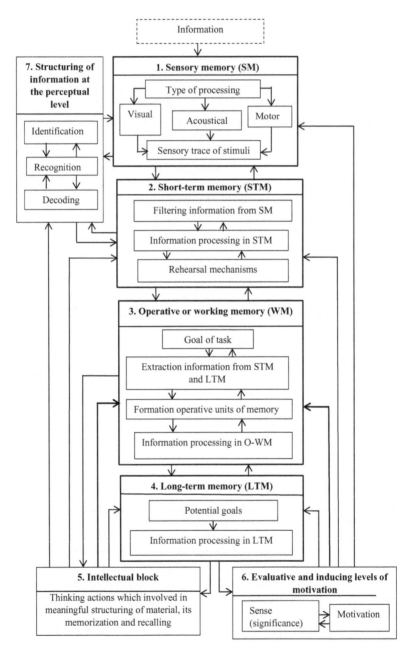

FIGURE 4.2
Self-regulative model of memory.

memory. Constant revision of expectations using some logical rules during perception, making decisions about received signals, and assignment of a certain meaning to such signals suggests the existence of mechanisms that are responsible for these functions.

In cognitive psychology, such mechanisms include processes for monitoring and control that govern the distribution of the flow of information between different memory storages. Various authors assign different names to these mechanisms. These mechanisms can function at conscious and/or unconscious levels of processing. The functioning of these mechanisms indicates that they belong to thinking that processes information in our memory. So the memory model should demonstrate the relationship of memory with perception and thinking. Thus, in our model we have two function blocks: block 7 – structuring of information at the perceptual level and block 5 – intellectual block. Block 7 is tightly connected with block 5, which is involved in the structuring of information. Intellectual block 5 directly interconnects with blocks 7 (structuring information at the perceptual level) and block 2 (short-term memory). At the same time, intellectual block 5 is connected with block 1 (sensory memory) indirectly through blocks 7 and 2. Thus, the model of memory presented demonstrates that even at the first stage of memory functioning, intellectual block 5 still plays some role in memorization.

In our model, there are also short-term memory block 2 and operative or working memory block 3. Due to the interaction of block 1 and block 2 with intellectual and perceptual blocks (blocks 5 and 7), the functioning of sensory and short-term memory is goal directed. This influences the selection and storage of information at the early stages of memory functioning. Intellectual block 5 in its reduced form can affect automated operations that dominate in sensory memory.

Block 2 has a leading role in purely mnemonic tasks. In contrast, block 3 plays a main role in performing tasks in a production environment that cannot be considered as purely mnemonic.

Interaction of intellectual block 5, perceptual block 7, and emotionally motivational block 6 with sensory memory block 1 and short-term memory block 2 demonstrates that these two types of memory cannot be considered as having rigidly defined properties or characteristics. When studying the visual short-term memory, Zinchenko et al. (1980) proved how consciousness, motivation, thinking, and so on can affect these types of memory. These processes are not hard-programmed and invariant as it has been assumed in the models of visual memory suggested by Sperling (1960). The amount of information that can be reproduced turned out to depend on physical conditions (the time of exposure to a stimuli), and the individual characteristics of a subject (levels of motivation, etc.).

Zinchenko et al. (1980) also demonstrated how verbal repetition and the duration of concentration affects the retention of a visual stimulus in visual memory. Intellectual block 5 also interacts with block 3 (operative or working memory). Designated by a bold line, connection of block 5 and block 3 is critically important. The intellectual block 5 can also interact with block 4.

Interaction of blocks 5, 3, and 4 has an important role in the creation of semantic memory (Lindsay and Norman, 1977) which, in turn, is critical for the functioning of long-term memory.

Cognitive psychology also presents evidence that various cognitive processes are involved in functioning of other types of memory. For example, the collected data proves that information is stored in short-term memory acoustically and semantically.

Thinking is especially important for functioning of short-term memory and working memory. The most active processing of information takes place in working memory. Thinking facilitates awareness of what information is stored in this memory. A possibility to storing information semantically or in terms of its meaning clearly demonstrates the role of thinking in the functioning of memory. The level of processing model (Lindsay and Norman, 1977) of human memory also emphasizes the importance of thinking for memory functioning and interpretation of meaning of the memorized information. Deep processing means significant involvement of thinking in memory functioning. Interaction of mnemonic, perceptual, and intellectual functions made up images of the memorized items and therefore increasing the probability of memorization and recalling various information.

At the same time, other factors also play a specific role in memorization. For example, significance (emotionally motivational factors) of information plays a specific role in memorization. Various factors influence each other during memorization. The analysis of block 5's interaction with other blocks of memory demonstrates how intellectual block 5 includes various mechanisms that monitor other described function blocks of memory as well as how the flow of information between different memory storages is distributed. In general, this block plays a significant role in the functioning of operative or working memory and long-term memory.

Now, let's summarize the already-discussed data for a better understanding of memory functioning as a goal-directed, self-regulative system, which becomes handy when this model is utilized in task analysis. Information in long-term memory is organized into a complex structure that interconnects units of information and creates past experiences. Remembering and extracting information from long-term memory involves thinking operations and utilizing mental strategies necessary for task performance. Hence, the interaction of intellectual block (block 5) and long-term memory (block 4) is achieved through the working memory block (block 3). The last block is vital for purely *mnemonic* tasks and tasks *with increasing memory workload* in a production environment. This type of task, with significant memory workload, is specifically important for ergonomists, because an operator has to select information from the environment, compare it with information that is retrieved from memory for a specific purpose, put information in specific order, and keep this order in memory until a task or its stage is completed. This is especially difficult when an operator makes numerous decisions during task performance. The study of memory functions in such cases involves analysis of mental strategies and mechanisms of thinking that guide working memory and long-term memory. At the same time, the thinking block does not affect sensory memory directly. This block influences block 1 indirectly through function block 7.

Tasks that are performed in stressful conditions. Thus, block 6 (evaluative and inducing levels of motivation) becomes critical in such situations, that involve emotional stress. Block 6 has two sub-blocks. One involves the evaluation of a situation's significance or its separate characteristics (sense or subjective significance). The other is an inducing or motivational sub-block that affects all function blocks of memory. Thus, cognitive, emotionally evaluative, and motivational mechanisms are involved in selecting, adding, and storing information in memory. Feed-forward and feedback connections between various blocks demonstrate that our memory function is a goal-directed, self-regulative system. Thanks to self-regulation, a subject can utilize various strategies of selecting, storing in memory, and recalling information. Such strategies depend on a subject's past experience, individual differences, and the specifics of a task. In each block, information can be processed sequentially or in parallel.

The presented model emphasizes the special role operative or working memory plays in task analysis. In SSAT, operative or working memory is distinguished from short-term memory. Working memory is understood here as mnemonic processes that are serving activity that is directly involved in a specific task performance. The goal is very important for this type of memory and for consciousness or awareness of memorized data. Long-term memory carries potential goals that can be extracted and then transferred into operative or working memory. The interconnection of blocks 5 and 6 with blocks 3 demonstrates the importance of intellectual and emotionally motivational mechanisms for memory functioning and specifically for working memory during task performance. Blocks 3 and 5 are the main mechanisms for developing a *mental model of a task or a situation.*

For a more detailed analysis of a mnemonic mental model, operative thinking should be analyzed. Intellectual block 5 is involved in identifying and structuring information on the perceptual level, and the interaction with operative or working memory. In particular, these blocks are used for classifying data and organizing it in operative or working memory. This block is involved in developing a strategy of memorizing and reproducing information, and in constructing the execution program of mnemonic actions in block 3. Verbally logical components of activity are involved at this stage of memory functioning. Block 5 is also involved in decision-making related to the implementation of verbal or motor actions. This aspect of task performance is not considered here. Instead, we choose to focus on the internalization aspects of information processing. In the model considered, intellectual block 5 has various feed-forward and feedback connections with block 3 (WM) and block 4 (LTM).

According to Norman (Norman, 1976), there are three main components of memory: the data base, the interpretive system, and the monitor. The monitor guides the interpretive system and examines the database. The interaction of the monitor and the database sets up the strategies used to evaluate information. The interpretive system examines the database structure.

According to our model, the *interpretive system and the monitor are thinking mechanisms*. Thus, memory cannot be considered as an independent system. It includes other mechanisms associated with various cognitive processes that are integrated into a self-regulated system.

Function blocks represent basic mechanisms that include various psychological operations. Main units of analysis for each function block are cognitive actions, operations, and specifically mnemonic actions.

The concept *"operative units of memory"* has profound significance when studying memory. Operative units of memory are the semantically holistic entities used by mnemonic actions during task performance. Working memory is a key mechanism for formatting operative units of mnemonic actions. In cognitive psychology, the term "chunks" is used instead of the term "operative units". These two terms have a similar meaning. However, there are some differences in understanding of these terms. Operative units of memory are tightly connected with concepts of action and the goal of action. In SSAT, there are also such terms as operative units of thinking and operative units of perception. These units are generally labeled as *operative units of information* (OUI) that are formed during the acquisition of a specific activity. Such units can be understood as contextually defined entities or items (image, concept, statement, etc.) that are formed through experiences that enables a subject to manipulate them mentally using various cognitive actions. External motor actions manipulate material objects while cognitive actions manipulate mental items (operative units of information) in accordance with the goal of an action. When compared with short-term memory, information can be stored longer in working memory through the interaction of short-term and long-term memory. Temporary storage of operational information requires performing a specific task that is facilitated by continuous interaction of short- and long-term memory and the transformation of data from one type of memory to another. Working memory provides chunking of material, rehearses it, relates new information to information that is already stored in memory, structures information, and so on. This mnemonic activity is performed by using conscious mnemonic or other cognitive actions and unconscious mental operations. Operational memorization (keeping required information until a task is completed), can be presented as an interaction of working and long-term memory. It includes not just voluntary but also involuntary reproduction of new information delivered by operative or working memory.

Long-term storage of information is possible at the level of operative or working memory. Voluntary and involuntary memorization and reproduction at the level of operative or working memory depends largely on the goal of the activity undertaken. In addition, the significance of incoming information for a subject, strategies of activity, and the individual characteristics of a subject are of importance.

The proposed model shows why memory should be considered a self-regulative system as well and why special attention should be paid to various strategies of memory functioning during task performance. The suggested memory model offers very important information about how memory functions when specific tasks are performed. It depicts memory as a structurally organized self-regulative system and facilitates efficient application of available psychology data about functioning of the operator's memory during task performance.

4.3 Applying Experimental Methods to Study of Attention

The process of receiving information is a selective one. Subjects don't receive all incoming information equally; they handle it actively, accepting some information and ignoring other information. Such selectivity is facilitated by the mechanism of attention. The more complex the task is, the higher the level of attention or mental effort is required. The subject can perform time-sharing tasks differently depending on the complexity of each task.

There are various methods of studying attention. One method involves performance of two tasks simultaneously to figure out what is being attended to in one task and ignored in the other. In this chapter, we also study how a subject performs two tasks. However, tasks are performed in sequence. Each task consists of performing complex choice reactions. A more detailed analysis of the experimental data is presented in Bedny et al. (2015).

Before a brief review of our experimental data, it is necessary to consider some theoretical data regarding the studies of attention. Currently, there are two classes of attention models. One considers attention as mental effort (Kahneman, 1973); the other treats attention as an information-processing system (Lindsay and Norman, 1977). The model suggested by Norman is interesting, but it ignores energetic components of activity. The mental effort models of attention can be related to either *single-resource theory* or *multiple-resource theory* (Wickens and McGarley, 2008). There are *data-driven processing* and *conceptually driven processing*. Data-driven processing starts with sensory-perceptual data and conceptually driven processing starts with general knowledge of the events and specific expectations generated by subjects' knowledge. Data-driven processing is more automatic and depends on input information. It is to some extent similar to the concept of involuntarily attention, while conceptually driven processing is close to the concept of voluntary attention in AT.

The other theoretical aspect of studying attention is the relationship between peripheral and internal processes. One of the deficiencies of *multiple-resource theory* is its inability to determine whether the advantage of cross-modality tasks over intra-modality tasks are attributable to central or peripheral processes. Time-sharing may not in fact be a result of central resources but rather the result of peripheral factors that constitute the two intra-model tasks.

The above-presented material shows that there are two basic methods of studying attention. One method suggest that subjects perform two tasks simultaneously, and the other assumes that two tasks are performed in sequence. For our analysis, the method suggested by Pashler (1998) is the preferred one. In their studies, the first task included two acoustical stimuli, and the second one involved three visual stimuli. The interval between these two tasks changed from 50 to 450 milliseconds. It has been discovered that the shorter the time interval between these two tasks, the longer the response time is for the second task.

In our study, two tasks were also performed in sequence, but the second task was performed immediately after the first task was completed. However, the complexity of the two tasks varied. We analyzed how the complexity of the first task influenced the performance of the second task. We also considered how this influence was altered when the complexity of the second task also changed.

It is also important to consider how the modality of the stimuli can influence the performance of two tasks. There are inter-modal and intra-modal procedures that are utilized in the studies of the time-sharing tasks. For example, in the intra-modality study, a message is given through one earphone to the left ear and to the right ear through another earphone. In the intra-modal procedure, messages are of the same modality (two auditory or two visual). In this case, they can mask one another; in other words, the peripheral factor influences task performance because these messages require the *same attentional resource*. In inter-modal or cross-modal time-sharing tasks, one message can be visual and another can be acoustical. Therefore, such tasks require *separate attentional resources*. Therefore, in a cross-modal time-sharing task, the influence of peripheral processes is eliminated. Due to this factor, cross-modal time-sharing can be more efficient. Such conclusion is an important argument for using the *multi-store model* of attention. In order to demonstrate the invalidity of such a conclusion and to prove the advantage of the *single-store model* of attention, one should conduct an experiment in which peripheral factors are excluded. If, in a time-sharing task where peripheral factors are excluded, and both tasks would still interfere with each other depending on the complexity of the task, only single *single-resource theory* can explain such interference.

Many single-channel theories of attention are based on the research of the psychological refractory period (Meyer and Kieras, 1997; Pashler and Johnston, 1998, etc.). However, the psychological refractory period is only one mechanism of attention. Attention also depends on consciously regulated strategies. For example, in our study (Bedny, 1987) it has been discovered that subjects can start the formation of a second action program while executing the first action. Such strategy depends on the complexity of the first and second motor actions. If the first action is very complex, or is associated with danger, subjects can utilize a sequential strategy. As we will demonstrate in this work, an ability to perform elements of activity in parallel depends not only on mechanisms of the psychological refractory period but also on strategies of self-regulation of activity in general.

Thus, currently, there is no single theory of attention that would completely explain the phenomenon of attention. Here, we want to present some new data that can be useful for the understanding of attention mechanisms. In this study, the subjects performed sequential tasks of various complexity (Bedny, Karwowski and Bedny, 2015). The experimental curves presented show the performance time of complex reactions performed by the left and the right hands. The model of attention has been developed as a result of this study. The model of attention presented in this chapter is an enhanced one.

Let us consider shortly our experiment that involved interaction of choice reactions. The experimental bench that had a subject panel on one side and a research panel on the other side had been designed for this study. The subject panel had a digital gauge that displayed these signals: lighted numbers and a sound device that produced a clear tone sound. There were two start positions on the panel for left and right hands' fingers of the subject. The switches for left and right hands were located radially from the start positions. Four switches were used for the left hand and eight for the right hand. Two meters on the researcher's panel measured reaction time for the left and right hands. University students had been selected to participate in this experiment. For two days, they were trained (1 hour session per day) to perform required tasks. The number of sound stimuli for the left hand could be increased from one to four. The selected sound stimuli could be clearly discriminated by subjects without any difficulty. The number of visual stimuli for the right hand could be increased from 1 to 8. The second reaction was performed by the right hand as a response to the visual stimulus that was presented only after performance of the first reaction by the left hand to the acoustical stimulus. There were also scenarios when subjects should react only to visual stimuli with the right hand. Thus, a researcher could present a variety of combinations of acoustical and visual stimuli. The experiment included seven sets with different combinations of sound and visual signals.

It is important to emphasize that in classical studies of attention, two streams of acoustic information are presented simultaneously. One stream is presented to the right ear and the other to the left ear (Lindsey and Norman, 1977). In our study, two streams of information are presented in sequence

TABLE 4.1

The General Plan of the Experiment

Day of Experiment	The Number of the Set of Experiment	Program (Relationship between Sound and Visual Signals)
The first	1	0–1; 0–2; 0–4; 0–6; 0–8;
	2	1–1; 2–1; 3–1; 4–1;
	3	1–2; 1–4; 1–6; 1–8;
The second and the third	4	2–2; 2–4; 2–6; 2–8;
	5	3–2; 3–4; 3–6; 3–8;
	6	4–2; 4–4; 4–6; 4–8;
The forth	7	$2' - 2; 2' - 4; 2' - 6; 2' - 8;$

and have different modality. The general plan of the experiment is depicted in Table 4.1.

Let us consider some programs. In column three, the first number reflects the number of sound signals and the second number shows the number of visual signals. For example, in the first day the number of sound signals was zero and the number of visual stimuli varied from 1 to 8. The subjects had to react only to the corresponding number of visual signals by pushing the switch by the right hand. In the second set of experiments, the subjects had to react to the sound signal by pushing the switch by the left hand. There were 4 different sounds signals and each signal had a corresponding switch on the left side of the panel. After that, subjects react only to the same visual signal with number 1. At the third set of experiments, only one sound signal and visual signals that changed from 2 to 8 were presented. At the fourth set of experiments, 2 sound signals and visual signals that changed from 2 to 8 were presented. In the fifth and sixth set of experiments, the 3rd and 4th sound signals were presented. The number of visual signals varied from 2 to 8.

At the last day of this experiment, the program was the same as on day 4. However, two sound signals were selected with similar tones and their discrimination was evaluated by the subjects as more challenging. In our studies, it was important to demonstrate that the position of the logarithmic curve for the visual stimulus can vary depending on the complexity of a previous reaction to a sound signal.

The task consisted of the following sequence of events:

1. Subjects kept their sight on the digital indicator. When one of four sound signals was delivered through a headphone to the left ear, they had to react by pushing the corresponding switch with the left hand. During the execution of this action, subjects had to keep looking at the digital indicator.

2. Immediately after the response to the sound signal, one of eight numbers lit up on the digital indicator and the subject had to react

with the right hand by pushing the corresponding switch. Each reaction time has been measured separately. The number of alternatives in each set of experiments was the same.

3. A warning signal was given before the start of the next sequence of signals.

The within-subjects two-way ANOVA statistical analysis has been performed to test the effects of the number of visual stimuli, acoustical stimuli, and their interaction with reaction time to the visual stimulus. The statistical data is not presented in this abbreviated description of our study. The average result of all measures for each set of experiment was calculated based on 40 reactions of each subject to the corresponding signals. Erroneous reactions were not considered.

The result of the experiment is presented in Figure 4.3.

In the first set of our experiment, only visual signals were presented. The number of signals was increased from 1 to 8. This set was marked from 0–1 up to 0–8 indicating there was no previous reaction with the left hand to the sound stimulus. Reaction with right hand to visual stimulus is performed when number of stimuli is changed from 1 to 8. The interdependence of reaction time and the number of stimuli can be depicted by logarithmic curve (see Figure 4.3 with the symbol ▲ – no acoustical signals). This function is called Hick's law (Hick, 1952) which is well known in psychology. He was the first to discover the linear relationship between the average information (I) and reaction time (RT).

$$RT = a + bI \qquad (4.1)$$

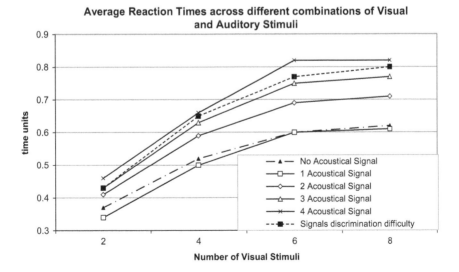

FIGURE 4.3
Reaction time to the visual stimuli performed by the right hand.

where a and b are constant coefficients that are dependent on the conditions of the experiment. This relationship between the information conveyed by the stimulus and the choice reaction time to the stimulus contributed to considering a subject primarily as an information-processing mechanism. The concept of "information" became an important one in psychology. According to this law, people have a relatively constant bandwidth of information processing that is reflected by the inverse slope of the Hick law function (bits/sec).

In the second set of our experiment, the number of sound signals for the left-hand reaction was varied (1 out of 4) and only 1 visual stimulus was given for right-hand reaction. The second reaction to simple visual signal (only number 1 was presented by digital indicator) followed immediately after the first sound signal. It has been discovered that reaction time to the second visual signal was 0.12 seconds. However, in our preliminary study, a simple right-hand reaction, when only one visual signal was presented, took 0.23 seconds. Thus, preliminary reaction to sound signals of different complexity reduced reaction time to the second simple visual signal. The result of the second set of experiments are presented on the Figure 4.3.

The result of the third set of our experiment, when only 1 sound signal was used and a number of visual signals for the right-hand reaction was changed from 2 to 8, is depicted by curve with symbol □ in Figure 4.3. It was discovered that the reaction time of the right hand after a simple left-hand reaction to the sound stimulus has a tendency to decrease in comparison to the reaction time of the right hand without previous reaction of the left hand. This difference is not statistically significant. Nevertheless, this result is interesting because curves ▲ and □ do not intersect.

On the second and third days, the fourth, fifth and sixth sets of the experiment were conducted. Curves with symbols ◇, Δ, and x are presented in Figure 4.3. The result of this experiment demonstrates that the more complex the first reaction to the sound signals is, the higher is the position of the curve for visual signals.

The results of the fourth day of testing, or set 7 (when subjective discrimination of two sound signals were evaluated as difficult), demonstrate (see curve with symbol ■) that discriminative features of an acoustical stimulus have approximately the same effect as the number of acoustical stimuli. One can see that curve ■ is positioned between curves Δ and x. The position of this curve is much higher than the position of curve ◇, when subjects preliminarily react to two well-distinguished sound stimuli.

The data demonstrate that the more complex the previous reaction to the sound signal is, the higher is the curve position which reflects the reaction time of the visual stimulus. We compared the significance of the difference between reaction time for when the reaction was performed only by the right hand (curve ▲), and by the right hand with previous left-hand reactions (curves ◇, Δ, ■, and x).

Hence, a rise in the complexity of the previous reaction influences the performance of the following one. The level of complexity of the second reaction is very important as well. The more complex the previous and the following reactions are, the more they influence each other. The data obtained contradict Hick's law. This outcome can be interpreted based on attention theory. According to cognitive psychology, there were two streams of information that were presented during the time-sharing task performance. One source of information was presented to the right ear and the other one was presented to the left ear. In our experiment, two streams of information were presented sequentially, so a subject dealt with only one source of information at each time period. The more complicated the previous and the following portions of information and response selection processes in both tasks were, the more difficult it was to shift active attention from one source of information to another. This has been observed despite the fact that two such steams of information are traditionally considered to be independent. The result of the experiments conducted proves that they cannot be considered as independent.

The debriefing of the subjects uncovered an interesting fact. During the performance of the most complicated task, where 4 acoustical signals and 8 visual signals were used, there was the subjectively noticeable break between identifying the digit and making a decision about the performance of the second reaction. (The subject said, "I see the digit, but I can't make a decision and move my hand. My hand is glued to the start position").

The data obtained offered an explanation for how mechanisms of attention influence the strategies of information processing. Recognition of stimulus is made by a passive automatic process using a low level of attention, but making a decision is linked to the active processes with a high level of attention. The active processes reorganize slower than the passive automatic processes. The reorganization of attention mechanisms, when unconnected sequential portions of information are presented, is the same as when a subject shifts attention from one portion to another portion of simultaneously presented information. In our experiment, a subject cannot keep all information about two reactions in short-term memory. It becomes necessary to use information from long-term memory. The search for information in long-term memory by a scanning device can start at any node in a structure of the database of long-term memory (Norman, 1976). Hence, during the process of extracting the information from long-term memory, the alphabet used by a subject constantly changed. This alphabet is dynamic. As a result, the speed of information processing changed as well.

Let us look again at the results of the experiments conducted on the fourth day, when two acoustical signals subjectively difficult for differentiation were presented. The similarity of curves Δ (3 acoustical signals are presented) and curve ■ (acoustical signals that are difficult to differentiate were used) in Figure 4.3 allows us to conclude that the deterioration in differentiation of the sound signals influences the reaction time on the visual signal the same way the increase in the number of alternative sound signals does. It has been discovered that the reaction time on the visual stimulus performed by the right hand increased as a result of the deterioration in differentiation of the sound signal in the first reaction performed by the left hand when it became more difficult to decide which sound signal was presented. The first task (react with the left hand to a sound stimulus) became more complex and therefore more difficult for the subjects. As a result, the process of adjusting the mechanisms of attention from the first to the second task became more complicated. Hence, the second reaction time depends not only on the amount of information presented for the first reaction but also on the difference between signals. This demonstrates that the complexity of two considered subtasks determines the strategies of shifting attention from one task to another.

Comparing the results of the observation, debriefing of the subjects, and analysis of experimental data shows that individuals do not react to various stimuli but actively select and interact with the information. Depending on the results obtained from activity, individuals reformulate their goals and strategies. This results in the transformation of the conscious content of activity into unconscious and vice versa. Voluntary attention is the mechanism through which consciousness is attained. Our observation showed that during the experiment, subjects shifted their attention between two tasks, allocating their attention and efforts in attempt to perform one task quicker, slowing down other tasks, or vice versa, correcting errors, and attempting to enhance their strategy. As a result, individual actions were integrated into a holistic structure based on self-regulative mechanisms. Thus, describing an individual behavior in terms of stimulus-response is inaccurate. Individual behavior can't be explained as the sum of independent reactions to a series of independent stimuli, as described by Hick's law. When developing a model of attention, it's necessary to take into consideration the self-regulation process and voluntary and involuntary attention mechanisms. Experimental data demonstrate that the model of attention should incorporate the self-regulation process and take into account various voluntary and involuntary attention mechanisms.

Figure 4.3 clearly demonstrates that reaction time by the right hand depends on the complexity of the reaction by the left hand. The more complex the reactions of both left and right hands are, the more time is required for the second reaction performed by the right hand (the higher the position of right-hand curve is on the graph). When the subjects react only by the right hand, it takes less time. Reactions performed in sequence cannot be considered as independent. They influence each other. Subjects developed various strategies to optimize their performance. Existing ergonomics or cognitive psychology studies consider only isolate reaction time (see, for example, Hick-Hyman law; Hick's law; (Hick, 1952; Hyman, 1953)). The maximum speed of response to a single stimulus cannot be the basis for determining the task performance time. A subject is not a reactive system but a voluntarily acting individual. Our experiment demonstrates that in order to determine the time for task performance one needs to consider the *possible pace utilized by a subject.*

Our study showed that an increase in complexity of each task increased its effect on other tasks even when the tasks were performed in sequence and the subjects used two modality information processing channels. This contradicts the multiple-resource theory, in which people have several different capacities in terms of the resource properties. According to this theory, two considered tasks are independent and should not influence one another. However, our study showed that they did influence each other. Thus, despite the fact that two tasks utilized information of different modalities and were not performed simultaneously, they still influenced each other. Our study clearly demonstrates that the model of attention should be based on the *single-resource theory.*

In our experiment, we presented two modality-different stimuli, where one was acoustical and another one was visual. According to the multiple-resource theory, this should make it easier to perform considered tasks simultaneously. The second factor that should reduce interference of these tasks is that they were performed in sequence. The subjects should perform the second task only after completing the first one. However, our study showed that an increase in complexity of each task increased the effect of its interference even when the tasks were performed in sequence and the subjects used two modality information processing channels. This contradicts multiple-resource theory, in which people have several different capacities in terms of the resource properties. The mechanism that is responsible for the investment of required resources of attention is called "available level of arousal". Physiological studies consider specific and nonspecific arousal. When we evaluate the task difficulty, nonspecific arousal is especially important (Aladjanova et al., 1979). The more complex the task is, the higher the probability is that this task would be difficult for a subject. Hence, the higher the task complexity is, the higher the degree of limited energy resources it requires.

4.4 Self-Regulative Model of Attention

In this chapter, as in Section 4.3, we are going to demonstrate the usefulness of integrating ideas of cognitive psychology with the ideas of SSAT when studying attention. More specifically, in Section 4.3 the cognitive approach is combined with functional analysis developed in SSAT, where attention is described as a goal-directed, self-regulative system. Consideration of psychological functions of attention from the self-regulation perspective is viewed as functional analysis of attention. In this chapter, we briefly consider the modified model of attention that is based on the experimental data presented earlier. In the model presented, two new function blocks are introduced, and some additional changes are made. This modified model of attention is discussed in this section (see Figure 4.4).

The experimental basis for the creation of the model of attention is the study of the interaction of complex choice reactions described in Section 4.3. The basic units of analysis of the model of attention are function blocks that represent the specific stages of information of processing and have feed-forward and feedback connections with other stages. Each stage has a

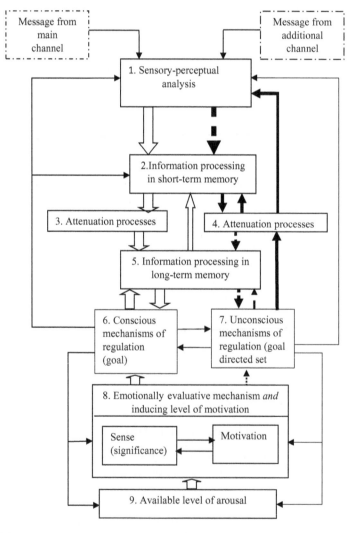

FIGURE 4.4
Self-regulation model of attention.

particular purpose in attention regulation. Depending on the task specifics, each block can interact differently with other stages of the attention process. In our model, there are two information processing channels. One of them is auditory and the other one is visual. According to multiple-resources theory (Wickens and McCarley, 2008), this factor alleviates the process of simultaneously performing two tasks. The second factor that can reduce interference between the tasks is that the subjects perform two tasks in sequence. However, our results demonstrate that in spite of these factors, the tasks interfere with each other. This contradicts multiple-resource theory, wherein people have several different capacities in terms of resource properties. According to this theory, the two tasks studied by us are independent and should not influence one another, but our data show that these tasks influence each other.

Our data clearly demonstrate that the model of attention should be based on *single-resource theory*. Complicated tasks make access to the resources more difficult and diminish the ability to allocate the resources.

There are peripheral and central processing stages of attention. Peripheral analysis starts in the first function block (block 1 – sensory-perceptual analysis). This stage of analysis can be performed automatically without awareness. This information proceeds further, to block 2 (information processing in short-term memory), where it is transmitted from two channels. One channel transmits information from the main task and the other from the additional task. Hence, in this model, we have the main and additional channels of transmitting information. Information in block 2 is conscious due to the feedback from function block 6 (conscious mechanism of regulation) that includes the goal. The conscious thinking operations are particularly important at this stage. The goal is the main mechanism that is responsible for allocation of information and energy resources between two considered channels. Due to taking into account the feedback from the goal to block 1, the sensory-perceptual analysis is more adequately to the goal of task. At the same time, automatic processes dominate in this block. Therefore, selectivity, to some extent, takes place at the sensory-perceptual level. In contrast to Broadbent's (1958) filter model, Treisman (1969) introduced the concept of attenuator as an important mechanism of attention, according to which attenuator as a filter weakens, rather than entirely rejects, unattended information. These inductive relationships between attenuators have been described by Norman (1976). If one becomes more active, the other one becomes less active, which means that different psychological and physiological processes are involved in this mechanism. This has been proved by studies in neuropsychology (Fafrowicz and Marek, 2008; Sokolov, 1969). According to them, activation of some neural attention network of the brain can inhibit another neural structure of attention. Hence, the attenuation blocks simply demonstrate that there are complex relationships between activation and inhibition processes in the neural structure of the brain. Attenuation mechanisms provide, first of all, unconscious influence on others blocks.

The goal is involved in conscious regulation of the main channel of information processing. The goal-directed set (block 7) that has close connections with the orienting reflex plays the leading role in unconscious and involuntary regulation of attention. Each channel of updating information, and specifically additional channels, has limited capacity. Switching from one channel of attention to another is provided by connections between goal (block 6) and goal-directed set (block 7). The more complex the main tasks are, the more difficult it is to switch from block 6 to block 7 and vice versa, and therefore to switch from one channel to another. If goal-directed set (block 7) is transformed into the conscious goal, the role of channels in the processing of information can changed. The main channel becomes the additional one, and the additional channel becomes the main channel of information processing.

Information in block 5 (information processing in long-term memory) is performed under supervision of block 6 or block 7. The message from the main channel is not attenuated. On the contrary, the flowing along the additional information channel is partly attenuated in short-term memory and mostly during its transformation into long-term memory. Block 6 (goal) can partially suppress functioning of the goal-directed set (block 7). Thus, the possibility of regulating information processing in the additional channel is restricted. This, in turn, is associated with activation of attenuation processes in block 4. All this influences the processing of information in blocks 1, 2, and 5. Feedback influences from blocks 6 and 7 to the above blocks are presented in Figure 4.4.

The incoming, usually unexpected, information activates the orienting reflex. The orienting reflex is conveyed by such responses as moving eyes, altering sensitivity of sense organs, and so on. For example, due to the feedback, the goal-directed set can involuntarily be tuning sensory-perceptual analysis or increasing the involuntary level of arousal into attention process. The goal-directed set is close to the concept of goal, but is not sufficiently conscious or can be even totally unconscious. Through the feedback, it gives a goal-directed *tendency* to unconscious regulative processes of attention. Hence, blocks 6 and 7 reflect two types of feedback regulation. One is associated with unconscious regulation, and another one is associated with conscious regulation of attention (see Figure 4.4). The goal-directed set can be transformed into the conscious goal of activity and vice versa. Therefore, there are two interdependent mechanisms. One is called "conscious mechanism of regulation (goal)" and the other is "unconscious mechanisms of regulation (goal-directed set)".

There should be a mechanism that regulates the allocation of resources. Let us consider block 8 (*emotionally evaluative mechanism and inducing level of motivation*) and block 9 (*available level of arousal*) that are involved in this process. Block 8 has two sub-blocks that influence each other. One of them is *emotionally evaluative sub-block* which is involved in assessment of "*sense or significance*" of tasks. The other sub-block determines *motivational or inducing* components of activity. They are involved in creation of the vector "motive → goal" which gives activity its goal-directedness.

The model of attention requires mechanisms that provide energy support not only at the psychological but also at the physiological level. Physiological studies consider specific and nonspecific arousals. The nonspecific arousal plays the leading role in evaluation of the task difficulty. The more complex the task is, the higher the probability that the task would be difficult for the subject. And the more difficult the task is for the subject, the more mental effort required, and block 9 starts to play a more important role.

In the model presented, multiple instances of feedback reflect interaction of the peripheral and the central processes.

We've just discussed how information is received, processed, and stored in memory. Two self-regulative models of cognitive processes have been presented. In this chapter, we also present the self-regulative model of attention. All models previously listed are instrumental in task analysis because they allow prediction of preferable strategies for task performance when receiving information and when mnemonic functions are especially important. SSAT uses a functional block as the basic mechanism of self-regulation models. Such blocks interact with other mechanisms by utilizing feed-forward and feedback connections, which symbolize specific stages of information processing. Each block presents a coordinated system of functions that are most important at a particular processing stage. Each function block determines the scope of questions to be considered at a particular stage of task analysis. The main units of analysis of each function block are cognitive actions and operations.

When studying perceptual processes, sensory actions such as perceptual actions play a leading role. In studies of memory, mnemonic actions come into focus. The information obtained can be compared with data collected by analyzing other function blocks. It is not necessary to use all the function blocks for analysis of a cognitive process. A specialist should pay special attention to the blocks that play leading roles in each task performance.

For instance, if there is a weak stimulus that requires detection of a stimulus in the presence of background noise, then this stage of task performance should be considered, and the analysis of the first function block (detection) is required. Moreover, the detection stage of the perceptual process is so complex in some situations that it appears as an independent task, the purpose of which is detection of a signal in the presence of noise. However, if the signal presented can be easily detected by an operator, function block "detection" can be omitted from the task analysis.

Similar to perception, a mnemonic task is considered as a type of activity that includes goals, emotionally evaluative components, motives, means of memorization, and mnemonic and other cognitive actions. Each block of memory model performs certain functions in task performance. The feed-forward and feedback connections and emotionally motivational blocks allow this model to depict activity as a goal-directed, self-regulative system.

Attention in this chapter is also described as a complex goal-directed, self-regulative system. The goal is the main mechanism that is responsible for switching of attention. The goal is involved in conscious regulation of the main channel of information processing. The more complex the main task is, the more difficult it is to switch attention to another task. The goal-directed set plays a leading role at the unconscious level of attention regulation. The energetic mechanisms of attention (emotionally evaluative mechanism and inducing level of motivation) and available level of arousal also take part in allocation of resources of attention. They are involved in creation of the vector "motive → goal" that gives attention its goal-directedness. Finally, study of attention clearly demonstrates that the model of attention should be based on *single-resource theory.*

The models presented are adequate for the analysis of stages of receiving information and mnemonic functions during performance of various tasks. Psychological processes, including receiving information and processing it in memory, depend on goals, motives, and specificity of tasks, and the models offered here reveal strategies of task performance.

Each function block in the models presented is associated with a set of questions that should be considered by a specialist at each stage of cognitive processes during task analysis.

In the models presented, each block should be primarily analyzed from the theoretical perspectives.

Studies of the cognitive processes considered clearly show the interconnection between cognitive psychology and SSAT. At the same time, cognitive psychology does not pay sufficient attention to various cognitive processes as components of a subject's activity and does not clearly describe the role of goal, motivation, self-regulation, cognitive and motor actions, and so on in task performance. In contrast, our models concentrate on these aspects of perception and memory functioning. The models presented here, as well as other models of self-regulation developed in SSAT, clearly demonstrate how SSAT is closely linked to cognitive psychology and why the suggested models are instrumental in task analysis.

4.5 The Role of Thinking in the Study of Human Performance

The review of the literature in cognitive psychology, and in ergonomics and engineering psychology, demonstrates that human performance studies do not pay sufficient attention to the analysis of thinking processes. For example, thinking and decision-making are presented as one block of an information-processing system (Wickens, Gordon, and Liu, 1998, p. 147). However, only decision-making is discussed further. In Wickens, Gordon, and Liu book, which is very influential, the authors restricted their discussion about thinking to just that one word, "thinking." In the later published by Wickens and Hollands (2000) book, an excellent analysis of data obtained in experiments in cognitive and experimental psychology and its relation to the study of human performance is presented. The information processing models described in this book cover all cognitive processes, excluding thinking (Wickens and Hollands, 2000, p. 11 and p. 295). The authors discuss decision-making without relating it to the thinking process.

In general, there is a tendency in cognitive psychology to pay attention to memory instead of thinking. However, even when a subject performs a task based on information stored in memory, the main mechanisms that facilitate a set of operations with this information are the thinking mechanisms. The involvement of thinking in performance of the problem-solving task enables the information-processing system to get information into memory, restructure it according to the required goal of the task, and so on. In general, thinking is directed to the discovery of new properties, the relationship between phenomena, and the objects of reality that are not directly given in the perceived situation. Thinking plays a leading role in performance of problem-solving tasks. The task-problem very often is not given to a subject in a ready-made form and should be formulated by him/her independently. A subject has to transform objectively given requirements into the subjectively accepted goal, which should be adequate to the objectively presented data.

In the study of thinking as well as in the study of other cognitive processes, different approaches can be identified. In this chapter, we discuss the key principles of studying thinking in cognitive psychology and in AT and SSAT. From a cognitive psychology perspective, the information-processing approach is a dominant one in the studies of thinking. This approach is the cornerstone of contemporary cognitive psychology in general. It should be compared with various approaches of studying thinking in AT.

Cognition unfolded in time as a process that involves thinking mechanisms. Thinking can be presented as a sequence of steps, each of which is associated with the implementation of the set of operations involving information that is coming from the outside environment or from memory. Thinking can be involved even at the earliest stages of information processing.

According to the information-processing approach, thinking is a subsystem of cognition that manipulates symbols in the same way that a computer does. However, the assumption that thinking can be presented as a symbols manipulating system is incorrect. The thinking process involves operating with not just the symbols but also with various mental and material objects and their images that can be presented by the symbolic system. Such symbols have a specific meaning for the human subject. The concept of meaning should also be taken into consideration. The assumption that thinking can be reduced to the process of manipulation of symbols is another limitation of the approaches to thinking considered.

In general, human information processing is a very useful approach in the study of thinking. At the same time, it has some weaknesses. There is no clear terminology for the description of basic psychological concepts. For example, Gilhooly (1988) defines thinking as the process that is directed toward solving a problem and can be regarded as exploration of a mental model of the task to determine a course of action that should be the best one. The author does not have the concept of a goal of the thinking process and does not explain how he understands such concepts as the mental model of actions. When studying thinking, scientists often focus on consideration of solutions of various puzzles which can be perceived as well-defined problems. The possible sequence of moves is represented as a state-action tree (Lindsey and Norman, 1992). However, one or a number of motor moves is preceded by a number of cognitive actions to decide what motor movements to undertake. These cognitive actions are not considered in such studies. Moreover, the terms movement and motor action are not the same. Such studies of thinking are useful because they demonstrate that analysis of external behavior and the obtained intermediate results of the task solution can help in understanding some aspects of the thinking process when analyzing specific problems. The starting state, goal state, and a set of processes that are directed to reaching the goal state are outlined to define the problem (Reitman (1965). However, what is the meaning of such terms as starting state, goal state, and processes that are directed to achieve a goal state? Such definitions do not say anything about consciousness of the goal and its relation to motives, the goal of actions and the goal of a task, the mental model that includes verbally logical and imaginative components, and so on.

According to AT, such concepts as cognitive and motor action and the concept of internalization play an important role in the study of thinking. Genetically, motor actions with material objects are considered to be the primary actions In the process of internalization, they are transformed into mental actions (Piaget, 1952; Gal'perin, 1969; Leont'ev, 1978). However, SSAT does not consider internalization as a process of transferring the external into the internal plane (Bedny and Karwowski, 2007). According to the theory of self-regulation, we see internalization as a process of constant transformation of activity structure. External material activity contains cognitive components and serves as the basis for the formation of internal mental actions. Internal and external components of activity regulate and check each other based on self-regulation and feed-forward and feedback influences. During the acquisition process, mental actions can often be performed only with the support of external motor or verbal actions, and only later can they be performed independently. Mental activity is guided by external activity. The motor actions emerge as external tools for mental actions. Verbalization of motor activity facilitates better concentration of various components of activity. When mental components are acquired, the need for verbalization decreases, and mental components become less conscious. In this case, we are not talking about the transformation of the external into the internal plane but about comparison of various actions based on the feed-forward and feedback connections and formation of the internal thinking actions.

Via the process of thinking, the object is included in all new connections and relationships and appears with all new properties and qualities. This mechanism of thinking is called analysis-by-synthesis (Rubinshtein, 1946). Isolation of certain properties of objects (analysis) and their transformation are accomplished through comparison with other features of an object or with the other objects (synthesis). During genetic development, practical analysis and synthesis are carried out by motor actions. This method precedes the development of theoretical analysis and synthesis. Analysis and synthesis in the external form can be used in combination with the mental analysis and synthesis as a supplementary method.

In cognitive psychology, conversion of a single complex problem into a number of smaller and easier solved sub-problems is known as the *problem-reduction approach*. The method for developing sub-problems is known as *means-end analysis*. This approach has been used by Newel and Simon's (1972) General Problem-Solving program (GPS).

Comparison of the program steps utilized by the GPS and the thinking aloud protocol used by subjects when performing a symbolic logical task demonstrates some similarities between problem solving in GPS and the study of the problem-solving tasks in activity theory.

However, there are some differences. First of all, the concept of a goal in SSAT differs from the one in cognitive psychology. In SSAT, thinking is considered as a self-regulative process that includes multiple external and internal mental feedback. Not only verbally logical but also imaginative non-verbalized components are important in this analysis. Therefore, the thinking aloud method is not sufficient for the study of thinking.

Such psychological concepts as actions or operators that are selected from memory, reducing the difference between the current state and the new state, and so on are not clearly defined in the problem-reduction approach. Human goal-directed actions cannot be considered as movements or operators, as has been stated in this approach. The term operator can be applied in the study of information processing in computer systems, and there is a need to redefine this term when it is applied to human activity. There is a fundamental difference in the processing of information by people and by computers. The development of standardized psychological terminology is of fundamental importance for this purpose.

The problem-reduction approach ignores the fact that subjects operate not with symbols and signs but with mental or material objects and signs' systems that have different meanings. Human thinking has a conscious goal-directed character and cannot be reduced to the iterative trial-and-error process when only probabilistic connections between events are taken into account. Emotionally motivational components play an important role in the thinking process. They determine strategies of selection and interpretation of information and the possible program of task performance or finding a solution to a problem. In general, computers can reproduce only some individual functions of human thinking.

Gilhooly (1988, p. 25) described the problem-reduction approach from a cognitive psychology perspective by using the following example:

If the problem is to travel from London to New York, the difference between the starting state and the goal is a large distance. Large distances can be reduced by using airplane (the first operator). But to use a plane one needs to have a ticket and to be at the airport. That is, the preconditions of the first operator have to be met.

This fragment of the text shows that the term operator is not an action in psychological terms here. This is simply an example of a stage of a task performance that can be of various sizes. It demonstrates again that the concept of action does not have any psychological meaning in cognitive psychology. We also cannot agree with the notion that thinking actions are simply selected from memory.

It is often the case that subjects do not have required actions for solving a new problem in their memory. Such actions should be developed during problem-solving stages. We do not discard the problem-reduction approach developed in cognitive psychology (Newell and Simon, 1972). However, the term action does not have an accurate psychological meaning in their concept of thinking. It is also obvious that the concepts of cognitive and motor actions do not exist outside of activity theory. These concepts play an important role in general AT, but they are not sufficiently developed there for practical use.

The concepts of cognitive and motor actions are precisely defined in SSAT. We now present the model of thinking that describes the problem-reduction approach by using SSAT terminology (see Figure 4.5).

This model also demonstrates that there are feed-forward and feedback connections between various stages of the problem-solving process. Each stage in the model is called a functional mechanism or function block. It is important to point out that each function block of thinking is connected with mechanisms of memory and specifically with the working memory functions. Adding the description of connections of each block with memory significantly complicates the model of thinking. This is the reason such connections are not depicted in this model. It is important to underline that thinking does not just provide extraction of required information but also transforms information in memory and restructures it.

In this model, the program of performance is understood as a logically organized system of cognitive and behavioral actions utilized by a subject. The program of action performance cannot just be selected from memory but might also be developed by a subject during the problem-solving process. The model presented can be useful in task analysis. SSAT provides very specific and standardized description for all terminology utilized in this model. The basic idea that a subject converts a single complex task or problem into a number of easily solved sub-problems (subtasks) is preserved in our model. This model demonstrates that the solution of the problem has certain stages.

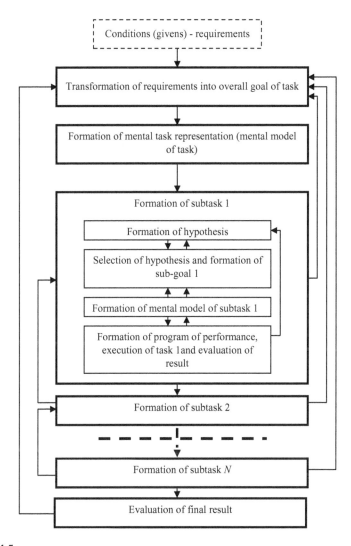

FIGURE 4.5
Model of thinking that describes the problem-reduction approach.

These stages have feed-forward and feedback connections. At the heart of the solution lies the continuous reformulation of the problem until an adequate solution can be found. The subtask's solution is compared with the overall goal of a task. The result of the following subtask is also compared with the result of the preceding subtask. A subject pays special attention to comparing the final result with the overall goal of a task. If such comparison is viewed as satisfactory, the solution of the problem is completed. The more effective the solution of the problem is, the fewer intermediate stages of problem solving are required.

Let us consider from the SSAT perspective how a subject compares the goal with the starting state using the *means-end* analysis steps. In SSAT, these steps are described as an evaluation of intermediate or final result (see Figure 4.5). As it has been described in the model of self-regulation of orienting activity, the result is not always the same as a goal. Even when a subject's result exactly matches the goal, a subject can still evaluate this result negatively depending on the selected criteria of *"successful result"*. For example, a subject has the goal of producing 100 parts during the shift. However, during the task performance he/she discovered that he/she can produce 110 parts and earn more money. In this case, a subject can immediately change his/her strategy and continue to work until he/she produces 110 parts.

The other cognitive psychology approach to the study of thinking is Anderson's (1985) theory of acquisition of problem-solving skills. As do Newell and Simon (1972), Anderson describes the thinking process as a means-ends analysis. Such analysis involves dividing a problem into a series of smaller sub-problems, each of which is then solved. Anderson (1993b) has elaborated on Newell and Simon's ideas, introducing such important concepts as declarative and procedural knowledge. According to Anderson, thinking takes place within a *means-ends problem-solving process*. Problem solving can be understood as a transition from an initial state of a problem to intermediate state and finally arriving to a state that satisfies the final goal. Anderson's Theory of Cognitive Architecture (ACT*) differentiates between *'knowing that'* and *'knowing how'* (Anderson, 1983). The former is *declarative knowledge* and the last is *procedural knowledge*. Thus, the first type of knowledge encodes our factual knowledge, and the second one encodes cognitive skills, including problem-solving skills.

According to Anderson, the production system framework can be useful in analysis of cognitive tasks that include thinking components. Anderson (1993a) described origins and nature of problem-solving operators that are utilized in the means-end analysis. The initial stage is called the *interpretive stage* and the second stage is the *knowledge compilation stage*. The last stage is the process of transiting from the interpretive stage to the procedural stage. Transiting from the interpretive stage to the end stage is facilitated by using *problem-solving operators* that are the main engine of means-ends analysis. Such operators are actions or operations of the thinking process that are described in terms of production. Each operator leads to a new state of affairs, and the last one can triggers new rules. In Anderson's theory, the concept of *production system* is a set of conditional-action pairs called *production*. They are used as units of analysis and presented in the following form: If condition B applies, take action A. Anderson (1993a) demonstrates application of this system using an example from geometry: IF conditional statement specifies some condition that must be met for the THEN conditional statement are to be executed.

IF the goal is to prove two triangles are congruent,
THEN try to prove corresponding parts are congruent.

IF segment AB is congruent to segment DE, and segment BC
Is congruent to segment EF, and segment AC is congruent to DF,
THEN conclude: triangle ABC is congruent to triangle DEF
because of the side-side-side postulate.

Let us consider some basic terminology employed by Anderson's theory and then analyze the example presented above from the SSAT perspective. The interpretive stage is known in AT as the orienting stage, after which the executive stage of activity can follow. These stages, concepts of goal, cognitive and motor or behavioral actions, operator, and so on are clearly defined in SSAT. In ACT theory, such terms as goal, operator, actions, activity, and so on do not have a clear description. The concept of self-regulation is not utilized in ACT theory, and therefore possible strategies of task performance cannot be considered accurately.

Considering the origin and nature of the problem-solving operators that are utilized in Anderson's means-ends analysis, it can be seen that they include three stages: the interpretive stage, the knowledge compilation stage, and the tuning stage. These stages are well known in cognitive psychology. We, as an example, consider the knowledge compilation stage. The procedural stage is a system of production rules consisting of condition-action pairs. At this stage, logical rules are activated in memory and various actions are performed based on these rules. The result of such actions adds new information to working memory and a chain of production rules is created. However, human thinking cannot be reduced to computer-like If-Then rules and to their automatic actualizations in memory. A subject might know rules but is not able to perform the required thinking actions. Emotionally motivational aspects of the thinking processes are also ignored. Selection of information and the way it's interpreted depends on these aspects of human thinking. The meaningful interpretation of a sign system depends on the significance of this system for a subject and on a subject's motivational state. If emotionally motivational state changes within the same situation, the objective meaning is transformed into the subjective sense (personal meaning for a subject who is working on solving the problem).

A subject is not a computer, and her/his thinking cannot be reduced to manipulation with the symbols stored in memory. Thinking also includes manipulation of mental objects such as meanings, images that correspond to some real objects, or events, and so on. Such manipulation can be performed by cognitive and motor actions at the conscious level. At the unconscious level, such manipulation employs mental operations that usually are automatic and are considered in AT as components of conscious actions. Thinking actions play the main role in development of human knowledge.

This role of actions is not discussed in cognitive psychology. In Chapter 2 we considered the relationship between concepts of cognitive actions and knowledge. It has been demonstrated that the objects of action are not just material ones but also images and knowledge along with their associated concepts and meanings. Moreover, procedural knowledge is often not enough for practical application and should be transformed into the skills for their practical use. One can see this in gymnastics, where the gymnast knows the technique of the gymnastic element but cannot perform it without extensive practice. Therefore, knowledge should be transformed into skill in order to be applied in task performance. Skill acquisition is tightly connected with the concepts of cognitive and behavioral actions. In this process, knowledge is transformed into actions that are transformed into skills through training. This process involves gaining automaticity of performance of cognitive and behavioral actions and their combinations.

It is also necessary to consider such concepts as an object of study (do not confuse this with an object of activity) and a subject of study. An object of study is a phenomenon or an object that exists independent of our knowledge about it and calls for some theoretical or empirical methods of analysis. A subject of study is the aspects of analysis of a considered object chosen for extraction of specific knowledge about a considered object. Our knowledge about the object is not identical to the object itself. This knowledge is determined not only by the imperfection of the chosen method of study but also by the purpose of studying it.

The same object of study can be related to various subjects of study and vice versa. Let us consider an example. Suppose, a person needs 10 nails to fix a fence at a summer house.

She/he found a few nails in one place and took several nails from another place. Then she/he decided to count the nails found in both places to determine if she/he has a sufficient quantity of nails. Five have been collected in one place (counting 1, 2, 3, 4, 5) and four in the other (counting 1, 2, 3, 4). The performed summation gave nine nails as a result. After that, a person deducts nine from ten and obtains the number one, concluding that one more nail is needed. In this example, the objects of study are nails. They have different features like size, material, their conditions (new, old, shape, size, etc.). Suppose that all nails are not exactly the same, but they can be used for their intended purpose. After checking their quality, a person determined that one nail did not meet the qualitative criteria. For simplification of our discussion we will not consider the qualitative aspects of nails' evaluation and only consider the stage of assessing the quantity of nails.

When counting the nails, a subject operates primarily with numbers and not with nails. A number is not a simple visual icon or a specific sound. We utilize numbers because they reflect the specific quality of various objects and particularly their quantity. Such quality as numbers can exist independently from various objects. We can perform special actions with numbers and as a result measure the quantity of various objects. In our example, such actions as *addition and subtraction* by using numbers has been performed. The subject acts with nails in a specific way that is different from acting with numbers. It is important to underline that a subject performs actions not with signs but with numbers which have various quantitative characteristics. Evaluation of the nails' quantity is the subject of study in this case, which is different from an object of study – nails.

There is no subject of study in the object itself. One can choose a subject of study as a specific matter by utilizing cognitive and practical actions with a considered object. As can be seen, calculation of nails involves verbally logical thinking and associated with it, thinking actions. Thus, thinking and thinking actions play a leading role in the knowledge acquisition process.

Anderson's "knowing that" or "declarative knowledge" and "knowing how" or "procedural knowledge" are results of human cognitive (internal) and practical (external) actions. Declarative and procedural knowledge are independent. A subject can have declarative knowledge but no required procedural knowledge, and the relationship between this knowledge should be learned.

The material presented in Chapter 3 demonstrates that meaning and sense are important mechanisms of activity self-regulation. Here, we will briefly analyze these mechanisms in the contexts of the study of thinking. Sense is the emotionally evaluative mechanism in the model of self-regulation (Bedny, 2015; Bedny, Karwowski and Bedny, 2015). In general psychology, sense is described as a subjective context of the meaning (Leont'ev, 1978). The verbal meaning is particularly important for thinking. Thinking involves sensual, volitional, emotional, and other aspects of human activity. Hence, thinking is broader than just the logical operations with meanings. The psychological studies of meaning are rooted in psycholinguistics and verbal learning (Ausubel, 1968; Piaget and Inhelder, 1966, etc.). Vygotsky is the founder of psycholinguistics in AT (1978). During the process of mental development, the individual internalizes various sign systems and uses them as internal tools for thought. Hence, there are two kinds of signs, one of which exists in the external world and the other in the subject's mind. The signs in the mind of the subject fulfill the role of an *internal psychological tool*. One of the most important aspects of the study of the sign is the elucidation of the relationship between the sign and its meaning. Signs do not exist in isolation but are integrated into language systems. Here, we refer to language in a broad sense as a system of socially fixed signs, gestures, sounds, written images, and so on that facilitate communication and interpretation of various real phenomena. From this perspective, language includes not only words but also mathematical symbols, formulas, geometrical figures, and so on. Signs can be manipulated in the same way that other objects are. However, in order to obtain the knowledge of a given sign system, it is not enough to manipulate the material form of its signs. The most important aspect in subjects' interaction with signs is their meaning. Subjects cannot manipulate signs the way they manipulate regular objects. A sign can be utilized in human activity in accordance with the laws applied to the meaning of the signs. In order to understand the sign, it is important to consider it in relation not only to its referent but also to the activity of which it is a part and which grants it its meaning and sense (Shchedrovisky, 1995). A symbol is a sign only because people can interpret its meaning. At the same times, interpretation of a sign is not a purely subjective process. The meaning of a sign has an objective character in that it is the result of a sociocultural development, which gives a sign a standardized method of its interpretation. The fact that people can interpret signs in the same way is a proof of the objective existence of meaning, which is independent of the subject interpreting the sign. Objective meaning is formed in the process of development of human activity that has a cultural-historical nature. From the AT perspective, the meaning of a sign should be studied not only in relation to the object or other signs but also in the context of human activity. Meaning is determined through the relations of action to a situation, and from this point of view it exists only during the performance of a particular action (Genisaretsky, 1975).

There are different types of meaning, such as object meaning, category meaning, non-verbalized meaning, and so on. Object meaning can be seen as a network of feelings and experiences associated with a particular object (Rubinshtein, 1957). It derives from individual practical experience. Category meaning or idealized meaning is part of the verbal categories that one masters. This type of meaning has a stable character and is independent of the situation. During acquisition of knowledge, object meaning can be transferred into category meaning. The constancy of meaning and its relationship to culture allows us to view culture as a semiotic system, or a net of meanings, which is superimposed by the individual on the surrounding natural environment and artifacts (Sokolov, 1974).

There are conscious and unconscious meaning in the thinking process (Pushkin, 1978). Thinking actions are the conscious elements, and thinking operations (components of actions) are the unconscious elements of the thinking process. Due to unconscious mental operations, a subject can extract unconscious meaning in various situations. With the help of non-verbalized meaning, a subject can extract distinct and essential characteristics of the situation that are germane to the solution of a particular problem. Thinking actions are the tools that are responsible for extraction of the conscious meaning from various elements of a situation.

The meaning of an object, sign, or word (a verbal sign) is an objective phenomenon, which can be transformed into the personal, subjective sense. Thus, the concept of sense has two interpretations. In SSAT this term is considered only as an emotionally evaluative mechanism. It is involved in the subjective evaluation of the significance of events or a phenomenon by a subject. In general psychology, sense is considered as a subjective interpretation of an objective meaning. Such a distinction should be taken into account in the context of each specific study. The emotionally motivational components of activity play a particular role in transformation of the objective meaning into the subjective interpretation or sense. This process depends not only on the emotionally motivational components of activity but also on past experience and current goals of activity. Sense is personal and depends on general characteristics of activity. It helps an individual to adjust to a specific situation or a problem. The difference between the objective meaning and the subjective sense is clearly exposed when people try to interpret the same political event, such as a presidential election. The objective meaning and subjective significance of events result in the specific interpretation of such events. When considering the notion of sense, we focus on the aspects of meaning that are specific to a given subject. Meaning determines the position or role of an object among other objects. Sense, on the other hand, determines the relationship between objects and the needs of an individual (Gal'perin, 1973).

Based on actions performed, a person can discover relationship between the perceived object and the elements of the situation and is able to identify this object. Interpretation of meaning of the situation depends on past experience and the utilized strategies of performance. Therefore, self-regulation of activity plays a critical role in meaningful interpretation of the situation. Meaning is responsible not only for orientation in a situation but also for regulation of the executive actions of a subject.

Meaning and personal sense are important in thinking. But associations are also vital mechanisms. Some scientists consider thinking not just as logical operations but mainly as based on the associations between related cases from past experience or precedents (Eisenstadt and Simon, 1997). The approach proposed by these authors describes thinking as performed in memory automated mental operations. These mental operations or associations are viewed as automatically triggered mental rules. In AT, mental associations are considered to be a result of learning. They can be transformed into mental associations through multiple solutions of specific problems and automation of mental actions and should be considered as thinking operations. Such associations or thinking operations are not conscious. Thinking is a combination of intuitive mental operations that are combined with internal mental and external behavioral goal-directed actions.

Thinking is the cognitive process of fundamental importance for various problem-solving tasks. The problem-solving task arises from a problem situation but is different from it. A problem situation is usually vague and not entirely conscious. An operator begins to realize that something is wrong in functioning of a technical device, but the specific cause of the problem is unknown to him/her. Therefore, he/she does not know what type of actions should be taken. At the first stage, the task conditions are given, but requirements for what should be achieved are not precisely formulated. At the second stage, a subject formulates a goal of tasks based on givens and requirements. Thus, the task-problem is often not given to a subject in a ready-made form and should be formulated by a subject independently. Such tasks usually have their origin in a problem situation. Thinking, according to

AT involves active manipulation of the internal representation of the external world according to a goal of activity. Our intuition is always in play in the conscious goal-directed activity.

We can conclude that thinking actions play a leading role in thinking. They can be of various types: 1) thinking actions that are involved in analysis of a situation and development of its mental model; 2) verbal actions that are involved in the thinking process; 3) decision-making actions that are used at intermediate and final stages of the thinking process; 4) other types of cognitive actions, primarily perceptual and mnemonic actions, play an important role in thinking. The analysis of consequences of various motor actions is also a critical factor in thinking. We would like to mention that a goal of task and goals of separate thinking actions are essential for thinking. Our further analysis of thinking is based on the above-stated theoretical platform.

Although decision-making plays a critical role in the thinking process, it should not be studied independently but should be considered as a specific stage of the thinking process. Of course, it's important to study decision-making in unity with other psychic processes and specifically with memory. There are automatic "perception-action decisions" and a more complex analytical decisions. The first one is derived based on the perceptual data. For example, if the red light is on, then press the red button. If the green light turns on, press the green button. The analytical decision requires a higher level of involvement of the thinking process.

Decision-making tasks include a number of stages such as analysis of the received information and creation of a mental model of the situation; formation of a hypotheses and meaningful interpretation of the situation; evaluation of the hypothesis and selection the most preferable one; the selected most adequate hypothesis is utilized for formation and selection of the alternative course of actions.

Promotion of a hypothesis can be done at the conscious or unconscious level. At the conscious stage, a subject promotes a *conscious hypothesis*, and at the unconscious stage, a subject generates a non-verbalized, *unconscious hypothesis* in a problem solution. It is should also be pointed out that not all described steps of decision-making are always necessary. Thinking is the key cognitive process that is required for identifying the most preferable choice.

Let us consider how some authors view the final stage of decision-making. According to Wickens, Gordon and Liu (1998), the final stage of decision-making involves *generation and selection of actions*. A decision maker generates one or more alternative actions by *retrieving* options from memory. As an example, these authors describe the situation when after diagnosing acute appendicitis, the surgeon generates several alternative actions, *including waiting, conducting additional tests, and performing surgery.*

Waiting, conducting additional tests, and performing surgery are not alternative actions but rather three alternative courses of medical treatment (*possible medical treatment processes*). A medical treatment process can be divided into medical tasks, which include various courses of actions by personnel.

For example, waiting involves monitoring a patient's physical condition during a certain period of time. Therefore, waiting, conducting additional tests, and performing surgery are not the decision maker's alternative actions. He/she performs complex diagnostic tasks that can be divided into subtasks. Based on the data obtained, a surgeon decides which course of action from the three available ones is the mostly appropriate for the patient. Other medical personal can perform various medical procedures based on this decision. Making the final decision about the method for medical treatment is a complex problem-solving task that might follow a number of preliminary decisions that are made by medical experts. Therefore, decision-making about the acute appendicitis diagnosis is a multistage medical treatment process where outcomes of one decision influence subsequent decisions. This process includes a number of medical tasks which can be performed by different personnel; each task includes a logically organized course of cognitive and behavioral actions. This is another case that proves that without standardized terminology, the analysis of human performance cannot be conducted properly.

4.6 Self-Regulative Model of Thinking

In section 4.5 we presented the model of thinking that depicted the problem-reduction approach (see Figure 4.5). However, this model covers only some aspects of thinking. In this chapter, we describe thinking as a complex self-regulative system. The model presented here is useful for studying thinking in the context of task analysis. It allows description of a variety of strategies for solving the same task-problem. The self-regulative process of thinking includes various stages that can be presented as function blocks with feedforward and feedback connections.

When analyzing a specific task, not all function blocks are equally important. Some function blocks can be even omitted during the analysis of thinking. The model presented in Figure 4.6 is the enhanced model of self-regulation of the thinking process that has been described initially in Bedny, Karwowski and Bedny (2015). The contents of some blocks have been modified, and additional interconnections between blocks have been introduced. This model can be beneficial for analysis of complex rule-based tasks, problem-solving tasks, and decision-making tasks. Each function block includes various cognitive processes. However, various thinking operations are vital in each function block. Analysis of thinking in terms of function blocks helps to concentrate our attention on certain stages of the thinking process.

Let us now discuss the model presented below.

A subject perceives information from the external environment and from different instruments on a panel or from a computer screen. This information

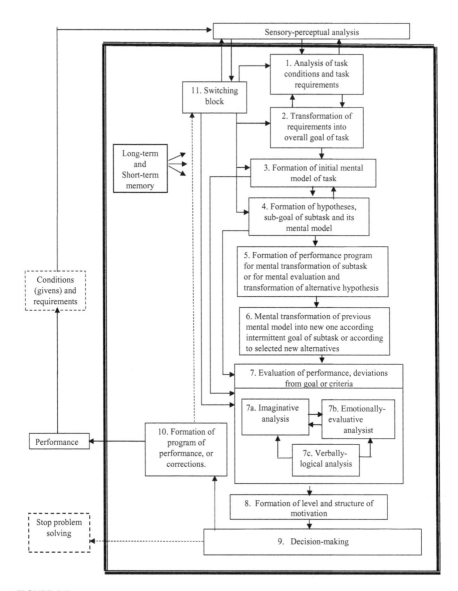

FIGURE 4.6
Self-regulative model of thinking.

can be viewed as an informational model. A subject then selects information that is subjectively most important in a presented situation based on
sensory-perceptual analysis. The selection of information also depends on
the past experience and on her/his emotionally motivational state. This is
the first stage of task performance, in which sensation and perception play a
leading role. This stage of receiving information is closely related to thinking
that is involved in understanding of the nature of the perceived phenomena.

At this stage, attention is continuously shifted from one object or element of the situation to another. Receiving information involves sensation and perception and is closely linked with past experience or memory, attention, and thinking. The content of perception also depends on goals and motives of activity (see feedback from blocks 1, 2, and 11 to sensory-perceptual analysis block, Figure 4.6). At the sensory-perceptual stage of task performance, a subject does not meaningfully evaluate the relationship between elements of a situation, because at this stage thinking is not yet directly involved in receiving information. At this stage, voluntary and involuntary mechanisms of attention play a specific role in analysis of the information presented. In some cases, the success of subsequent stages of thinking depends on how information is organized and perceived. This functional block is depicted outside of the boundaries of the thinking process because this is a stage of activity that precedes this process.

However, mechanisms of the thinking process can significantly change strategies of selecting information at the sensory-perceptual analysis stage. It has been discovered that during meaningful analysis of visually presented information, eye movement strategies can significantly change. Eye movement that is involved in the perceptual process is significantly different from the eye movement that is a part of the thinking process. Changes in strategies of gathering visual and other types of information under the influence of thinking mechanisms are called the gnostic dynamic. Pushkin (1978), who was the first one to introduced the concept of gnostic dynamic, understood it as the ability of a person to constantly change his/her mental picture of a situation in spite of its external constancy. This in turn influences the eye movement strategy. "Non-verbalized operational meaning" or "situational concept of thinking" are important notions of gnostic dynamic. Gnostic dynamic is a result of conscious thinking actions and unconscious mental operations of manipulation with externally given unchanged information. For example, when an operator receives visual information, different distinct essential characteristics can be extracted from identical situations that are important for the solution of a particular problem by using external fixation as well as internal mental scanning. These strategies are not always conscious and cannot always be verbalized. According to SSAT, the self-regulation process lies at the heart of gnostic dynamics. Gnostic dynamics facilitates mental restructuring of the same situation, and based on evaluation of the obtained result, facilitates restructuring of the situation again until a useful result is achieved.

For example, when playing chess, a player can evaluate the relationship within one group of figures and then select and evaluate the relationship within another group of figures despite the fact that the position of the figures physically remains the same. This is an example of thinking actions that are performed based on visual information. Thus, our model depicts feedback from various functional mechanisms or function blocks of thinking to the sensory-perceptual analysis block that is shown as an external one. When

function blocks that are directly involved in the thinking process are activated, the role of thinking in receiving information and therefore the role of
the gnostic dynamic significantly increases. In our model of self-regulation,
the gnostic dynamic is activated through the feedback from various blocks
of the thinking process to the "sensory-perceptual analysis" block. When
information about the situation is presented to an operator adequately, the
gnostic dynamic is more effective.

Self-regulation can also be revealed in the functioning of separate sense
organs. It is provided by feed-forward and feedback connections between
receptors and the central part of sense organs in the brain. For instance,
nerve impulses from the eyes are delivered to the central part of the sense
organ, and after specific processing based on the feedback, information is
transmitted back to the eyes (Bedny, 2015).

Sometimes in an emergency situation people need to stop to perform
motor actions because the situation is changed. In other words, people have
to change their initial decision within a very short period of time. For example, a driver decides to press the gas pedal to move the car. However, in a
fraction of a second the situation has changed and in order to avoid a collision, a driver has to change her/his decision and push the brake pedal. It
has been discovered that in a fraction of a second after the first decision, it is
very difficult to make an opposite decision. Scientists at The Johns Hopkins
University (Xu, Anderson, et al., 2017) demonstrated that different areas of
the brain are responsible for this contradicting decision-making. One area of
the brain began to carry out the original plan to accelerate the car. Stopping a
plan that is under way is rather hard and takes some time because it is initiated by another region of the brain. A lot of brain power from other areas of
the brain is required to suppress the previous plan to carry out actions. Based
on neuropsychological studies, scientists proved that it is not just a problem
of regulating our movements. The delay is due to the interaction of different
areas of the brain that are involved in changes in the original plan of actions.
People have to be more cautious in making initial decisions in risky situations. The more cautious people are when they make the first decision, the
better they can voluntarily regulate their motor actions in a given situation.
It is obvious that studying the brain's stop system could be helpful for understanding the success of performance in emergency and risky situations.

The meaning of the situation can be comprehended through either conscious mental actions or through unconscious mental operations. The first
stage involves promotion of a conscious hypothesis and problem solution.
This is the conscious stage of thinking. The second stage of thinking is
unconscious and involves the non-verbalized operational meaning.

The first functional block that is directly involved in the thinking process
is block 1 (*analysis of task conditions and requirements*). At this stage of task
performance, a subject defines task conditions and task requirements that
are the most relevant ones to solving a problem. The subject restructures
the situation presented and assesses the results obtained according to the

subjective significance of the elements of a situation. At all stages of restructuring a problem-solving situation, such thinking operations as analysis
and synthesis are most clearly revealed. Restructuring of a problem-solving
situation can be performed not just practically but also mentally. The inner
(internal) speech plays an important role at these stages of thinking regulation. The inner speech enables planning and regulation of practical and
mental actions. Sokolov (1963) studied inner speech using electromyography
(EMG) of the muscles of the throat and lips as people solved various mental
problems. It has been discovered that an increase in the complexity of the
thinking tasks is accompanied by an increase of the EMG level. This activation of EMG is specific for verbal-logical thinking. The data considered demonstrate that speech and verbal-logical thinking are tightly interconnected.

The main operative units of thinking actions are concepts, propositions,
and mental images. Thinking is considered as an object-oriented process
and cannot be reduced to manipulation with symbols based on the internal
rules of the brain. Rules are not actions. One can know the rules but not be
able to utilize them. A subject should learn how to act according to the rules.

Function block 1 demonstrates that at this stage of the thinking process,
the objectively presented *requirements* of a task are not yet accepted by a subject as a goal of a task. This becomes possible only after block 2 (*transformation of requirement into overall goal of task*) is activated and a goal of a task is
formed and accepted. This is a critically important step of problem solving
because acceptance or formation of the overall goal of task-problem is the
critical stage of thinking. Blocks 1 and 2 have forward and backward connections, and therefore these blocks are mutually adjusted (see Figure 4.6).
Interconnection of block 2 with block 1 allows reformulation of the preliminary accepted overall goal. The goal can be gradually specified and become
precise. In thinking, the overall goal of a task can be formulated in a very general form in the beginning. And then gradually this goal becomes more and
more specific. Hence, a goal is not a ready-made standard to which human
activity is directed, as stated by some scientists. In SSAT, understanding of
a goal and a process of its formation is different from its understanding in
cognitive psychology. For example, a goal is not the end state of a task. It does
not include the motivational component, but together with the motives it creates the vector motives → goal, and so on.

The critical stage of a problem-solving task performance is *formation of an
initial mental model of a task* (block 3). A mental model is dynamic and can be
corrected during task performance. However, it should be adequate to the
formulated goal of the task. Both the overall formulated goal of a task and
its initial mental model are the most difficult and important stages of the
problem-solving task performance. Only when the initial mental model of a
task is created can the solution stage of the thinking process begins.

The mental model includes verbal-logical and imaginative components
that, in turn, consist of conscious and unconscious components depending
on the specificity of their combination. The initial mental model of a task is

constructed by a subject and can be modified if a goal of a task is changed or if a subject formulates a sub-goal of a task and a mental model of this subtask is created (see interaction of block 3 with block 4 – *formation of hypotheses, sub-goal of subtask and its mental model*). The interaction of blocks 3 and 4 as the stages of the thinking process provides a better understanding of a task by a subject. As a result of such interaction, a subject becomes aware of the task as a purely problem-solving task or as a decision-making task that includes problem-solving components. These two types of tasks have their specifics. For example, when a task is interpreted or formulated as a decision-making task, a formation of various hypotheses plays an important role in selection of the alternative course of actions at the final stage of decision-making.

The thinking process is the cyclical one. The outcome of each cycle and its intermediate stages can be evaluated and corrected based on the feedback. The results of the thinking process can be reflected in changes of the external situation or be accompanied by mental restructuring of the situation.

The problem solving can be performed by restructuring and by the following evaluation of the externally presented situation, or by mental restructuring and evaluation of the situation when the external situation is unchangeable. These strategies can be combined. When evaluating the first strategy, a subject utilizes external feedback. The second strategy can be evaluated only by using mental feedback. Hence, thinking is a self-regulative cyclical process during which strategies of thinking are modified.

Our model includes these types of feedback in evaluation of the thinking process.

Blocks 7 through 11 play an especially important role in each cycle of the self-regulation of the thinking process (see, for example, feed-forward connections of blocks 3 and 4 with block 7 and further connection with blocks 8, 9, 10, and 11). Blocks 9, 10, and then block 11 through the mental feedback (see dashed line) depict the closed-loop structure system that enables corrections of all blocks during the thinking process. It is important to point out that contents of block 7 (*evaluation of performance, deviation from goal or criteria*) can also be corrected (see connection of block 11 with block 7). Similarly, block 4, 5, and 6 can be also mentally corrected.

There is also a possibility of a real transformation of the situation as a result of the thinking process (see connection of block 10 with the external block "performance"). In such cases, a subject introduces some modification to the problem-solving situation and reevaluates it based on changes in sensory-perceptual data and following modification of block 1. Thus, there are three ways of evaluation of the thinking process.

One way involves modification of the external situation only and then evaluation of the thinking process based on the external feedback. Another way includes modification of the situation just mentally with the following evaluation of the thinking process based on mental feedback. The third strategy is a combination of the first two. The process of evaluation of sensory-perceptual data under influences of the thinking mechanisms without real modification of the externally presented information is known as gnostic dynamic. Thus, feed-forward and feedback connections with evaluative block 7 and the following blocks describe evaluation of all stages of thinking based on external or mental feedback or their specific combinations.

The interaction of blocks 3 and 4 demonstrates that when the initial mental model of a task is created the task's solution stage begins. Hence, after understanding a task at hand and realizing that the initial mental model is not adequate, or cannot be helpful (see block 3) in finding an adequate solution, a subject divides a task into subtasks (see block 4). A subject begins formation of subtasks by formulating various hypotheses. Each hypothesis has its own potential goal. Based on comparison and evaluation of such hypothesizes, a subject selects one of them and formulates the first sub-goal associated with that hypothesis. Comparison of a new sub-goal with an existing mental model of a task-problem allows, if necessary, transformation of an original mental model into a new one that is more adequate to a new sub-task's goal (see interaction of blocks 3 and block 4).

The mental restructuring of the situation involves intuitive and logical components of thinking. Inner speech (internal speech that is not spoken aloud) plays an important role at this stage of thinking regulation. External and internal speech are tightly connected with thinking actions that are utilized by a subject for mental transformation of an unchanged external situation. Sometimes practical (motor) actions also can be used for manipulating the element of the externally presented data. Hence, block 4 is involved in formation of a new hypothesizes associated with it and potential goals of subtasks and their mental models.

An analysis of the first developed subtask might turn out to be insufficient for an acceptable solution, and therefore a new subtask is developed similarly. Comparison of a new sub-goal with the previously created mental model allows transformation of this mental model of a sub-goal into a new one that is adequate for the new subtask. This stage of the thinking process is associated with block 5 (*formation of program of performance for mental transformation of subtask*) and block 6 (*mental transformation of previous mental model into new one according to intermediate goal of subtask*).

This stage of the thinking regulative process is adequate for developing a new mental model of a subtask. Based on the program developed in block 5, block 6 executes a mental transformation of the previous mental model into a new one according to the intermediate goal of a subtask. This block also carries out thinking operations that are involved in generation and selection of new hypotheses, which are especially important for performing the decision-making tasks.

The result obtained in block 6 is evaluated in block 7 (*evaluation of performance, deviations from goal or criteria*) based on its adequacy to a *goal* or *sub-goal* of a task and *subjective criteria of success*. The last one can deviate from the objective goal, and therefore, a subject can evaluate her/his own result as a successful one even if it is lesser than the required standard.

For example, a blue-collar worker can lower the quality of her/his performance in order to increase quantity.

Therefore, we have to distinguish a goal from the subjective criteria of success. A goal is formulated in advance, but the subjective criteria of success can change during task performance. For example, there are two methods of problem solving. One method is easier and less precise, and the other one is more difficult but more precise. The last method fits the requirements better, but a subject can select the first method of problem solving in order to reduce fatigue and evaluate the result obtained as subjectively acceptable. The more significant a problem-solving task is for a subject, the more rigorous are the subjective criteria of success. If a task is especially significant for a subject, he/she can introduce subjective criteria of success that are even higher than the given requirements or a goal. The subjective criteria of successful completion of the problem-solving task can coincide with the goal. *Verbal-logical analysis* (sub-block 7c) and *imaginative analysis* (sub-block 7a) are involved in *cognitive evaluation* of the task performance and of the deviations from the goal of a task or subjective criteria of success.

Interactions of such blocks are also involved in evaluation of the *difficulty* of task performance. Difficulty is a subjective characteristic of a task, and its complexity is its objective characteristic. The more complex a problem-solving task is, the more cognitive demands it imposes on a subject. In some cases, a subject might overestimate the complexity of a problem-solving task and, as a result, discard it or select a subjectively simpler strategy for its performance by utilizing various heuristics for its solution that might not even be sufficiently accurate. An individual can also underestimate the objective complexity of a problem-solving task, which can influence the efficiency of such task's solution or even result in failure to solve it.

We would like to stress that the *imaginative analysis* (sub-block 7a) is not entirely conscious and therefore is not adequately precise as an evaluative process. This process of evaluating the problem-solving tasks is often not verbalized and sometimes even unconscious.

Emotionally evaluative (sub-block 7b) is involved in determining the *significance* of a task and the *significance* of discovered deviations from the standard established by a subject. Under- or overestimation of complexity of the problem-solving tasks and its acceptance or rejection depend first of all on relationships between *emotionally evaluative analysis* (sub-block 7b) and *verbal-logical analysis* (sub-blocks 7c). Interaction of these sub-blocks influences *motivational* block 8. This process can be explained using the following example. If *imaginative and verbal-logical analysis* sub-blocks (sub-blocks 7a and 7c) indicate that a task is difficult, and an *emotionally evaluative* sub-block 7b shows that the task is not significant for a subject, then this influences *block 8 (formation of level of motivation)*, and a subject can reject the task altogether. In general, block 7 influences the structure and level of motivation.

Block 8, in turn, influences block 9 (decision-making). Depending on the specifics of the interaction between blocks 8 and 9, a subject can make a decision to *enhance an already developed strategy* (the first version of a deci-sion). In such cases, feedback goes from block 9 to upper-level blocks (see, for example, the left side of Figure 4.6 and connections between blocks 10 and 11 and from block 11 to other blocks of the thinking model). There is another option. If the preliminary stages of performance are evaluated posi-tively, *decision-making* block 9 can trigger block 10 (*formation of a program of performance or correction*) and then activate the 'performance' block outside of the model (practical restructuring of problem-solving situation). A subject restructures the problem-solving situation not just mentally but also practi-cally. All of the above-described material pertains to the self-regulation of the thinking process during performance of the problem-solving tasks. The model presented shows that thinking has a loop-structure organization.

Block 7 also has another critical function in the self-regulation of the think-ing process. Together verbal-logical sub-block 7c and imaginative sub-block 7a provide evaluation of task *difficulty*. *Emotionally evaluative* sub-block 7b assesses *subjective significance* of a task-problem for a subject. The relationship between *subjective difficulty* and *subjective significance* of a task determines the level of motivation (block 8). The content of motivation can be changed in a wide range due to the complex relationship between these variables. For example, if the subjective evaluation of the task difficulty is very high and evaluation of the task significance is also very high, then a subject is still motivated to complete the task. At the same time, if the subjective evaluation of the problem-solving task difficulty is very high and its subjective signifi-cance is very low, a subject rejects such a task. The relationship between ver-bal-logical and imaginative analysis influences the contents of motivation. This means that block 7 influences the formation of the level and structure of motivation in block 8.

There is also another possibility in the performance of a problem-solving task. Subjects employ numerous problem-solving strategies that effect the successful solution. Some strategies can be successful, and others do not lead to success. Hence, problem-solving is a cyclical process that does not always guarantee successful solution. The subject needs to evaluate the result obtained and decide whether the solution is successful, and if it is, the subject must stop the solution cycle. If the solution is not satisfactory, the subject has to decide either to continue the problem-solving process or to stop the attempts to solve the problem because it is too complex for him/her. Therefore, decision-making block 9 is also involved in making a decision to continue or to stop the problem-solving cycle (see the dashed box outside of the model). For example, if a problem-solving task is evaluated as very difficult and not significant, this would result in the decision to stop its performance.

Thus, the connections between blocks 9, 10, and 11 or the external block *performance* reflect a decision to continue to solve the problem, while the con-nection between block 9 and the external *dashed box* (stop problem solving) demonstrates the case when a subject decides to stop the thinking cycle.

The connection of blocks 10 and 11 demonstrates that a subject can manipulate the problem purely mentally, because block 10 also influences all other function blocks through block 11.

If a subject evaluates the selected strategy of performance as acceptable, and based on relationships between *imaginative* and *verbal-logical analysis* (block 7) discovers drawbacks in the utilized strategy, he/she can correct this strategy based on the feedback from *decision-making* (block 9) to block 1 or the external block (*sensory-perceptual analysis*). Finally, if the selected strategy is adequate and does not require correction, then the *decision-making* block triggers block 10 (*formation of a program of performance, or correction*), and a subject can introduce required changes in the external situation.

The model presented demonstrates that the emotionally evaluative analysis is important not only for the decision-making stage but also for the thinking process in general. For example, each of us experiences and/or observes that when discussing the political events in any country, people emotionally reflect and interpret the political situation. In general, emotions can influence selection and interpretation of information when solving various problems.

Simon (1957) was the first who paid attention to the fact that people often do not follow a goal of making the optimal or the best decision. Instead, they try to choose a "good enough" decision. Similarly, Tversky and Kahneman (1981) demonstrate that people frequently violate assumptions of the normative theory of decisions. In our model of thinking, such a violation can be determined by the relationship between blocks 7, 8, and 9. For example, if the decision-making process prescribed according to the normative theory of decisions is evaluated by a decision maker as a difficult and not sufficiently significant one, the probability of the violation mentioned by the authors increases. If the subject estimates decisions as not difficult ones and at the same time significant, he/she often chooses the normative decisions. There is a block, "*long-term and short-term memory*", at the left side of Figure 4.6. This block is not directly related to thinking. However, this block can interact with any thinking block during problem-solving task performance.

Let us summarize our analysis of the model considered. The function blocks of the thinking process form a closed loop system that allows for multiple mental transformations of the situation and its evaluation via various strategies of mental problem solving. However, the real situation during such transformations can remain unchanged. Hence, at the mental stage of the thinking process, one strategy can be abundant and a new strategy of solution and decision-making can be mentally developed. Only after completion of the mental analysis, can a subject develop her/his program of performance that can be utilized not only for mental but also for the *real transformation* of the problem-solving situation (see block 10, *formation of performance program or corrections*) and connection with external block "performance." When internal stages of thinking regulation are completed, the performance block is activated, the external situation can be changed, and the new mental analysis can start again. Therefore, external and internal activity are inter-dependent. The performance block is located outside of thinking stages of activity regulation because it depicts an executive stage, when a subject modifies a problem-solving task and adds newly obtained data. At this stage, changes in the problem-solving situation are made not just mentally but are also made in practice. Such changes involve human external behavior and the sensory-perceptual block. At the same time, problem-solving strategies involve, first of all, mental formation of various subtasks and their evaluation and performance.

Based on the feedback, the result of the problem solving and decision-making can be used in two ways. One involves transmitting information into block 1, and then the whole cycle of thinking and decision-making process can be repeated. The other one involves switching to block 11, where the result of sensory-perceptual analysis can be used by blocks 2, 3, or 4. For example, if data is transmitted to block 4, then this information bypasses blocks 1 through 3. This means that after performance, a subject starts to evaluate the subtask considered before. Switching block 11 can also transmit information directly to block 7 (*evaluation of performance, deviation from goal or criteria*). The actual result of a problem-solving task is evaluated in block 7 and the data obtained through blocks 8 and 9 can be used for corrections of performance in block 10.

Switching block 11 and block 9 can selectively influence various blocks. Connections between blocks demonstrate that the model can describe vari-ous strategies of the thinking and decision- making process. Therefore, the circle of thinking regulation can be shorter or longer, depending on the specificity of a problem-solving and decision-making task and the thinking process strategies selected by the subject.

The *ensory-perceptual* block can periodically interact with blocks 3 and 4 through the switching block 11 and the feedback of those blocks to *sensory-perceptual*. Similarly, the sensory-perceptual block can interact with block 7 (through switching block 11) when internal mental activity needs to be sup-ported by the externally presented data. Due to the interconnection of con-sidered functional blocks or stages of thinking regulation, a problem-solving task is continuously modified actually or mentally until an adequate solution can be obtained. The model presented demonstrates that a subject, when solving a problem, breaks it down into subtasks and then works with those subtasks because they are simpler to solve. Self-regulation of the thinking process is the basis for selection of such strategies.

Analysis of the self-regulative model of the thinking process shows that there are two types of feedback. One of them involves the presence of *external feedback* used after actual practical implementation of actions that change the external situation takes place. This type of feedback is associated with the completion of some stage of the thinking task or final evaluation of its result. Another type of feedback is mental, when a subject manipulates elements of the situation in her/his mind and evaluates the result of such manipulation with ideal objects. In such a strategy, a subject utilizes *internal mental feedback,* which allows *prevention of real errors.*

It is important to understand the difference between the interpretation of feedback in cognitive psychology and SSAT. In cognitive psychology, feedback is possible only when a subject receives information about a result of her/his external behavioral actions' execution. In SSAT, mental feedback is possible during the execution of mental actions. The description of functional blocks in the above-presented model demonstrates the range of issues that needs to be considered at various stages of analysis of the thinking process.

Analysis of blocks 1 through 6 demonstrates that the problem-solver process operates with various hypotheses and therefore can involve a number intermediate mental decisions. The final decision in block 9 is made only after completion of processes in block 7 and 8. This decision-making is especially important. For example, it determines if a subject follows the normatively prescribed strategy or uses shortcuts that immediately come to mind. The later strategy can be subjectively easier but at the same time less accurate. A subject's decision to correct the current strategy, or even reject continuing to solve the problem, also depends on such a decision.

Each block of the thinking process requires its own specific methods of analysis. Therefore, analysis of thinking as a self-regulative process also requires development of interdependent and supplemental methods of analysis that should be organized into stages in accordance with selection of the analyzed functional blocks. Forward and backward connections between such blocks show relationships between stages of thinking and allow conducting an efficient analysis of the strategies of the thinking process. Depending on the nature of a task-problem, a practitioner can choose the most relevant blocks for its analysis.

Analysis of the self-regulative process of thinking demonstrates that each functional block can be considered as a window. We open each window one at a time and consider the same object from different perspectives. Such analysis allows consideration of various aspects of the thinking process. All aspects of this analysis can be compared, and therefore this method of study can be qualified as a qualitative systemic analysis.

The material presented demonstrates that the described concept of thinking is important for conducting task analysis where thinking and decision-making play essential roles.

The above-presented concept of thinking can be used for analysis of computer-based and computerized tasks. These tasks should often be considered as problem-solving tasks. Some of these tasks have well-defined attributes: an overall goal or task, a method of task performance, and so on. Some computer-based tasks can be ill-defined. Studying mechanisms of human thought is critical for an understanding of the nature of computer-based tasks and strategies of their performance. Computer-based tasks have a lot of similarities with the tasks that are performed based on dynamic visual information and involve operative thinking. Information is presented on the screen in visual form. Data on the screen consists of a number of discrete elements and can be considered as a structure with static and dynamic elements. A user continuously changes this structure. Based on the result obtained, this data is immediately corrected. Performance of a Human Computer Interaction (HCI) task can be treated as a sequence of cyclical processes, the purpose of which is obtaining a desired result that should be adequate to the accepted or formulated goal of the task. Through various stages of task performance, a subject can formulate the subjective criteria of success that is adequate to each stage of task performance. Not just verbal-logical but also imaginative components are important in this analysis. Feedback is a result of not only material or motor actions but also of the thinking mental actions (mental feedback). An overall goal of task and sub-goals of subtasks are considered as conscious cognitive mechanisms. These mechanisms perform predictive and regulative functions at each stage of task performance. Thinking evolves as a goal-directed, self-regulative process. An important role in this process is played by decision-making.

Summing up the analysis of self-regulation of cognitive processes, some general conclusions are in order. Cognitive processes often play a critical role in human performance. The self-regulation models of separate cognitive processes are very useful because they allow prediction of preferable strategies of task performance. The main units of analysis of such models are function blocks. Each function block in the models presented is associated with a set of questions that should be considered during task analysis for each stage of cognitive processes' regulation. Interactions of the blocks considered is an important aspect of such analysis.

References

Aladjanova, N. A., Slotintseva, T. V. and Khomskaya, E. D. (1979). Relationship between voluntary attention and evoked potentials of brain. In E. D. Khomskaya (Ed.), *Neuropsychological Mechanisms of Attention* (168–173). Moscow, Russia: Science Publisher.

Anderson, J. R. (1983). *The Architecture of Cognition*. Cambridge, MA: Harvard University Press.

Anderson, J. R. (1984). Spreading activation. In J. R. Anderson and S. Kossylyn (Eds.), *Tutorials in Learning and Memory*. San Francisco, CA: Freeman.

Anderson, J. R. (1985). *Cognitive Psychology and its Application*. 2nd ed. New York, NY: Freemen.

Anderson, J. R. (1993a). Problem solving and learning. *American Psychologist*, 48(1), 35–44.

Anderson, J. R. (1993b). *Rules of the Mind*. Hillsdale, NJ: Lawrence Erlbaum Associates, Publishers.

Ausubel D. P. (1968) *Educational Psychology: A Cognitive View*. New York, NY: Holt Rinehart and Winston.

Bardin, K. V. (1982). The observer's performance in a threshold area. *Psychological Journal*, 1, 52–59.

Bedny, G. Z. (1987). *The Psychological Foundations of Analyzing and Designing Work Processes*. Kiev, Ukraine: Higher Education Publishers.

Bedny, G. Z. (2015). *Application of Systemic-Structural Activity Theory to Design and Training*. Boca Raton, FL and London, UK: CRC and Taylor & Francis.

Bedny, G. Z., Karwowski, W. and Bedny, I. (2015). *Applying Systemic-Structural Activity Theory to Design of Human-Computer Interaction Systems*. Boca Raton, FL and London, UK: CRC and Taylor & Francis .

Broadbent, D. E. (1958). *Perception and Communication*. London: Pergamon.

Craik, F. I. and Lockhart, R. S. (1972). Levels of processing: A framework for memory research. *Journal of Verbal Learning and Verbal Behavior*, 11, 671–684.

Dmitrieva, M. A. (1964). Speed and accuracy of information processing and their dependence on signal discrimination. In B. F. Lomov (Ed.), *Problems of Engineering Psychology*. (pp. 121–126). Saint Petersburg, Russia: Leningrad Association of Psychology Publishers.

Eisenstadt, S. A. and Simon, H. A. (1997). Logic and thought. *Minds Mach*, 7, 365–385.

Fafrowicz, M. and Marek, T. (2008). Attention, selection for action, error processing, and safety. In O. Y. Chebykin, G. Z. Bedny and W. Karwowski (Eds.), *Ergonomics and Psychology. Development in Theory and Practice* (pp. 203–220). Boca Raton, FL; London, England; and New York, NY: Taylor & Francis.

Gal'perin, P. Y. (1969). Stages in the development of mental acts. In M. Cole and Maltzman (Eds.), *A Handbook of Contemporary Soviet Psychology* (pp. 249–273). New York, NY: Basic Books.

Gal'perin, P. Y. (1973). Experience of development basic notions in psychology. *Questions of Psychology*, 2, 146–152.

Genisaretsky, O. I. (1975). Methodology of organization of activity system. In E. G. Yudin (Ed.), *Development of Automatic Systems in Design*. Moscow, Russia: Science Publisher.

Gilhooly, K. J. (1988). *Thinking: Directed, Undirected and Creative*. London, UK and San Diego, CA: Academic Press.

Hick, W. E. (1952). On the role of gain of information. *Quarter Journal Experimental Psychology*, 4, 11–26.

Hyman, R. (1953). Stimulus information: A determinant of reaction time. *Journal of Experimental Psychology*, 45, 423–432.

Leont'ev, A. N. (1978). *Activity, Consciousness and Personality*. Englewood Cliffs, NJ: Prentice Hall.

Lindsay, P. H. and Norman, D. A. (1977). *Human Information Processing. An Introduction to Psychology*. 2nd ed. San Diego, CA; New York, NY; and London, England: Harcourt Brace Jovanovich, Publishers.

Kahneman, D. (1973). *Attention and Effort*. Englewood Cliffs, NJ: Prentice Hall.

Kahneman (1981). The framing of decisions and the psychology of choice. *Science*, 211, 453–458

Meyer, D. E. and Kieras, D. E. (1997). A computational theory of executive cognitive processes and multiple-task performance: Part 1. Basic mechanisms. *Psychological Review*, 104, 3–65.

Newell, A. and Simon, H. A. (1972). *Human Problem Solving*. Englewood Cliffs, NJ: Prentice Hall.

Norman, D. A. (1976). *Memory and Attention: An Introduction to Human Information Processing*. 2nd ed. New York, NY: Wiley.

Pashler, H. E. (1998). *The Psychology of Attention*. Cambridge, MA: MIT Press.

Pashler, H. and Johnston, J. C. (1998). Attention limitations in dual–task performance. In H. Pashler (Ed.), *Attention* (pp. 155–190). East Sussex, UK: Psychology Press.

Piaget, J. (1952). *The Origins of Intelligence in Children*. New York, NY: International University Press.

Piaget, J. and Inhelder, B. (1966). *The Child's Conception of Space*. London, UK: Routledge and Kegan Paul.

Pushkin, V. V. (1978). Construction of situational concepts in activity structure. In A. A. Smirnov (Ed.), *Problem of General and Educational Psychology* (pp. 106–120). Moscow, Russia: Pedagogy.

Reitman, W. R. (1965). *Cognition and Thought: An Information-Processing Approach*. Hoboken, NJ: Wiley.

Rubakhin, V. F. (1974). *Psychological Foundation of Human Information Processing*. Moscow, Russia: Science Publishers.

Rubinshtein, S. L. (1957). *Existence and Consciousness*. Moscow, Russia: Academy of Science.

Schedrovitsky, G. P. (1995), *Selective Works*. Moscow, Russia: Cultural Publisher.

Shekhter, M. S. (1967). *Psychological Problems of Recognition*. Moscow, Russia: Pedagogical Publishers.

Sokolov, E. N. (1963). *Perception and Conditioned Reflex*. New York, NY: Macmillan.

Sokolov, E.N. (1969). The modeling properties of the nervous system. In M. Cole and I. Maltzman (Eds.), *Handbook of Contemporary Soviet Psychology* (pp. 671–704). New York, NY, and London, UK: Basic Books, Publishers.

Sokolov, E. V. (1974). *Culture and Personality*. Moscow, Russia: Science Publisher.

Sperling, G. (1960). The information available in brief visual presentations. *Psychological Monographs*, 74(1-29).

Treisman, A. (1969). Strategies and models of selective attention. *Psychological Review*, 76, 282–299.

Vygotsky, L. S. (1978). *Mind in Society. The Development of Higher Psychological Processes*. Cambridge, MA: Harvard University Press.

Wickens, C. D. and Hollands, J. G. (2000). *Engineering Psychology and Human Performance*. 3rd Ed. New York, NY: Harper-Collins.

Wickens, C. D. and McGarley, J. S. (2008). *Applied Attention Theory*. Boca Raton, FL; London, England; and New York, NY: Taylor & Francis.

Wickens, C. D., Gordon, S. E. and Liu, Y. (1998). *An Introduction to Human Factors Engineering*. London, UK: Longman Publisher.

Zaporozhets, A. V., Venger, L. A., Zinchenko, V. P. and Ruzskaya, A. G. (1967). *Perception and Action*. Moscow, Russia: Moscow State University.

Zinchenko, P. I. (1961). *Involuntary Memorization*. Moscow, Russia: Pedagogy.

Zinchenko, V. P., Velichkovsky, B. M. and Vuchetich, G. G. (1980). *Functional Structure of Visual Memory*. Moskow, Russia: Moscow State University Publishers.

5

Tasks with Complex Logical and Probabilistic Structure and Assessment of Probability of Decision-Making Outcomes

5.1 Problem-Solving and Decision-Making Tasks

Behavioral economists are mainly interested in studying the decision-making tasks that include problem-solving aspects and risky decisions. Specific attention to these types of tasks is also paid in work psychology and ergonomics. There are decision-making tasks for which a subject can formulate alternatives independently or such alternatives are offered to him/her in a ready form. A subject then has to select the most valuable course of action. However, there are also problem-solving tasks where necessary information is not available for the formation of hypotheses about possible alternatives or potential decisions. Some information that is required for formulation of alternatives is lacking, and there is no prospect for a subject to make a correct decision. Such scenario transforms a decision-making task into a problem-solving task that includes at the final stage formulation of alternatives and decision-making. At the first stage, a subject defines the problem based on analysis of the situation, at the next stage she/he formulates possible alternatives, then evaluates and selects a hypothesis, and at the final stage possible actions are determined and performed. These are decision-making tasks that include problem-solving aspects. In SSAT, this category of tasks is related to the group of *non-algorithmic decision-making tasks*, and we are going to discuss them later in greater detail.

Decision-making tasks are studied not just in psychology but also in human factors and economics. As has been discussed in the previous chapter, thinking plays a leading role in performance of problem-solving tasks.

The focus of our further analysis is decision-making involved in performance of the rule-based tasks. Let us consider the similarities and differences between *problem-solving* and *non-algorithmic decision-making tasks*. Problem-solving tasks are involved in the independent search for the unidentified methods and rules of solving the problem, and/or applying the known methods to new situations, and so on. Each problem has givens or conditions that are presented to a subject and he/she should independently select the relevant information. There are also requirements for what should be achieved as a result of problem solving that should be transformed into the subjectively accepted goal of the task. The goal of a task can be also formulated independently. Therefore, the goal formation stage is especially significant for the problem-solving tasks. The goal should not be considered as the end state of the task performance. The end state is often unknown in creative tasks. The final goal gradually is specified based on the intermediate formulated goals. A subject breaks a task down into subtasks, each of which has a sub-goal. Problem solving includes various strategies such as working backward, drawing analogies, utilizing various heuristics, and so on. The difference in the understanding of these processes in cognitive psychology and in SSAT is that in the latter these strategies are considered as a result of the process of self-regulation. Thinking and human activity in general are considered as self-regulative processes. An objective goal is presented as requirements and givens of the considered situation. Transformation of such objective goal into the mental model of task is a very important stage of the problem-solving task performance. Due to the fact that thinking plays a leading role in problem solving, the basic units of analysis here are thinking actions. Concepts, propositions, and images are the basic idealized objects that are utilized by the thinking actions. External motor actions have an auxiliary role in thinking. This type of action is especially important for the explorative activity. The result of such actions gives a subject information about the situation. Analysis of problem-solving tasks demonstrates that in most cases such tasks have only one acceptable solution. A task is divided into stages that have a cyclical organization. A subject continuously evaluates intermediate and final stages of the solution, and based on this process decides to select a specific hypothesis and associated with it alternatives for correcting the method of solving the problem. Hence, decision-making still plays an important role in solving the problems, even it has only one right solution. These findings have implications for understanding the difference between *problem-solving tasks*, *non-algorithmic decision-making tasks* and *rule-based tasks*. Other mental processes and specifically short-term and long-term memory are also involved in solving various problems. *Emotionally motivational processes* are playing a critical role in this process as well. For example, emotions play a key role in assessing the significance of the information in problem-solving situation, meaning that selection of required information from long-term memory, from short-term memory, or from the external environment can be determined by this factor.

The *non-algorithmic decision-making tasks* include problem-solving aspects in their performance. These tasks are studied in various applied fields such as economics, work psychology, and human factors or ergonomics. Hence, we are going to discuss these types of tasks in greater details.

When performing this type of task, a subject has to choose one out of several alternatives. These tasks have much in common with the problem-solving tasks. The formulation of alternatives and choosing one of them is the main distinctive feature of the decision-making tasks. Two types of decision-making theories have emerged over the years. The first one is called *rational decision theory*, the central concept of which is the expected value (Edwards, 1987). The expected value depends on the assigned generally agreed value to various outcomes of a choice. However, the value for the same events can be subjectively different for different subjects. This is the foremost cause for the development of the *psychological decision-making theory* (Newel and Simon, 1972; Tversky and Kahneman, 1974; Kahneman and Tversky, 1979; Tversky and Kahneman, 1973). The mathematicians and economists made the main contribution to the rational decision-making theory. This theoretical approach attempts to discover how people make decisions in a highly structured situation. The psychological decision-making theory strives to find out how people make decision in reality, when problem situations are less structured and often ambiguous. Decision-making tasks may be static and dynamic, well or poorly defined, and risky or not risky.

Rational and psychological approaches to the study of decision-making influence each other. As an example, let us consider such important concept of the rational decision-making theory as the subjectively expected value (von Neumann and Morgenstern, 2007).

$$\text{SEV} = \sum_{J=1}^{m} P\,S_J\,V_J,$$ \hfill (5.1)

where:

   $P\,S_J$ is the probability of events
   V_J is the expected value

Probability values range from zero to one. The expected value for a gambler is the expected winning or losing in the long ran. However, the outcome has a subjective value. For example, for a rich person the subjective value of money is different in comparison with its value for a poor one. Therefore, these scientists suggested using the concept of *utility*. Utility is determined as the subjective evaluation of value.

Subjectively expected utility SEU is determined as:

$$\text{SEU} = \sum_{J=1}^{m} P\,S_J\,U_J,$$ \hfill (5.2)

where:

   $P\,S_J$ is the probability of events
   U_J is its expected utility

According to this formula, the optimal strategy of choice is based on maximization of the subjectively expected utility of alternatives a_r. Alternatives a_r is optimal if

$$\text{SEU}(a_r) \geq \text{SEU}(a_i) \quad \text{for all } i = 1, \ldots, n \tag{5.3}$$

The psychological methods of utility evaluation are different. Some of them are based on the analysis of human behavior; the others are based on the judgments about the usefulness of a certain object. In the latter case, scientists often suggest employing the conditional scale of utility for various objects which has the numbers and an explanation of their meaning. A subject has to assign a certain number to a particular object by using the suggested scale. According to SSAT, a subject can relate an object or an outcome in different ways depending on the significance of this object or outcome for this subject. The factor of significance is always connected with the emotionally motivational components of activity. One of the serious shortcomings of the above-mentioned type of scales is that they don't take into account this factor. Not only cognitive but also emotionally motivational factors are important in evaluation of utility of various objects or outcomes. Neumann–Morgenstern utility theory allowed the formulation of principles of analysis of decision-making in risky situations. However, in fact, people's risky choices do not always match the behavior that is described by rational decision-making theory. In most cases, real tasks are much more complicated than those that are described by this theory because they can be multidimensional and dynamic. Evaluation of outcomes of such tasks can turn out to be very difficult. Despite the shortcomings of this theory, it is a very important one because it triggered development of the psychological decision-making theory. Due to this theory, it became possible to compare normatively prescribed decisions with real solutions. Psychological decision-making theory focuses on the heuristic processes. Newell and Simon (1972) had pioneered the studies in this area of research. Their objects of study were heuristic programs, also called heuristic strategies, that can be presented as rules, instructions, or intuitive solutions. They are not reliable, and in some cases may lead to mistakes. Their advantage is that they reduce the difficulty of the task and the decision-making time. These heuristic strategies play a special role in solving creative and complex problems. The essence of this method is that a subject describes aloud how he/she solves the problem. Unlike in introspection, where people describe their emotions and feelings, here, a subject describes her/his steps to solve the problem. Based on this description, the report is created and later analyzed by a researcher who tries to uncover heuristic strategies and compare them with the theoretically possible strategies. These studies brought interesting data in psychological theory of decision-making. The collected data is of great interest to economics, engineering, and human factors, specifically when studying dynamic, risky, and poorly defined (open-ended) tasks-problems that are typical in the real world.

As we have discussed above, the expected value theory has been utilized in economics for analysis of decision-making, when people use money or their value as its currency. However, in reality, when people make decisions, they do not view money as a linear function of its worth. It has been discovered that people try to maximize an expected utility (Edwards, 1987). Utility is considered as a subjective value of various outcomes.

Kahneman and Tversky (1984) introduced a function that depicts the relationship between an objective value and a subjective utility. Analysis of this curve proves that consequences of losses are subjectively evaluated negatively quicker than positive evaluation of the consequences of gains. This, in turn, influences the decision-making process subjects. Moreover, equal changes in value produce progressively smaller changes in utility. This factor also depends on subjects' individual features such as risk addiction and financial status. According to SSAT, motivation to perform the task that involves gains and losses depends on the relationship between the above-considered positive and negative significance of the task.

The relation between value and utility clearly demonstrates that the decision-making choice cannot be precisely evaluated as value. Conditions or givens and requirements or goals of tasks cannot be clearly defined for risky and open-ended tasks. A subject does not know all alternatives and information about their consequences. The situation is very dynamic, and the goal is formulated in general terms. The last can be modified and specified during the process of solving the problem. This type of task is very difficult to formalize. Therefore, the detailed qualitative analysis is critical when studying such tasks and especially the goal of a task which is not clearly formulated. One interesting way of studying the goal formation process is considering the aspiration level (a*), which interacts with the objective scale of achievement. For example, if the weight-lifter lifted 100 kilogram weights at the previous competitions, the goal would be to add another ten kilograms to the weights at the next one. According to this concept, the aspiration level sets the standard used by people to evaluate the outcome of task performance (Kozeleski, 1979). According to SSAT, when a subject initially attempts to solve a problem, she/he creates a level of aspiration. This aspiration level is conscious and can be considered as the initially formulated goal. This goal to some degree depends on the understanding of task and on the self-assessment. From the activity self-regulation point of view, at the first stage of task performance, a person often executes exploratory cognitive, and behavioral actions. The purpose is to use a sequence of trial-and-error actions to analyze their consequences. After that, the individual creates a hypothesis about the situation and then formulates an intermediate goal of the task. This goal depends on the self-assessment and on the understanding of the meaning of results about exploratory actions. In this situation, the personal significance of the formulated goal is also critical. Thus, cognitive and emotionally motivational factors are vital for this process. The aspirational level (a*), as it is related to the goal and decision about how to choose the goal, can be analyzed from the utility or u(a*) stand point.

If a decision is considered to have a low probability of success, then u(a*) > u(a) and subjects discard u(a). If a task is perceived by a subject as a very difficult one, then the aspirational level can decrease. Therefore, the potential success or failure can alter the aspiration level. From the AT perspective, and taking into consideration the process of self-regulation, the aspirational level is just one important component of *goal formation*. We considered the level of aspiration as a distinguished value that a subject wants to achieve. However, there are situations when a subject has to choose a goal when the level of aspiration is a multidimensional phenomenon. In such cases, the level of aspiration as a goal can have several meaningful values. For example, a manager formulates a goal to increase coal production to 100 tons and to reduce the fuel consumption by 10%. The usefulness of the aspiration level should be determined on multiple scales. Thus, a combination of such concepts as *goal and aspiration* can be beneficial.

According to AT, there are two classes of decision-making. The first one involves *decision-making on the sensory-perceptual level* and the second one involves *decision-making at a verbal thinking level*. The first type of decision-making is well known in psychophysics in relation to the detection of a signal in noise. These studies demonstrate that detection of a signal on noisy background depends not only on the sensitivity of the sense organs but also on the criteria for "yes" or "no" response used by a subject. Swets (1964) utilized just one horizontal x-axis for "yes" or "no" response, and the criteria can move along the x-axis. Bardin (1982, 1988) has demonstrated that an observer, while performing the signal detection task, can use additional criteria corresponding to other axes of measure. For instance, when two signals are very difficult to discriminate by their loudness, subjects began to use another criterion. They started perceiving other previously unnoticed qualities of acoustical signals such as resonant, dull, dimmed. If a subject uses more than one type of signal quality, this means that he/she uses an additional axis transferring from x-axis to y-axis. In SSAT, this process of changing the utilized criteria is considered as changes in the strategy of task performance utilized by a subject. Thus, even a simple psychophysics task includes decision-making processes. More complex decision-making processes are performed at the verbally thinking level. Complex cognitive processes such as thinking, short- and long- term memory are involved in such decision-making. It is necessary to distinguish between decision-making that is involved in problem-solving tasks, in decision-making tasks, and in rule-based tasks. Moreover, any decision-making includes emotionally motivational components and volitional processes. For example, if a subject has to perform just one but critical or dangerous action (act or do not act), such action is usually accompanied by emotional stress.

We cannot restrict our analysis of decision-making by considering just two types of theories that are known as rational and psychological approaches.

In the following section we discuss rule-based decision-making that is critically important for studying the efficiency of the production process and therefore is significant not only for human factors but also for economics.

Below we briefly give some critical comments on the existing psychological decision-making theory. This analysis is useful from both theoretical and practical points of view. From the SSAT perspective, analysis of this theory demonstrates that psychological decision-making theory to a significant degree is reduced to the study of cognitive processes and specifically functioning of memory. When considering such concepts as objective value or subjective value (utility), there is no clear understanding of how such concepts correlated with emotionally motivational components of activity. The useful and important *utility* concept should be studied in unity with such concepts as objective meaning (verbalized or non-verbalized), subjective sense and associated with it the concept of significance (emotionally evaluative components of activity), and a goal of a decision-making task. For example, a subjectively accepted or formulated goal can influence formulation and assessment of various alternatives.

In SSAT, the concept of sense has two aspects of study. One aspect is associated with the emotionally evaluative mechanism of sense (factor of significance) and the other one considers its relation to the objective meaning. Meaning is related to individual consciousness as an image or concept and is embodied in language. Thanks to meaning, a subject understands various environmental phenomena.

Let us consider again the relationship between objective value and utility. This relationship can be explained not only by cognitive mechanisms but also by significance (emotionally evaluative factor) of gain and losses to a subject. For example, if a person is very rich and the possibility of losing a considerable amount of money does not really matter, she/he can neglect losses and accept a risky strategy to experience positive feelings and satisfaction from possibly winning. We can conclude that different evaluation of losses and gains by different subjects can be explained based on the analysis of objective meaning and the personal sense of the considered events. Losses and gains can have the same objective meaning. However, they can have different personal sense for different subjects because sense integrates cognitive and emotionally evaluative components. Therefore, a subject evaluates losses and gains not only cognitively in an objective manner (objective meaning) but also in a personal sense (what it means for me). A subject might be more concerned about possible losses because they usually are more significantly (as a negative event) than expected gains (as positive event).

Sense refers to the emotionally evaluative aspects of activity where cognitive components play a subordinate role. Sense that has a value for a subject and is accompanied by positive emotions during achievement of the goal of a task should be considered as having *positive significance*. Sense that is accompanied by various obstacles, by danger, and by a negative emotional state should be considered as having *negative significance*. If the goal of a task includes only positive significance, the block *sense* in the model of self-regulation of orienting activity (see Figure 3.1) has a *homogeneous* structure and has positive significance. However, according to our model of self-regulation, the goal of a task can not only have positive value for a subject but can also include some negative aspects or some negative value. In such cases, block *sense* has a *heterogeneous* structure and reflects both positive and negative significance. The proportion of these types of significance determines the integrative character of evaluation of the task significance. The mechanism of significance plays an important role in the formation of motivation that is considered to be an inducing mechanism of activity. Vector motive-goal gives activity its goal-directedness. In the situations, when block *sense* includes only negative significance, a subject can avoid such task performance because it has no sense for a subject. There are also other function blocks that can influence the subject's motivation. Here, we only briefly consider function blocks *"assessment of task difficulty"* and *"criteria of evaluation"* (see Figure 3.1). Block *assessment of task difficulty* interacts with block *assessment of task sense*. If for example, a task would be evaluated as very difficult and not significant, a subject would reject the task. At the same time, if a task is difficult but very significant for a subject, he/she would be motivated to perform the task. If a subject acquired some strategies of task performance in the past and they were successful, she/he would have a tendency to use these strategies in similar situations. This decreases the mental effort needed for the task performance, and a subject becomes more confident that he/she can obtain a satisfactory result.

It is interesting to consider the historical fact that occurred in 1724 and involving Isaac Newton, who not only was a brilliant mathematician but also had a strong financial background (King, 2017). In 1724, a finance company was created that according to today's terminology operated on the principle of the financial pyramid. So the new investors' money went to pay the high percentage profit to the old investors. Upon careful consideration, it would become clear that within a particular period of time, the number of payouts would exceed the amount the company had. Newton did not pay attention to this fact. He invested a lot of money in this company and after a while lost all of it. As a great mathematician, he could easily figure it out, but he made a purely emotional decision, the same as many others did. He made a purely emotional decision without spending any cognitive effort on determining if it was a rational one. Thus, not only cognitive but also emotionally motivational mechanisms of activity regulation play an important role in evaluation of losses and gains.

The outcome of decision-making should also be considered in our discussions. A positive outcome that is expected by a subject should be considered as the goal of a decision-making task when choosing an alternative. An unexpected outcome that is perceived by subject as a positive result is not a goal of a decision-making task and should be considered as a supplementary result. In risky tasks, a possible outcome can conflict with the goal of a decision-making task. The goal of a task can influence the selection of alternatives.

For example, choosing an adequate military strategy can depend on the goal formulated by the senior officers in specific conditions. Both memory and thinking are critical factors in the studies involving decision-making tasks. However, analysis of different publications demonstrates that attention is mostly paid to short-term and long-term memory and attention, and the role of the thinking process usually is underestimated. The objectively presented task conditions or givens are not simply perceived by a subject but information is filtered and subjective representation of a task is actively developed consciously or even unconsciously. Cognition as well as emotionally motivational mechanisms are involved in this process. The subjective representation of a task depends not just on the ability to manipulate the perceptual data or concepts and their meaning but also on the ability to translate this data into a subject's idiosyncratic impression. This is the transformation of an objective meaning into an internal personal sense. This process is important for understanding of the concept of utility. Performance of problem-solving or decision-making tasks generates new motives and emotional states. The cognitive and emotionally motivational processes during decision-making task performance interact dynamically and influence each other. Thus, we cannot agree that decision-making is achieved only by cognitive mechanisms. It is also important to stress that subjective statements about the probability of events can also be influenced by the emotionally motivational state of a person. People have a tendency to evaluate the probability positive and negative events in different ways. Intuitive processes have an important role in decision-making tasks.

Let us consider how risk-addictive people behave when they perform risky tasks. It is possible to observe the situation when an inexperienced driver has a tendency to gradually increase the speed of the car until an accident occurs (Kotik, 1978). Applying the SSAT concept of self-regulation, this situation can be explained in the following way. If the dangerous task is repeated multiple times, the difficulty of the task decreases, and this in turn diminishes the factor of significance and motivational level of its performance (see model of self-regulation, Figure 3.1, function blocks *"assessment of the task difficulty"*, *"assessment of sense of task"*, *"formation of the level of motivation"*). This, in turn, depresses a subject's interest in this task performance. The subject tries to maintain his or her interest in the task performance by increasing the task complexity and therefore by increasing its difficulty, which leads to increasing significance of the task and the subject's motivational level. The cycle can recur again during the skill acquisition. A subject might repeat the cycle until the failure can take place.

In conclusion, we outline several stages in performance of decision-making tasks.

1. Informational search involves discovering relevant elements of information and comparing them with information that is stored in memory.

2. Based on selected information, a subject creates a subjective representation of a task. He/she creates a mental model of the situation that can be stable or dynamic.

3. Based on the created mental model, a subject formulates various hypotheses and possible alternatives.

4. Based on analysis of existing alternatives and their possible outcomes, a subject makes a required decision.

5.2 Decision-Making in Rule-Based or Instruction-Based Tasks

In the previous chapters, we considered the concept of task and suggested the new system of classification of tasks. We also briefly considered the concept of task analysis and design. This material is important for our further discussions. This task classification allows defining the area of task analysis where the concept of design can be used. In a production environment, a subject, in most cases, performs the task based on the given instructions or rules. While such tasks might be complex, they can still be described algorithmically. This is not a regular algorithm but a human algorithm that describes human activity during task performance. It depicts the logical sequence of cognitive and motor actions that are performed by a subject when he/she executes a specific task. The specific nature of the production environment is that a subject performs a set of operationally related tasks or repeats the same task multiple times. Development of efficient methods of performance of such tasks is critical for both economics and human factors. Analysis and design of rule-based tasks are associated with aspects of human performance studies such as design of equipment, safety mode, training, reliability, complexity, professional selection, time analysis, compensation, and so on. Thus, in this chapter we consider the role of decision-making processes in performance of algorithmic or rule-based tasks because this is a critically important type of task in a production environment. Such tasks are performed according to given instructions that derive from technological requirements. Rule-based or algorithmic tasks can be divided into *"deterministic-algorithmic tasks"* and *"probabilistic-algorithmic tasks."* In the first type of tasks, a subject makes decisions that have two outputs with equal probability ("if-then" type). For example, "if the temperature exceeds the permissible level, turn toggle switch off", "if the temperature is lower than permissible level, turn toggle switch on." In probabilistic-algorithmic tasks, subjects make decisions with not just two but also with three or more outputs, each of which can happen with varying probability. Any algorithmic task can include several decision-making steps.

Here, we present basic distinguishing characteristics of rule-based or algorithmic tasks.

In the production environment, tasks can include a number of relatively simple decisions, each of which can be relatively simple to perform. The simplest types of decisions are the complex choice reactions studied in the laboratory conditions. Such variables as number of alternatives, stimulus modality, intensity of stimulus, temporal uncertainty, and so on are considered to affect reaction time. In the context of a task, the speed of reactions is different, and such characteristics as expectation of a specific type of reaction, an ability to remember the rules for the performed reactions, and other factors should be taken into consideration. When performing a task at hand, a subject reacts but makes expedient choices for the correct action of varying complexity. Here, we are not talking about reactions but rather about simple decisions in the context of a task. Such decisions are commonly utilized in the skill-based tasks. There are usually a small number of the simplest decisions that are involved in the skill-based task performance. They have fewer outputs, and because of repetitive multiple performance, such tasks are performed in a habitual way. In skill-based tasks, cognitive and behavioral actions have a high level of automaticity. The complexity of skill-based tasks depends, first of all, on the complexity of the motor components of the activity and the specificity of their organization and regulation.

The rule-based tasks involve more complex types of decisions. Decisions have more than two outputs that occur with different probabilities and include complex executive actions that involve deductive reasoning. Such decisions are not complex when they are made out of the context of the task and there are no intensive thinking processes involved. However, in the context of the task, such decisions become sufficiently complex to make. One of the critical factors is how many decisions are involved in the task performance. The more such decisions are in the context of the tasks, the more rules a subject has to remember. Performance of a task many times during the shift and the necessity to make various kinds of decisions require constant attention and use of deductive reasoning. The information required for decision-making should be correctly allocated. External signals that are adequate to the given technical instructions and rules of task performance should be allocated properly at the design stage. The term *signal-instruction* means that a sign has to remind a subject which rule should be applied in a given condition. With an increase in the pace of performance, even the simple decisions become complex ones for a subject to perform repeatedly. In this regard it is necessary to externally present information when decisions are carried out and not to depend on memory. Variability of a task and the necessity to remember the rules of task performance lead to the excessive concentration of attention and as a result, increased fatigue.

Let us consider a simple example that has already been discussed.

If an even number is displayed, a subject has to push the red button, and if an uneven number is displayed, he/she has to push the green button. A subject has to remember these simple rules during the task performance. Moreover, the task should be performed within a certain time frame. Repetition of the same task or performance of this task a number of times among other tasks is a particular important condition of the decision-making. In such a situation, a subject has to remember the given rules. When information needed for making a decision is presented to a subject, he/she has to recall adequate rules and select adequate actions. In the study of reaction time, a subject perceives a small set of well-structured stimuli and simply has to quickly recognize one of them. In the production situation, a subject often has to actively select a corresponding stimulus. Such recognition is often transformed into a complex successive process of recognitions. Some decisions involve deductive syllogistic reasoning. In real life, erroneous choices might lead to undesirable consequences. Due to the fact that a subject performs a number of tasks and each task might include a number of decisions, there is a chance that a particular decision is made incorrectly. Such decisions overload a subject's memory. Based on the *signal-instruction* principle instead of indicators with even and uneven numbers, a designer should introduce red or green light bulbs as more adequate indicators when a subject has to choose to push the red or green button. It is then much easier to select red or green button. This is a hypothetical example. In the real production environment, it is not that easy to follow the signal-instruction principle. This is particularly relevant when a subject has to choose from more than two alternatives that appear with various probabilities. Moreover, in a production environment, in complex man-machine or human-computer interaction systems, decisions can be much more complex and require deductive thinking. In an unusual situation, a subject might have to independently select adequate decisions. Such tasks are not designed in advance. These tasks are called non-algorithmic task. Only some components of this type of task can include a standard method of performance.

An overwhelming number of decisions in a production environment make tasks variable and complex and can cause errors and a reduced quality of performance and productivity in general. Some tasks might become very variable and complex. In some cases, in order to make the right decision based on the given rules, it's not enough to simply perceive the externally presented signals; it is also necessary to interpret the situation by comparing its elements and using deductive reasoning. The correct decision when performing complex actions often requires extracting information from long-term memory and keeping information in working memory. So reducing memory workload becomes a critical factor. Extensive recurrence of the task is also a critical factor. For example, assembly line workers sometimes perform the same task more than 1,000 times during the shift. Continuous decision-making and alternation of the sequence of actions can become challenging when a subject is trying to maintain productivity, which can elevate the level of fatigue. With an increase of the pace of task performance, even making simple choices can increase the fatigue of a subject.

All this suggests that decision-making should not be studied just for problem-solving tasks or decision-making tasks that include problem-solving components. There is a wide class of normatively prescribed tasks where the subject acts in accordance given rules and instructions. The decisions are made within the structure of tasks and are not performed in isolation. There is a need to design an adequate structure for such tasks. Designing an adequate structure of the production tasks helps to reduce their complexity. One of the key factors for reducing the complexity of the tasks is a design that makes decision-making more efficient. It is also important not only for the design of performance methods but also for the design of equipment and software to increase productivity and efficiency of performance in general. Rule-based or instruction-based tasks are the main type of tasks in the production processes. For example, all tasks should be performed according to technological requirements. It is similar for complex man-machine systems. For instance, pilots in most cases perform tasks according to the prescribed rules. These rules can be very complex. Only in atypical situations would a pilot have to solve a problem that would include unusual creative components. The role of the rule-based tasks is often underestimated. The assumption that such tasks can be totally automatized is not always accurate. For example, total computerization of the main types of pilots' tasks is undesirable because a computer can fail. The pilot should be able to take over the control of the aircraft. Furthermore, full computerization and automation might not be economically feasible. Therefore, computerization does not exclude manual components of work. We consider computerized and computer-based tasks in the following chapters. In a production environment, full automatization can be very expensive or even impossible.

The material presented indicates that analysis of decision-making cannot be reduced to the study of just the problem-solving tasks, or the tasks with problem-solving components and decisions that are made at the final stage of task performance. There is a wide class of rule-based tasks where a subject makes decisions based on rules specified in advance. Analytical methods should be applied for the analysis of such tasks, especially during the design process. In economics, these methods are applied in order to increase productivity and efficiency of work and for job evaluation. An adequate design of such tasks reduces their complexity and as a result reduces cognitive demands for task performance when a subject is making various rule-based decisions. This area of study is also critical in human factors or ergonomics. Currently there are no analytical methods of task analysis outside of SSAT when the tasks under consideration are variable. The studies are commonly reduced to the observation of real tasks or creation of physical models for new tasks. Some psychologists and human factor specialists reject even the possibility of such analysis. Vicente (1999, p. 69) promoted the idea of *constraint-based* approaches to task analysis. He wrote:

> With constraint-based approaches, on the other hand, none of decisions are made up front by the analyst. Only guidance about the goal state and the constraints on actions are provided, not about how the task should be accomplished.

Consequently, workers are given some discretion to make decisions online about how to perform the task. This contrast shows that constraint-based approaches to task analysis provide more worker discretion than do instruction-based approaches.

Vicente has stated that the only method of studying variable tasks is the input-output analysis. This method simply identifies the goal state of the task, the input that is required to get to that point, and constraints that should not be violated when a subject tries to achieve the goal state. It is not specified how a subject should reach the goal. In economics and psychology, decision-making is also not considered for such tasks.

The study of isolated decision-making during performance of tasks with a complex probabilistic structure is not of particular interest. There are no methods of describing variable human activity when a subject performs complex rule-based tasks that include multiple decisions in ergonomics, work psychology, or economics. The method of studying such tasks is offered by SSAT (Bedny and Meister, 1997; Bedny and Karwowski, 2007; Bedny, 2015). However, analytical principles of analyzing the probabilistic structure of complex tasks that include multiple decisions with various number of outputs and various probabilities were not discussed in prior publications. These aspects of analysis also have never been discussed in other sources. Moreover, there is a tendency to reject even the possibility of such analysis. In the following chapters, the analytical methods to conduct such task analysis will be presented. Our approach will be demonstrated using the study of complex and variable computerized and computer-based tasks. This approach can be applied to the real work settings when the performance is done under challenging conditions such as uncertainty, time pressure, vague goals, and so on. The importance of studying such decisions is considered by such authors as Klein (2008) and Orasanu and Fischer (1997).

5.3 New Method of Assessing Probability of Decision-Making Outcomes in Task Analysis

The logical and probabilistic structure of activity depends on probabilities of outcomes of various decisions. There is a need to determine the probability of the events that influence decision-making outcomes. There are two basic methods of determining the probability of events: the objective method of evaluation and the subjective method of interpretation of the probability of events (Savage, 1954). The latter can be determined by subject matter experts (SME). In this chapter, we will give a brief analysis of objective and subjective methods of analyzing the probability of events. We also describe the method, developed in SSAT, of subjective evaluation of probability of events for rule-based tasks. This method has been important for studying variable tasks not only in ergonomics but also in economics.

There are a number of different approaches to the assessment and interpretation of the probability of events. The most widely used method in mathematics and statistics uses the frequency interpretation of events. The other main concept is the relative frequency. The latter refers to the ratio of the number of trials completed by occurrence of events to the total number of trials. In terms of statistics, the probability of occurrence of an event can be interpreted as the relative frequency of the event. The value around which the frequency of occurrence of an event fluctuates with an increasing number of trials to infinity is called the probability of the event. However, we can observe the frequency of occurrence of the events using as an example a limited number of trials. In addition, it is difficult to calculate the frequency of such events. Despite these limitations, the frequency interpretation of the probability method has an important role in statistics, psychology, economics, and other areas of science. The probability that is based on the evaluation of the frequency of events is also called *objective probability*. Another approach is called *subjective method of interpretation* of the probability of events (Savage, 1954). According to this approach, the probability can be considered as a belief of the subject, who can rationally estimate the probability of events. Probability obtained by such a method of assessment is called subjective probability. Another similar method of the subjective evaluation of probability is called *logical method of interpretation of probability* (Kozelecki, 1979). This approach evaluates the probability of correctness of judgment about the events. Such a judgment can be either true or false. Since the subject has not always possessed all the necessary information about considered events, one could only assume that the judgment is likely true. The probability that the judgment is correct can be based on comparison with other judgments and the drawing of correct conclusions or inferences from observation or facts. Methods of assessment of subjective probability are closely related and don't have a definite borderline. This can be explained by the fact that degrees of certainty or degrees of belief of a subject (*subjective interpretation*) affect the rational judgments about the likelihood of events (*logical interpretation*). The subjective probability is an important concept in psychology. Specifically, it is the critical factor in evaluation of risky events. Methods for assessing subjective probabilities are based on the fact that a subject forms judgments about the probability of events. However, the subjective assessment of the probability may be inaccurate. For example, a subject may not be competent enough to assess the probability. Anticipating various events, subjects not only use the laws of theory of probability but also some heuristic principles. The latter may result in errors in assessing the probability of events, but such strategies are simple and do not require significant mental effort.

As we already mentioned, one heuristic strategy that reduces mental effort is called *representativeness* (Tversky and Kahneman, 1974). This strategy is based on the degree of similarity between the sample and the population.

The sample based on which assessment of the probabilistic characteristics of population is performed has to reflect probabilistic features of real events. However, the strategy of representativeness does not guarantee correctness of evaluation of probability of considered events. Another strategy that increases the accuracy of the estimate of probability is the principle of *availability*. One of the factors that can reduce the accuracy of the estimation of the probability is that familiar events that are encountered by a subject more frequently can be evaluated differently in comparison to those that are encountered seldom. For instance, a boxing fan who knows that his country often wins Olympic medals for boxing can decide that men in his country are usually stronger than men in other countries. However, in reality, boxing might be very popular in his country and therefore there are more boxing clubs with good coaches. It has also been demonstrated that the properties of long-term and short-term memory are important for the evaluation of the subjective probability of events. Despite the fact that in some cases the subjective assessment of probability of events is not entirely accurate, and in some cases even erroneous, correctly selected experts can with high levels of accuracy assess the probability of events. We believe that in a production environment, properly chosen experts can estimate the probability of events that occur in specific production conditions with high accuracy. In some cases, 2–3 experts can be selected to estimate the probability of events. With significant discrepancies in evaluation of data, it's necessary to ascertain the reasons for their differences and repeat the assessment. Analysis of the data collected in the field has allowed us to enhance the accuracy of estimating the probability of the decision-making outcomes.

Below we describe basic principles of evaluation of decision-making outcomes developed in SSAT. As has been stated above, performance of the rule-based tasks requires making various decisions. Some of them are carried out based on the assessment of perceived information, while others involve extracting information from memory and using thinking operations. Simple decisions have only two outcomes, with a probability of 0 or 1. Additionally, probabilistic-algorithmic tasks include decisions with multiple outcomes, and their probability varies from 0 to 1. In algorithmic description of human performance, such decisions are called logical conditions, and they're depicted by a symbol *l*. Decision-making or logical conditions are major factors of rule-based task complexity. The probability of the outcomes of certain logical conditions or decisions determines the probability of performing the next steps of a task that follow such outcomes.

Probabilities of some decision-making (logical conditions) outcomes depend on the preceding external events that, in turn, can appear with certain probability. For instance, suppose there are two mutually exclusive events. One of them has the probability of 0.8 and another one has the probability 0.2. If event one occurs, then the corresponding decision-making outcome should have the same probability unless the operator made a mistake. The same is true for the second outcome. The probabilities of these outcomes determine the probabilities of the following step of task performance or members of the algorithm of the task performance until the next decision that can change the probabilities of the members of the algorithm is made (steps of task performance). Hence, the evaluation of the probability of outcomes of the decisions and the subsequent steps of the task performance is the critically important stage of the analysis of tasks with complex probabilistic structure. Experimental evaluation of such probabilities is often complex, time-consuming, and sometimes even impossible. Therefore, researchers attempt to use subjective measurement procedures. There are several good publications that compare objective versus subjective methods of measurement (Hennessy, 1990; Muckler, 1992). Studies demonstrate that the subjective measurements in some cases have an advantage in comparison to the objective ones. Below we describe the subjective measurement procedures that have been developed in the framework of SSAT.

In our study, the subject matter experts (SME) estimated the probability of events that influence decision-making outcomes. Developed methods involve utilizing special scales for evaluation of the subjective probability of various decision-making or logical condition outcomes. This method also involves the new event tree technique. In the developed event tree, the probability of the following events depends on the probability of preceding events. Thus, this event tree utilizes conditional probabilities.

The developed method also allows us to transfer qualitative data about outcomes of various decisions into the quantitative data as well as to transform the verbal subjective judgment of experts into quantitative data. Of course, such quantitative data is obtained with some approximation. Nevertheless, the considered method has precision that's adequate for its practical application.

The scale of the subjective probability evaluation of events has been utilized for the evaluation of probability of decision-making outcomes and the subsequent steps of the task performance (see Figure 5.1). The scale of the subjective probability evaluation of events has a line of equally probable events that corresponds to the numerical value of $P = 0.5$. Events with probabilities greater than $P = 0.5$ are above this line, and events with probabilities less than $P = 0.5$ are below it. There are two extreme ranges of probability: the extremely high-probability diapason corresponds to events with probability from 0.9 to 0.99 and the extremely low-probability diapason reflects probabilities from 0.1 to 0.01. The probability values on the vertical scale are divided into different ranges or diapasons of probability. In our scale, eight diapasons are suggested. Four of them are above and four are below the equally probable line.

The main characteristic of each diapason is its position relative to the equal probability line. The extremely high-probability diapason ($P = 0.9 - 0.99$) and the extremely low-probability diapason ($P = 0.1 - 0.01$) are considered as additional reference points. Because events might occur with different probabilities inside each diapason, there's a need to determine whether their

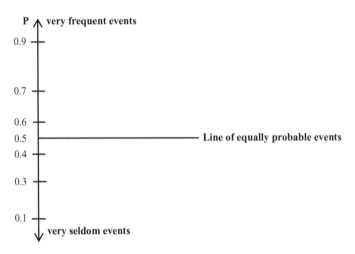

FIGURE 5.1
The scale of the subjective probability evaluation of events.

probability is closer to the top or the bottom of each considered diapason. Based on such a method, we can determine the probability of considered events with sufficient accuracy. Let us consider these two extreme diapasons. Figure 5.2 depicts the extremely low-probability diapason with probabilities $P=0.01–0.1$ (very seldom events), and Figure 5.3 reflects the extremely high-probability diapason with probabilities $P=0.9–0.99$ (very frequent events). The diapason with very-seldom events can be further divided into three areas with various probabilities: a) $P=0.01–0.03$; b) $P=0.04–0.06$; c) $P=0.07–0.1$. These diapasons are depicted graphically in Figure 5.2. If the considered events fall into one of these diapasons or into a considered range of probabilities, then based on qualitative analysis we can find the specific area within this diapason.

The extremely high-probability diapason for very frequent events can be presented similarly (see Figure 5.3).

Thus, the subjective assessment of an event's probability is accomplished in stages. First, using the scale, experts determine in what diapason a considered event falls, and then they define if the probability of this event is close to the top, middle, or bottom of this diapason. Based on this, it is possible to determine the probability of events within the considered range of

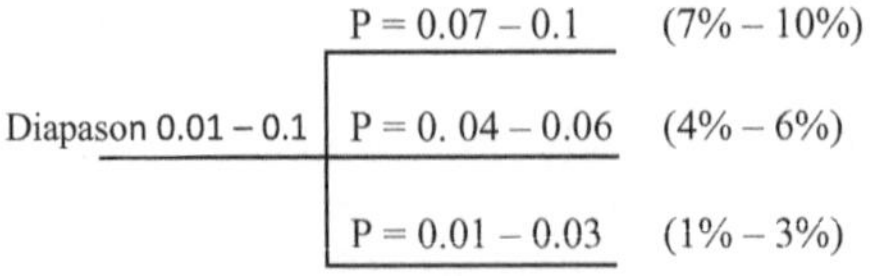

$$
\begin{array}{ll}
P = 0.07 - 0.1 & (7\% - 10\%) \\
\text{Diapason } 0.01 - 0.1 \quad P = 0.\,04 - 0.06 & (4\% - 6\%) \\
P = 0.01 - 0.03 & (1\% - 3\%)
\end{array}
$$

FIGURE 5.2
The extremely low-scale diapason covers very-seldom events with P from 0.01 to 0.1.

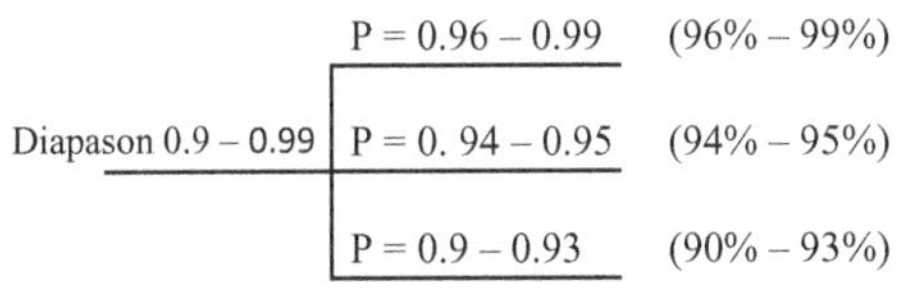

$$P = 0.96 - 0.99 \quad (96\% - 99\%)$$

Diapason 0.9 – 0.99 $\quad P = 0. 94 - 0.95 \quad (94\% - 95\%)$

$$P = 0.9 - 0.93 \quad (90\% - 93\%)$$

FIGURE 5.3
The extremely high-probability diapason covers very frequent events with *P* from 0.9 to 0.99.

probabilities. There are other types of diapasons such as diapason with probability $P=0.51–0.6$ that we designate as *"diapason for events with just noticeably higher than equal probability of occurrence"*. Diapason with $P=0.4–4.99$ is designated as *"diapason for events with just noticeable lower than equal probability of occurrence"*. We also distinguish such diapasons as: *noticeably more frequent events* with probability $P=0.60–0.69$; *rather frequent events* with probability $P=0.7–0.89$; *noticeably more seldom events* with probability $P=0.4–0.31$; *rather seldom events* with probability $P=0.3–0.1$; and *very seldom events* with probability $P=0.01–0.1$ (Figure 5.2).

Such a method allows experts to determine the probability of considered events with adequate precision. We would like to point out that the line of equally probable events and the diapasons used as reference points should be constant (extremely high diapason, extremely low diapason). Other diapasons may vary in size depending on the specifics of the task being analyzed.

As we already pointed out, the suggested method also utilizes event tree modeling. However, this method is used with some modifications. Below we demonstrate an example of hypothetical probabilistic event tree modeling that is built utilizing our method (see Figure 5.4).

The first two branches on the above model have the probabilities $P=0.15$ and $P=0.85$. The second two branches have probability $P=0.05$ and probability $P=0.1$. As it can be seen, these last two branches have a combined probability $P=0.15$. The combined probability of the last two branches is exactly the same as the probability of a branch that is designated by the bold line.

The proposed method shows that the probability of events on the child branches is determined providing that the probability of event at the first branch is $P=0.15$. This is the conditional probability of events within the considered process. In mathematics, the conditional probability of event B under condition that event A has occurred is defined as P_A (B). The suggested

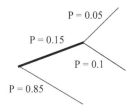

FIGURE 5.4
Hypothetical event tree model.

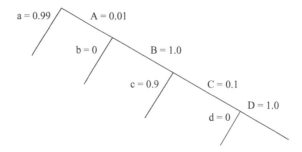

FIGURE 5.5
Example of human reliability analysis model (traditional method).

method of developing event tree models helps SMEs to determine the probability of events with higher accuracy.

Let us compare this event tree model with the event tree model developed utilizing the traditional method depicted in Figure 5.5. This method has been used in the human reliability assessment study (Kirwan, 1994). As can be seen in this figure, the probability of each subsequent branch does not depend on the probability of preceding branches, and each pair of branches has the accumulated probability of 1. Suggested in the SSAT method of developing an event tree model allows SMEs to determine the probability of considered events with high accuracy.

As can be seen, this method does not suggest an external tool that would help experts to evaluate the probabilities of the considered events varying in a narrow range, because according to this method the probability of events can change from $P=0$ to $P=1$. The only tool for determining the probability is the information extracted from memory and subjective opinions of the experts. This is a one-step procedure. In contrast, in SSAT, defining probability is a multiple-step procedure relying on externally presented tools and rigorously defined rules. Our method allows determining the absolute and conditional probability of events within the production process, not just their absolute probability as shown in Figure 5.5. Practical application of this method will be demonstrated in chapter 7.

References

Bardin, K. V. (1982). The observer's performance in a threshold area. *Psychological Journal*, 1, 52–59.

Bedny, G. Z. (2015). *Application of Systemic-Structural Activity Theory to Design and Training.* Boca Raton, FL and London, UK: CRC Press and Taylor & Francis.

Bedny, G. Z. and Karwowski, W. (2007). *A Systemic-Structural Theory of Activity. Application to Human Performance and Work Design.* Boca Raton, FL and London, UK: CRC and Taylor & Francis.

Bedny, G. and Meister, D. (1997). *The Russian Theory of Activity: Current Application to Design and Learning.* Mahwah, NJ: Lawrence Erlbaum Associates, Publishers.

Bardin, K. V. and Zabrodin, T. A. (1988). Changing sensory sensitivity when solving problems. *Problems of Psychology*, 1, 149–154.

Edwards, W. (1987). Decision making. In G. Salvendy (Ed.), *Handbook of Human Factors* (pp. 1061–1104). New York, NY: Wiley.

Hennessy, R. T. (1990). Practical human performance testing and evaluation. In H. R. Booher (Ed.), *MANPRINT: An Approach to Systems Integration* (pp. 433–479). New York, Van Nostrand Reinhold, A Division of International Thomson Publishing Inc.

Kahneman, D. and Tversky, A. (1979). Prospect theory: An analysis of decision under risk. *Econometrica*, 47, 313–327.

Kahneman, D. and Tversky, A. (1984). Choices, values and frames. *American Psychologist*, 39, 341–350.

King, T. (2017). *Isaac Newton: The True and Surprising Story of the Life of Sir Isaac Newton.* Palo Alto, California, Createspace Independent Publisher.

Kirwan, B. (1994). *A Guide to Practical Human Reliability Assessment.* London, UK: Taylor & Francis.

Klein, G. (2008). Naturalistic decision making. *Human Factors*, 50(3), 456–460.

Kotik, M. A. (1978). *Textbook of Engineering Psychology.* Tallin, Estonia: Valgus.

Kozeleski, J. (1979). *Psychological Theory of Decision.* Moscow, Russia: Progress Publisher.

Muckler, F. A. (1992). Selecting performance measures: "Objective" versus "subjective" measurement. *Human Factors*, 34(4), 441–455.

Newell, A. and Simon, H. A. (1972). *Human Problem Solving.* Englewood Cliffs, NJ: Prentice-Hall.

Orasanu, J. and Fischer, U. (1997). Finding decisions in natural environments: The view from the cockpit. In C. E. Zsambok and G. Klein (Eds.), *Naturalistic Decision Making* (pp. 343–358). Mahwah, NJ: Erlbaum.

Savage, L. J. (1954). *The Foundation of Statistics.* New York, NY: Wiley.

Swets, J. A. (1964). *Signal Detection and Recognition by Human Observers.* New York, NY: Wiley.

Tversky, A. and Kahneman, D. (1973). Availability: A heuristic for judging frequency and probability. *Cognitive Psychology*, 5, 207–232.

Tversky, A. and Kahneman, D. (1974). Judgment under uncertainty: Heuristics and biases. *Science*, 211, 453–458.

Vicente, K. J. (1999). *Cognitive Work Analysis: Toward Safe, Productive, and Healthy Computer-Based Work.* Mahwah, NJ: Lawrence Erlbaum Associates, Publishers.

von Neumann, J. and Morgenstern, O. (2007). *Theory of Games and Economic Behavior.* Princeton, NJ: Princeton University Press.

Part II

Design of Computerized and Computer-Based Tasks with Variable Structure

6

Morphological Analysis of Activity

6.1 Introduction to the Morphological Analysis of Task Performance

The term *morphological* comes from the Greek morph, which means shape or form. The morphological method of study describes arrangement of different elements of a holistic system.

The term *morphology* is used in anatomy, biology, and other fields of science for the description of constructive features of a complex system. A morphological analysis precedes quantitative analysis of the such a system. Morphological analysis as an independent approach to the study of the various types of complex systems in an abstract manner was developed by Zwicky (1969), who utilized this method in such diverse fields as the development of jet and rocket propulsion systems, classification of astrophysical objects, and other projects. More recently, this approach has been applied by a number of researchers in the United States and Europe in such fields as policy analysis, development of new products, and so on. This method is not quantitative but very useful for description of objects' structure (Ritchey, 1991). This formalized method facilitates further quantitative analysis.

The main purpose of morphological analysis is to transfer the ambiguous qualitative description of a complex system into a clearly structured one. Morphological analysis is a general method for non-quantified modeling of various objects. The morphological approach identifies the parameters or dimensions of a complex problem or system. Each parameter can have a range of relevant values or conditions. These data can be presented as a table or special matrix. Suppose, for example, that there is a system that consists of three dimensions that have different values. The number of values for various parameters might not be the same. Such data can be presented as a Zwicky box or as a matrix (see Table 6.1).

The corresponding values should be identified for each parameter or dimension. After that, "cross-consistency assessment" is performed (Ritchey, 1991). The purpose of this step is reducing the number of possible pairs of cells that can present potential contradictions. The technique of determining the pair-wise consistency looks for contradictions that can be induced by

TABLE 6.1

An Example of a Morphological Box

Parameters or Dimension Values or Conditions	Parameter 1	Parameter 2	Parameter 3
Value 1	1.1	....	n.1
— — —	— — —	— — —	— — —
Value N	1.N	....	n. N

causality, can be logical contradictions, or can be empirical inconsistencies. This technique helps to eliminate internally inconsistent configuration in a considered structure. As a result, a solution space is created for the farther analysis. It allows concentration on a manageable number of internally consistent configurations.

The original method of morphological analysis of human activity was developed in SSAT (Bedny, 1981; Bedny, 2015; Bedny et al., 2015). In SSAT, it is utilized for the description of the activity structure during task performance. At the first stage, a qualitative task analysis is performed, which is followed by the morphological analysis of activity. This approach, developed in the framework of SSAT for task analysis and design, is in agreement with the above-considered theoretical data. The morphological analysis offered by SSAT includes development of a human algorithm of activity and then the activity time structure. In this book, new aspects of morphological analysis are discussed. This data is presented in the following sections where analysis of the task with extremely complex probabilistic structure is conducted.

Our approach to morphological analysis is totally different from the method proposed by Zwicky. However, the purpose of our approach is the same, which is the qualitative description of a variable activity structure that unfolds in time as a process. This approach utilizes the standardized language of activity description that allows development of analytical models of activity. There are no approaches outside of SSAT that can create analytical models of extremely variable activity. After morphological analysis, quantitative methods of analysis such as task complexity and reliability evaluation can be performed. Based on such analysis, one can optimize task performance, evaluate equipment design, introduce new principles of software and equipment design, increase safety, and so on. Without analytical models, there is no design. Ergonomists usually utilize empirical experimental procedures that often lack external validity because ergonomists cannot develop analytical models that precede physical models of equipment. In contrast, any prototype that is evaluated by experimental procedures in engineering has analytical models that are developed at the earlier stages of design. In SSAT, the morphological analysis includes two basic stages. The first one is the algorithmic task description, and the second one is the time structure analysis.

The algorithmic task description is an important method of morphological analysis of human activity. There are purely mathematical understandings of algorithms and concepts of computer algorithms. Activity theory introduced the notion of a *human algorithm*. A human algorithm can be viewed as a system of logically organized mental and motor actions that are used for solving a specific class of problems or for the performing of various tasks. A computer, the same as a human algorithm, guides the process toward solving problems. However, a computer algorithm is not necessarily modeled the way in which humans solve problems.

The main units of analysis in an algorithmic description of activity are members of an algorithm that consist of cognitive and behavioral actions. Below we describe the human algorithm and show how it can be developed by using analytical procedures.

An algorithmic analysis allows description of very variable human behavior, including its cognitive components. It depicts logically organized elements of human activity verbally and symbolically. Each member of an algorithm has one goal that integrates several actions into a hierarchical subsystem of activity. A goal of a member of an algorithm is hierarchically higher than goals of individual actions that this member of an algorithm is comprised of. The number of actions that can be included in a member of an algorithm is restricted by the capacity of working memory. Subjectively, a member of an algorithm is perceived by a subject as a component of activity that has logical completeness. Each member of an algorithm can, in turn, be presented as an activity subsystem that includes cognitive or behavioral actions. The first can be described as cognitive operations (elements of cognitive actions), and the last can be presented as a system of motions.

Each member of an algorithm includes one to four interdependent homogeneous actions (only motor, only perceptual, only mnemonic, or only thinking or decision-making actions). Such actions are integrated into a subsystem by a high-order goal of a member of an algorithm. An algorithmic description of human activity is an important step in analysis of multiple decisions made by a subject during task performance. A member of an algorithm also has relatively constant material components of activity, such as objects and tolls of activity that are important distinguishing features of a member of an algorithm.

It is critical for an algorithmic description of activity and development of its time structure to determine which elements of activity can be performed simultaneously and which ones can be done only sequentially. Special rules have been developed in SSAT. They are presented further below.

Members of an algorithm are of two main types: operators and logical conditions. Operators represent motor or cognitive actions that depict transformation of objects, energy, and information. For instance, there can be members of an algorithm that includes perceptual action (receiving information), mnemonic actions (extraction of information from long-term memory, or keeping it in working memory), thinking actions that are involved in analyzing situation, and so on. Logical conditions are members of an algorithm that includes decision-making action and determines the logic of selection of the next member of an algorithm. In the deterministic algorithm, each decision (logical condition) has only two equal probability outputs 0 and 1. In the probabilistic algorithm, each decision (logical condition) can have two or more outputs that come with different probabilities. In the complex rule-based tasks, decision-making or logical conditions are the most critical components of task analysis. Outside of SSAT, there are no analytical methods for studying such tasks. Cognitive and behavioral actions as units of analysis are the most distinctive features of a human algorithm, compared to other types of flow charts that are widely used to present human performance.

Let us now consider the language of activity description that is utilized for development of models of a human algorithm. Each member of such an algorithm is depicted by a standardized symbol and called an operator. There are operators that are involved in receiving information, keeping information in memory, performing motor actions (O^{α}; O^{μ}; O^{ε}), and so on.

Operators that are involved in extracting information from long-term memory, utilize the symbol μ and depicted as O^{μ}. The symbol $O^{\mu w}$ is associated with keeping information in working memory. The symbol O^{ε} describes executive or motor components of activity, such as "move a hand to a red button and press it". All such elements of motor activity are also called efferent operators. Each symbol shows what type of activity is involved at each stage of task performance. Therefore, for example, O^{ε} cannot include any cognitive actions. Similarly, O^{α} can include only perceptual actions. If after accepting information (O^{α}) it is impossible to perform decision-making immediately (the logical condition cannot be performed), this means that after receiving information, a mental analysis (thinking action) of situation is necessary. A thinking action can be described by one or several thinking operators that are depicted by symbol O^{th}.

In a morphological analysis of activity, we distinguish between simultaneously performed actions and combined actions. Each simultaneously performed action has its own goal. For example, a subject moves an object using both hands. This is the case of simultaneously performed motor actions. An example of combined motor actions is when a subject moves an object and turns it into a horizontal position at the same time.

According to SSAT rules, cognitive actions cannot be performed simultaneously. However, cognitive action can be combined. A subject can perform a thinking action based on externally presented information or perform it based on keeping information in memory. Thinking actions can be executed based on externally provided information (mental manipulation of externally presented data), or they can be performed based on the information held in or retrieved from memory (manipulation of data in memory), or performance of such actions can require keeping intermediate information in memory. Examples of such combined thinking operators are $O^{\alpha th}$ or $O^{\mu th}$ (α means that the thinking operator is performed based on external, for instance, visual information, and μ means that such an operator requires complicated manipulation of information in memory). Such a symbolic description is used when a combination of cognitive processes are critical for performance of specific members of an algorithm.

Decision-making actions are depicted by logical conditions that are associated with the symbol $l\uparrow$ with a corresponding number of outputs. Combined action can be encountered when performing decision-making actions. When a subject makes a decision, it might be necessary to keep information in memory until the decision is completed or the decision-making requires extraction of information from memory. Such a combination of decision-making and memory function complicates decision-making actions which should be taken into consideration during task analysis. Symbol l^μ is used for description of these logical conditions, where μ designates a memory function that complicates decision-making. Simple logical conditions include "and", "ör", and "if- then" rules. They have two outputs "yes" or "no" with the probability of occurrence being zero or one. More complicated logical conditions have more than one output, with each output having various probabilities, and the sum of probabilities of all outputs is 1.

The symbols "*l*" for a logical condition include an arrow with a number on top of it that is the same as a number of a logical condition associated with it. For example, logical condition l_1 is accompanied by an arrow $\uparrow^1$. A reversed arrow with the same number ($\downarrow^1$) is placed in front of a member of the algorithm to which the task flow skips after the decision is made. So the syntax of the algorithm is based on a semantic denotation of a system of arrows and superscripted numbers. An upward-pointing arrow of the simple logical condition $l_1\uparrow^1$ requires skipping all following members of the algorithm until the next appearance of $\leftarrow\downarrow^1$. The operation with this arrow is the next one to be executed.

A complex logical condition has multiple outputs. For example, $l_1\uparrow^{1(1-6)}$ indicates that this is the first complicated logical condition that has six possible outputs: $\uparrow^{1(1)}$, $\uparrow^{1(2)}$, $\uparrow^{1(3)}$ ... $\uparrow^{1(6)}$. Arrows after logical conditions $\uparrow^{1(1)}$ demonstrate transition to another member of an algorithm with $\downarrow^{1(1)}$. Therefore, a human algorithm can be deterministic as well as probabilistic (Bedny, 2015). A deterministic algorithm has logical conditions with only two outputs with probability 0 or 1 of their occurrence, and a probabilistic algorithm has logical conditions with more than two outputs occurring with various probabilities or two outputs that have any probability between 0 and 1.

In some cases, complex logical conditions can be a combination of simple ones. These simple logical conditions are connected by "and", "or", "if-then", and so on rules. Such connections can be designated by symbols "&", "∧", "⟶", and so on. Complicated logical conditions that comprise simple ones may be designated as L_1 (l_1^1 & l_1^2 & l_1^3). The L_1 symbol depicts the first complex logical condition of the algorithm. Numbers 1 through 3 are used as superscripts to designate simple logical conditions that are included in the complex one. Complicated logical conditions are particularly important for diagnostic tasks. In the above example, a complicated logical condition L_1 comprises three simple ones that are joined by the logical conjunction *and*. This complicated logical condition can be used, for example, to determine whether a particular phenomenon belongs to a certain category, especially when a phenomenon's attributes are connected via the conjunctions. These three simple logical conditions depict determining: Is feature 1 present? Is feature 2 present? Is feature 3 present? Only when all three questions receive a positive answer can one conclude that a phenomenon belongs to a particular category. In contrast, if in our example simple logical conditions are combined via the logical disjunction *or*, it is sufficient when any one simple logical condition is satisfied. Some complicated logical conditions include different logical connections. The above-presented symbolic description of logical conditions demonstrates the formalized description of various instances of decision-making encountered in task performance.

In some cases members of an algorithm of the same type follow each other. For example, several afferent members of an algorithm such as O_1^α, O_2^α, and O_3^α or several efferent members of an algorithm such as O_1^ε, O_2^ε, and O_3^ε are performed one after the other. In these cases, if the sequence of the performed actions can be kept in working memory, then one member of an algorithm should include no more than three to four actions.

If actions are simple and performed sequentially, and their order does not need to be kept in working memory, then their integration into separate members of an algorithm is determined by the logical completeness of parts of the activity. Actions included in the same member of an algorithm are integrated by a higher-order goal and have a limited number of interdependent tools and objects. The limited capacity of short-term memory can also influence strategies of the grouping of these actions. Only after the morphological analysis is completed is it possible to evaluate the activity complexity, reliability, and so on quantitatively. Models of activity which are obtained at the stage of morphological analysis can be developed by utilizing analytical methods or simulation or a combination of both. Some simplified experimental procedures can be used. A segment of task performance can be reproduced in laboratory conditions to obtain the performance time of separate cognitive actions that are included in a specific member of the algorithm.

An algorithmic description of activity demonstrates that activity as a system has the following organization: hierarchical, sequential, or logical-probabilistic. This classification of a system to some extent is similar to Simon's (1999) classification of system organization and derives from analysis of principles of algorithmic task description. A member of an algorithm includes actions and their elements-operations. A goal of a member of the algorithm integrates subordinate actions. Actions, in turn, have their own goals that integrate their smaller elements, called operations. Usually, a member of an algorithm integrates no more than 4-5 actions. The simplest member of an algorithm includes a single action. The capacity of working memory determines a span of control of a high-order goal of a member of an algorithm. Members of an algorithm are the main subsystem of activity during task performance. They, in turn, have a sequential and logically probabilistic organization that depends on outputs of logical conditions.

Our algorithmic analysis demonstrates that it is incorrect to reduce the relationships between elements of activity to the hierarchical one. It has a logically sequential organization. At the same time, members of algorithms and their elements are hierarchically organized. According to Simon (1999), an important characteristic of a hierarchical system is a "span of control" when elements of the system are subordinates to a single boss or a high-order goal of the system. This idea is ignored in contemporary task analysis.

6.2 Description of Individual Members of an Algorithm

In this and the following sections we are going to describe algorithmic analysis of activity in greater detail than it has been done in previous chapters. An algorithmic description of activity is a critically important step in morphological analysis of rule-based tasks. It consists of the division of activity into hierarchically organized units and determination of the logic of their organization. Figure 6.1 presents a morphological decryption of activity into hierarchically organized psychological units of analysis.

Algorithmic task description also uses technological units of analysis that utilize technological terminology or common words, which with some approximation give a general understanding of what is involved at each stage of task performance.

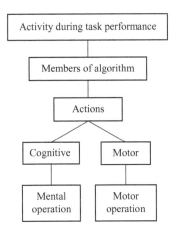

FIGURE 6.1
Morphological description of activity into hierarchically organized psychological units of analysis.

The following is an example of the actions' description using technological units of analysis:

(1) take a part; (2) install a part in a three-jaw chuck (in engine lathe); (3) look at the digital indicator 2; (3) decide to choose a part from the second bin. If one reads such a description of human actions without preliminary observation of the task performance, it is difficult to understand what elements of activity are involved in the task performance. This is a serious drawback of such a task description. In design process, a task often does not yet exist, and therefore it is not possible to observe a task performance. Design is, first of all, an analytical process and cannot be reduced to experimental procedures, as it is mostly done in ergonomics and work psychology.

Each motor action can be described using common language or using technological units of analysis. A motor action can also be described using psychological units of analysis. In SSAT, for this purpose, we use the MTM-1 system that offers the standardized description of movements that are elements of motor actions. The concept of a motor action does not exist in MTM-1, as it is introduced in SSAT. Thus, SSAT offers a new method of applying the MTM-1 system to the study of human activity. The abbreviated popular description of this system as it is given, for instance, by Barnes (1980) is not adequate for this purpose. The special MTM-1 system manuals such as UK MTMA, Ltd (2000) should be referred to instead.

Let us as an example consider a description of the motor action "move a hand and grasp a part". The distance for the part is 40 cm, and it is jumbled with others objects. So search and selection is conducted before grasping the part. This movement is identified by the MTM-1 system by a symbol *R-C* which specifies a movement *Reach to a single object jumbled with* other similar or dissimilar ones. This movement requires a high level of visual and muscular control and a mental decision when selecting one part out of a bunch of other objects. If we also consider the distance of this movement, then it is described as *R40C* that takes 0.6 sec to perform. This movement is followed by the movement *Grasp (G4)*. Suppose a part is jumbled with other parts ranging in size between $6 \times 6 \times 3$ mm and $25 \times 25 \times 25$ mm. In this case, we have to use *G4B* that takes 0.33 sec. MTM-1 provides the performance time of these motions. At this stage of analysis, we do not consider the performance time of motor actions. So the motor action "move a hand and grasp a part" is described as "*R40C + G4B*". In general, description of actions can be performed first by using technological units of analysis and then by using psychological units of analysis.

(1) "Move a hand and grasp a part" $\rightarrow$ (2) "*R40C + G4B*", $T = 0.93$ sec;

where:

1   is the description of the motor action utilizing technological units of analysis

2   is the description of the same motor action using psychological units of analysis

Usage of technological units of analysis at the first stage of description gives a general understanding of performed actions, and the standardized psychological units of analysis facilitate the precise understanding of what type of motor actions are performed.

The concepts of a tool and an object are also important for the description of motor and cognitive actions. These two concepts are described in Bedny and Karwowski (2007). There is no fixed relationship between these two terms. For example, a user performs motor actions (1) "Reach and grasp a mouse with the right hand" and (2) "Move cursor to the required position". In the first motor action, a mouse is an object engaged by a subject. The goal of this action is "to grasp a mouse". This is what a subject has to achieve as a result of this action. At the second action, when a subject wants to move the cursor into a specific position, a mouse becomes a tool of a motor action and the cursor becomes an object of this action (a subject changes the position of an object, or transforms its state or position according to a goal of an action). Thus, the concept of tool can be used differently depending on the role it plays in the analysis of activity during a task performance.

Cognitive actions also can be described by using technological and psychological units of analysis. For example, "look at the digital display 1" is an example of a description of a mental action using technological units of analysis. This action can be also described using standardized psychological terminology developed in SSAT (Bedny, 2015). It can be described as a "simultaneous perceptual action" with duration T = 0.4 sec. This information facilitates the clear understanding of the considered mental action. Thus, this action is described as:

$$(1) \text{"Look at the digital display 1"} \rightarrow (2) \text{"Simultaneous perceptual action"},$$

$$T = (0.4 \text{ sec}),$$

where:
1   is the description of the cognitive action utilizing technological units of analysis
2   is the description of the same cognitive action using psychological units of analysis

The above-presented principle of motor and cognitive actions' description is used for the morphological analysis of activity.

MTM-1 utilizes "eye focus" action (*EF*) and "eye travel" motion (*ET*) for defining relatively simple cognitive actions. According to SSAT, such actions are usually combined into one relatively simple cognitive action (*ET* + *EF*). For determining time for more complex cognitive actions, we utilize data gathered by the studies in aviation offered by handbooks and our own studies (Myasnikov and Petrov, 1976; Lomov, 1982; Zarakovsky et al., 1974; Bedny, 2015). In some cases, the chronometrical analysis can be applied.

Time studies cannot ignore the concept of pace as it is common in most cases in ergonomics where analysis of temporal data is reduced to the study of separate reactions. However, the performance time of separate reactions or actions is not the same in holistic activity (Bedny, 2015). Chronometrical studies of each cognitive or behavioral action should be performed not only in isolation but also in the context of holistic activity. In chronometrical analysis, subjects should not even be informed about what specific action's time is an object of analysis. Instructions should be focused on the optimal pace of performance. Thus, temporal data can be obtained from chronometrical analysis, different handbooks, and from the MTM-1 system. In cases when the time standard for performance of motor actions is taken from other sources, it is still necessary to preserve the MTM-1 system's symbolic description of motions.

For example, motion *M20C* in the MTM-1 system means "move an object 20 cm to an exact position." If we have more precise performance time for this motion in specific conditions that is obtained by a chronometrical study, we can utilize the italic description of the motion *M20C* to show it. Therefore, the standardized description of the motion is preserved and at the same time, the performance time of the specific motor action is obtained from other sources.

MTM-1 is useful for the description of simple perceptual action or the simplest decisions.

When studying computer-based or computerized tasks, the time for motor actions that are involved in typing should be determined. The time for one keystroke, as per MTM-1 system, is 0.38 sec. According to GOMS, the time standard for one keystroke is 0.28 sec. However, when typing, not only the fingers but also the hands makes slight movements. Depending on the location of the previous and the following letters or numbers, the distance of hand movement can vary, which is not considered in GOMS. For this reason, we have chosen to use the greater time for a keystroke. Therefore, time for typing "n" characteristics or signs is determined as $0.38 \times n$.

In addition, experimental methods can be used in order to obtain some chronometrical data. Simplified experimental procedures can be used for this purpose, even at the stage of analytical study when a real task is not yet developed. Such an experiment can imitate performance of separate members of an algorithm. For example, typing a seven-digit ID takes 2.3 sec based on a simple experimental study, and based on the MTM-1 system, it would take 2.66 sec; typing "restore communication" takes 7.6 sec based on MTM-1, but a simple chronometric measure records only 6.8 sec, and so on. So, chronometric data and calculation give similar results. In our chronometric study, we instructed the operator not to type with the maximum speed but rather use a comfortable typing tempo. The part of the activity that is an object of the chronometrical study should follow other components of activity rather than being considered in isolation. Usually, the psychological chronometrical data does not take into account the pace of performance, as is the case when Fitts' Low (Fitts, 1954) is applied. Critical analysis of this law is presented in our publications (Bedny et al., 2015).

Procedures for determining the performance time of members of an algorithm that are associated with cognitive components of activity are more complex. Some methods developed in cognitive psychology and activity theory can be adapted for this purpose. The chronometrical applied analysis should determine the duration of separate cognitive actions rather than the duration of the separate stages of information processing. The concept of reaction time is not adequate in such a study because the subject does not react to various isolated stimuli with maximum speed but performs goal-directed voluntarily cognitive and behavioral actions that are combined with other actions in the structure of activity (Bedny and Meister, 1997; Bedny and Karwowski, 2007).

Below, as an example, we present several members of the algorithm that are used further for description of the setup task 0.

O_1^α	Check for presence of inventory receiving screen menu before PO number is entered.	Simultaneous perceptual action $(ET+EF)$	$T=0.72$sec

O_1^α is an example of the standardized description of the cognitive member of an algorithm by using technological and psychological units of analysis, and such a description clearly demonstrates from a psychological perspective what type of activity is performed. An efferent (motor) operator O_2^ε that follows directly after O_1^α is depicted below.

O_2^ε	Type 1 and then press ENTER to choose ADD INVENTORY RECEIVING screen	$(R50B+AP1)+(R30B+AP1)$	$T=2.01$ sec

The members of the algorithm are described by the psychological unit of analysis in the left column in a symbolic manner and by technological units of analysis utilizing the common language in the next column. The third column also depicts the psychological units of analysis. For example, when we consider O_1^α, the verbal description as "simultaneous perceptual action $(ET+EF)$" is shown (ET - eye travel microelement and EF - eye focus microelement or eye fixation). Eye fixations reflect the time required for recognition of a simple signal or an object. In the MTM-1 system, there are no more microelements that would describe independent cognitive elements of activity. In SSAT, these two elements are considered as psychological operations that are integrated by a goal of a perceptual action. Such actions are called "simultaneous perceptual action". The right column displays the performance time of such action. Yarbus's (1965) method can be applied for analysis of eye movements. SSAT suggests a totally new method of interpreting this data (see Chapter 3). The time of the eye movement to α degrees can also be calculated using the following formula (Lomov, 1982):

$$T = 0.002 + 0.004\alpha \qquad (6.1)$$

Complex perceptual actions involve deliberate visual analysis of objects when a subject performs a complex perceptual action. A complex perceptual action includes no more than three to four simple perceptual actions that are integrated by a goal of such action due to the limitation of the capacity of working memory. Duration of an eye fixation for simultaneous perceptual actions on average is equal to 0.3–0.4 sec. However, this time can be significantly longer. For example, the average time for receiving information from an aviation display is 0.5 sec (Bedny et al., 2015). Suppose that duration of a pilot's fixation on a display is 3 sec. Then the duration of eye fixation for high-order mental components can be defined as:

$$T_{\text{ment}} = 3 - 0.5 = 2.5 \text{ sec} \qquad (6.2)$$

Based on a qualitative analysis, we can determine a specific cognitive action that is performed during this period of time. Additional data about the duration of mental components of activity can be obtained from various handbooks or experimentally.

In the second example for O_2^ξ, two motor actions are also described using psychological units of analysis in the third column of the table. The first action integrates $(R50B + AP1)$, and the second one consists of $(R30B + AP1)$. Here we can see that the first motor action includes $R50B$, and the second one has $R30B$. These are motions of the same type (Reach Case B). They only differ by the distance of movement (50 and 30 sm). The symbol $AP1$ depicts motion "Apply Pressure." There are two motions of this type: (1) $AP1$ and (2) $AP2$. According to the MTM-1 system, the second motion requires more time. SSAT considers these motions as being integrated by a goal of motor actions, and therefore this member of the algorithm includes two actions. So the third column displays psychological units of analysis with a detailed description of the activity element. The right column lists the performance time of this member of the algorithm. Such a description of each member of the algorithm facilitates a clear understanding of what is involved in the performance of each member of an algorithm. Hence, such concepts as psychological and technological units of analysis, their hierarchical description and their performance time allows the description of models of activity in a standardized manner.

Temporal data for individual motor actions can also be obtained experimentally by using an imitation of a fragment of a task that can be unexpectedly presented to a subject in the context of a performed activity. Although in such cases the temporal data for standardized motions is not taken from the MTM-1 system for determining performance time of considered motor actions, the standard symbolic description should still be carried out from it. For example, it has been experimentally determined that moving a hand 30 sm and grasping a very small object takes 0.9 sec, and this time is not the same as in the MTM-1 system where this motor action is described as $(R30C + G1B)$ with $T = 0.63$ sec. However, this motor action is performed in dangerous conditions, and a subject slows down his/her pace of action as a result. So based on the rules developed in the MTM-1 system and SSAT, the performance time for this motor action should be determined experimentally, but the symbolic description of motions and motor actions in general should be presented in a standardized manner. Such motor action should be described as "($R30C + G1B$), $T = 0.9$ sec because the experimentally obtained temporal data is more adequate for this specific motor action. However, its standardized description is beneficial for understanding of what specific motor action is performed.

Currently, some scientists demonstrate insufficient familiarity with the MTM-1 system and mix up this system with the work of Gilbreth and Gilbreth (1917). For example, in their authoritative work (Rodgers et al., 1986, pp. 118–119) wrote:

> The field of time and motion studies has long been associated with industrial engineering for setting job standards and estimating wage. The literature should be consulted for a detailed discussion of MTM techniques (Barnes, 1968). One such technique is to record the motions of the hands and body during a task or work cycle. By analyzing these motions, you may be able to define more efficient motion patterns and sometimes to identify motions that may contribute to worker's discomfort on a job.

(Gilbreth and Gilbreth, 1917)

This fragment shows that the authors do not clearly understand that the work of Gilbreth and Gilbreth is the precursor to the MTM-1 system, and that their work is not utilized in the contemporary task analysis. Moreover, Barnes (1980) presented only the popular description of the MTM-1 system. Analysis of the more well-known textbooks in industrial/organizational psychology (see as an example Landy and Conte, 2007; Schultz and Schultz, 1986) also demonstrates that the MTM-1 system is unknown in I/O psychology, and the citations are only made on the Gilbreths' work. They were outstanding scientists, but their time and motion study system has not been used for a while now. A much more powerful MTM-1 system has been developed by Maynard et al. (1948) based on the Gilbreths' ideas. In summary, analysis of the presented table form of the description of the members of the algorithm O_1^α and O_2^ε demonstrates that they are depicted in a standardized manner. The first column presents *psychological units* of analysis by using special symbols that classify a member of an algorithm. The second column contains *technological units* of analysis, and the third one uses *psychological units* of analysis describing members of an algorithm in a more detailed manner by utilizing various cognitive and motor actions. The right column contains the performance time of considered members of an algorithm. It is necessary to emphasize that the more standardized the conditions of actions that are described according to technological principles, the more often they are similar to standardized actions that are described based on psychological principles. This is due to the fact that the content of mental operations of these actions are also similar. Initially, actions are described in terms of technological units of analysis. Then they are described in terms of psychological units of activity. Introduction of such concepts as technological and psychological units of analysis are important for development of models of activity.

The above-presented material reveals the following basic principles of morphological analysis of activity. It includes psychological and technological units of analysis. Their comparison allows adequate description of the structure of activity during the design or redesign process. The hierarchical organization of units of analysis in the morphological description of activity can be presented as follows:

Motor activity: Member of algorithm (mode of motor activity) → motor action → motion or psychological motor operations;

Cognitive activity: Member of algorithm (mode of cognitive activity) → cognitive action → psychological mental operation.

6.3 Algorithmic Description of Activity as a Basis of Analysis of Variable Tasks

Considering that the current activity during task performance is extremely variable, it requires development of a special method for its description. For this purpose, a powerful method of describing extremely variable activity has been developed in SSAT. In the above sections, the separate steps of morphological analysis have been considered. We describe basic principles of algorithmic description of activity during task performance which allows demonstration of how this method can be utilized for analysis of variable tasks with complex probabilistic structure in the following sections.

Table 6.2 shows the simplest deterministic algorithm. Each logical condition in such an algorithm has only two outputs, with equal probability $P = 0.5$. An operator needs to check a digital indicator. If an even number is presented, then an operator should turn a two-positioned switch up. If an uneven number is shown, she/he should turn a two-positioned switch down. The left column depicts psychological units of analysis. Each member of the algorithm in the left column is described by using standardized symbols for psychologically different elements of activity. In the second column, each member of the algorithm is described using technological units of analysis. These units of analysis describe elements of tasks through common language. Symbol O_1^{α} is the first operator that designates a perceptual process of receiving presented information. The decision-making process is described by the logical conditions l_1.

TABLE 6.2

Description of the Algorithm for the Task "Check a Digital Indicator" in a Tabular Form

Members of Algorithm	Description of Members of the Algorithm
O_1^α	Take reading from digital indicator
$l_1^\uparrow$	If the even number is shown, then perform O_2^ε If odd number is displayed then perform O_3^ε
O_2^ε	Turn two-positioned switch up
$_1\downarrow O_3^\varepsilon$	Turn two-positioned switch down

If the probability of the output of l_1 is zero, then the second member of the algorithm O_2^ε is performed. If the probability of the output of l_1 equals one, then according to the arrow from the logical condition, the third member of the algorithm O_3^ε is performed. Hence, in the second case, O_2^ε is skipped and O_3^ε is performed. This is the simplest method of the algorithmic description of task performance.

Usually an algorithmic description is used for analysis of the rule-based tasks. However, this method can also be used for analysis and description of operative problem-solving tasks that are performed in time-restricted conditions and include problem situations whose plan of action is not known in advance. In such situations, a subject does not only perform actions based on prescribed rules but also performs his/her own independent cognitive actions which should be performed in order to achieve the required goal. In emergency situations, an operator strives to perform the task as quickly as possible. The pace of task performance is a critical factor in emergency situations. In cognitive psychology in all situations and specifically emergency situations, Hicks's law and Fitts's law are used for determining the performance time of elements of a task. These laws are based on a concept of reaction, not on a concept of action. However, the pace of reactions and the pace of goal-directed voluntary actions are different. A subject cannot be considered as a reactive system that responds to various unrelated stimuli. Therefore, in cognitive psychology, the concept of pace in task performance is totally ignored. Even in an emergency situation, a subject does not perform isolated reactions but executes a logically organized system of cognitive and behavioral actions. These logically organized actions are basic components of a task. The pace of performing such actions is not the same as the pace of isolated reactions. These aspects of task analysis are taken into consideration in our further discussion.

An example of algorithmic analysis of a problem-solving task performed in an emergency situation is present below. This task can be described as "driving in an unusual and unexpected emergency situation", when a driver tries to avoid a collision with a car that appears in front of her/him unexpectedly from the opposite direction. This example can be considered as the performance of an operative problem-solving task. This task can be related to a category of operative tasks because it is an unexpected problem-solving task that arose in time-restricted and dangerous conditions, and this situation represents a stable structure with embedded dynamic elements (two moving cars). According to the conducted analysis, the considered task should be related to the category of semi-algorithmic tasks that are performed in time-restricted conditions. This type of task can only be described by a human algorithm based on the conversations with the driver and further experimental and theoretical methods of determining the performance time of individual members of the algorithm, including various cognitive or behavioral actions.

Let us describe this task. After crossing a rural road which had a small slope down, the driver noticed that the brakes had failed in the car. This is not a wide road and one would have to perform a specific maneuver to avoid the cars moving in the opposite direction. Unexpectedly, the driver saw a car at a relatively close distance. The driver realized that she/he could not stop the car or go around the car that was moving toward him. The driver looked to the right and suddenly saw a heap of gravel on the roadside and decided to drive the car into this heap of gravel. As a result, the car stopped without a strong impact. Below we describe the above-considered task algorithmically according to the method developed in SSAT (see Table 6.3).

Analysis of members of the algorithm that are involved in task performance demonstrates that five out of its nine members represent the thinking process. There are also two logical conditions that include decision-making actions. The task is not performed based on preliminarily developed prescriptions or rules, and is executed only once in an unexpected circumstance. Therefore, this task can be related to problem-solving, operative tasks.

Let us analyze $O_1^{\alpha th}$. Based on a visual analysis of the external situation, the driver realized that the vehicle was moving too fast and decided to slow down. This member of the algorithm is based on the perception of visual information and its analysis that involves thinking. According to SSAT, this member of the algorithm includes simple verbally logical thinking action that is performed based on the analysis of visual information.

At the next step, the driver made a decision to slow down (see l_1). This decision has only one output because it includes a volitional component associated with a decision to push the brake to slow down (to act or not to act). The decision to turn or not to turn the car is made similarly (see l_2). These are simple decisions about what to do in the circumstances at hand.

Based on the made decision l_1, the driver performed O_2^ε. When describing O_2^ε, we have chosen a possible strategy of its performance. For example, the driver can push the brakes not two but three times. However, pushing the brakes two times in a time-limited condition is a more probable strategy. After the second pressing of the brake pedal, the driver realized that the brakes do not work. Instinctively, the driver continued to keep the pressure on the pedal but the car continued moving forward. It is very difficult to

TABLE 6.3

Algorithmic Description of the Task and Its Time Structure Description When a Driver Stops of the Car in Unexpected Emergency Conditions

Members of Algorithm	Description of Members of the Algorithm		
Psychological Units of Analysis	Technological Units of Analysis	Psychological Units of Analysis	Time in sec
$O_1^{\alpha th}$	Mental awareness "This rural road had a small slope and my car moves too fast"	Simple verbally logical thinking action based on visual information	$0.40+0.45=0.85$
$l_1^\uparrow$	I need to push the brake to slow down	Decision-making action at verbal logical level	0.3
$_1\downarrow O_2^\varepsilon$	Driver pushed the brake two times	(Action 1) $FM+FMP+FM$; (Action 2) $FM+FMP$;	$1.3+0.99=2.29$
O_3^{th}	Mental awareness that car keeps moving with the same speed	Simple verbally logical thinking action based on visual information	$0.40+35=0.75$
O_4^{th}	Mental awareness of situation "rural road had a small slope and my car has broken brakes"	Simple verbally logical thinking action	0.35
O_5^{th}	Mental awareness of situation "I need to move my car on the edge of the road and stop"	Simple verbally logical thinking action	0.35
O_6^α	Unexpectedly the driver saw the oncoming car at a relatively close distance	Simultaneous perceptual action	0.4
O_7^α	Driver looked to the right side of the road and saw heap of gravel ($\approx$5 feet wide)	Simultaneous perceptual action	0.4
O_8^{th}	Heap of gravel can stop my car without damage	Deductive reasoning (simple verbally logical actions)	5
$l_2^\uparrow$	The driver decided to turn to the heap of gravel and his car stops without a strong impact		0.3
$_2\downarrow O_9^\varepsilon$	Driver turns car's steering wheel to the right. (two short right turns)		4
Total time			15

determine how many times the driver pushed the brake pedal during performance of O_2^ε before she/he turned the car. With some approximation we have selected just the last two ones that would reflect the performance of two motor actions at the final stage of the case when the driver decided that something has to be done to prevent a collision.

The description of motor actions (O_2^ε and O_9^ε) and determining their performance time involves utilizing the MTM-1 system. It should be pointed out that there is no concept of motor action in the MTM-1 system. According to SSAT, motions of the MTM-1 system are elements (motor operations) of motor actions. These motions are integrated into motor actions by the goal of these actions. A highly skilled worker who performs the same task during a shift can maintain the MTM-1 system pace. The pace of motion performance in MTM-1 is slower than the performance time of the isolated reactions as it is stated in Hicks's law and Fitts's law. The MTM-1 system pace can be adequate when a subject performs motor actions in the structure of the task in emergency conditions (Bedny and Karwowski, 2007). In normal conditions, we utilize coefficients 1.1 – 1.2 and even higher if it is necessary.

In our example, the driver's task is performed in emergency conditions, and therefore we did not utilize considered coefficients. The considered motor actions are performed with the pace offered by the MTM-1 system. So in the described situation, pushing the brakes can be depicted by the following motions: *FM* (foot to pedal; raise foot with pedal) and *FMP* (press pedal). The last motion requires more time. SSAT uses not just the concept of a motion but also the concept of motor actions that integrate motions according to the goal of a motor action, to describe motor components of activity (Bedny, 2015). Thus, operator O_2^ε includes driver performance of two motor actions: $FM + FMP + FM$ (action 1) and $FM + FMP$ (action 2).

Let us now consider O_8^{th}. This member of the algorithm involves deductive reasoning, the process of reasoning from one or more statements (premises) to reach a logically certain conclusion. It is not easy to think deductively in time-limited and emotionally tense situations. However, in our case the driver handled the situation well. The driver's deductive reasoning in this problem-solving task can be presented as follows:

A - Gravel is soft (premise 1 is true) & B - Car hits a soft object (premise 2 is true) → C - car stops without extensive damage (conclusion C is true)

Performance time of O_8^{th} has been determined based on data presented in the engineering psychology handbook (Myasnikov and Petrov, 1976). Experimental data gave a similar result.

The performance time of O_9^ε has been determined experimentally.

Operators O_1^{th}; O_3^{th}; O_4^{th}; and O_5^{th} involved verbal-logical thinking that is performed by the driver in a time-limited condition. The driver used internal speech in his/her thinking process. Such speech is an abbreviated one. The performance time of these operators can be measured approximately by evaluating the changes in the bioelectric activity of the lower lip as a measure of internal mental activity utilizing the electromyogram (EMG). The bioelectric activity of the lower lip registered by EMG precedes the subject's verbal responses. However, in the considered example internal speech is not transformed into the external one. A verbal action should be considered as a single word or several interdependent words that convey one meaning (Bedny et al., 2015). This idea is in agreement with the study of Vygotsky (1978) about the meaning as a basic unit of activity analysis. It is in unison with ideas of cognitive psychologists Bainbridge and Sanderson (1991), who showed the possibility of segmentation of verbal activity in the verbal protocol analysis.

If we consider as an example O_4^{th} it is possible to extract with some approximation two verbal actions: "the road had a small downward slope" and "the car's brakes don't work." The performance time of these actions can be determined. The subject had to express these two sentences quickly several times, and the time of pronouncing them can be measure by using a stopwatch. The time of expressing each sentence mentally should be measured separately. The expressions should be formulated in an abbreviated manner and pronounced at a rapid pace. More precise data can be obtained utilizing EMG of the lower lip. However, in our example, for determining the performance time of thinking actions that are included in various members of the algorithm associated with thinking (excluding O_8^{th}), we utilized the engineering psychology handbook (Myasnikov and Petrov, 1976). The obtained data can be used when verbally logical thinking is the main form of thinking. When visual information and imaginative components of thinking dominate, eye movement data also should be utilized.

The presented example is interesting because this task has not been performed according to some prescriptions or instructions. It has been formulated and performed by the driver for the first time. This task has a low probability of being performed again. Despite these facts, we can describe such tasks in a formalized manner and even evaluate their complexity. The concept of task complexity will be considered in Chapter 8. The described method can be used in safety analysis and when there is a need to share experience during a learning or training process.

6.4 Development of the Activity Time Structure

The morphological analysis of activity includes two stages. The first one is algorithmic description of activity and the second one involves the time structure analysis. These two stages are tightly connected. Here we are going to discuss the second stage. The time structure of activity is a new concept that has been developed in SSAT. It includes the logical sequence of activity elements, their duration, and a possibility of their being performed simultaneously or sequentially. The time structure of activity also takes into consideration probabilistic characteristics of activity, or in other words describes the probability of the emergence of various activity elements in its structure. Activity time structure is described as a process that unfolds in time. As we have demonstrated above, there are two types of units of activity analysis: psychological (elements of activity described in a standardized manner) and technological units of analysis (element of a task that are described by common language or in technological terms). It is necessary to use both technological and psychological units of analysis in order to develop the time structure of activity.

Specialists who study human activity outside of SSAT do not utilize these units of analysis. As a result, they cannot distinguish the time-line analysis from the time structure of activity analysis. For example, in *A Guide to Task Analysis* (Kirwan and Ainsworth, 1992), the time-line analysis is discussed. In the time-line analysis only, the technological units of analysis are utilized. Algorithmic and time-structure analyses present activity as a structurally organized system that unfolds in time and can be further evaluated quantitatively by using objective measurement procedures. Only after transforming technological units of analysis into psychological units, decomposing them into actions or operations, and further determining their duration is it possible to describe the time structure of activity and quantitatively evaluate its complexity.

In Section 6.2, we have described two members of the algorithm O_1^α and O_2^ε. The right column presents the performance time of each member of the algorithm. If the whole task is described algorithmically, where each member of the algorithm is shown in the same manner, including its performance time, this would present not just the algorithmic description of the task but also its *time structure*. An example of the time-structure description of a task is presented below (see Table 6.4). This table shows not just members of the algorithm but also the time of their performance and utilizes both technological and psychological units of analysis. However, presented in Table 6.2 description of the task "check a digital indicator" does not show the time structure of activity because this table does not include the psychological units of analysis and performance time of each member of the algorithm.

Let us consider the fragment depicted in Table 6.4. The complete version of this table can be found in Bedny (2015).

Table 6.4 presents the algorithmic description of the task and information about the performance time of each member of this algorithm. It includes technological and psychological units of analysis. Hence, this is an example of the morphological analysis of activity during task performance.

In most cases, the time structure of activity is presented in tabular form. However, when the time structure is very complex, a graphical representation of the time structure can be utilized after a table is developed. The graphical method allows correction of an earlier developed table of the time structure.

TABLE 6.4

Algorithmic and Time Structure Description of the Military Task Test the State of Equipment

Member of the Algorithm	Technological Units of Analysis	Psychological Units of Analysis	Time (sec)
O_1^α	Visual perception of three positioning switches on the right-hand side panel	Successive perceptual action	0.55
O_2^ε	Turning of three positioning switches on the right-hand side of the panel (move an arm and grasp a switch)	$R50B + G1A + T40S + RL1$	0.95
---- -----	------- ------- ------	------- ------- --------	--- ----
$O_5^{\alpha th}$	Interpretation of information about the first subsystem	Thinking actions of categorization (the simplest thinking action performed on visual information)	0.30
$_1 l_1 \uparrow$	If subsystem 1 is blocked, decide to unblock the system (go to O_6^ε). If the subsystem is on decide, check the second subsystem (perform O_9^α)	The simplest "decision-making action" from two alternatives"	0.30
---- ----	----- ------ ------	------- ------- ------	--- ---

Below is presented a fragment of a very complex structural organization of motor and cognitive components of activity in time. Motor activity dominates in the considered task. The task's goal is installation of pins into the holes of the pin board. A subject moves two hands to the box, grasps pins, and moves them to the holes of the pin board according to certain rules:

A special physical model was developed for this experiment. It consisted of a pin board that contained 30 holes for metal pins. A box containing pins was placed behind the pin board. The latter was in front of the subject. Ten pins have clearly visible flutes. The subjects install the pins into the holes according to the following rules:

1. If the pins are regular (without a flute), they can be installed in any position.
2. If a fluted pin is picked by a subject's left hand, it must be placed in the hole with the flute inside the hole.
3. If a fluted pin is picked by a subject's right hand, it must be placed in the hole with the flute above the hole.

In this experiment, a subject fills a pin board with 30 pins according to the above rules. Subjects utilize the optimal strategy by using both hands to fill the center row first, starting with the holes closest to them, and then working upward and outward. Only 10 pins out of 30 have flutes. The experiment ended when subjects completed the task in time that matched the time standard that has been determined by the analytical method (time-structure development) in the last five trials. This is a manual manipulative task with a complex probabilistic structure. This type of task is performed a limited number of times in laboratory conditions. However, performing such tasks during the whole shift is challenging and cognitively demanding. As an illustration, we consider only three of the most complex members of the algorithm O_3^ε, O_4^ε or O_5^ε and O_6^ε (see Figure 6.2).

This is a complex task that has a probabilistic structure. It requires a careful perceptual control, and decisions are made based on the information that is extracted from memory.

Our purpose here is to illustrate the graphical method of developing the time structure of activity. A fragment of the graphical model of activity time structure for the task of installation of pins is presented below. The MTM-1 system is utilized here to describe motor actions and simple cognitive actions that include simultaneous receiving of information and derived from it a simple Yes/No decision. As has been already mentioned, the MTM-1 system does not utilize such concepts as motor or cognitive actions and concept of time structure of activity but instead concentrates on describing the isolated motions which violate the principle of systemic organization of activity structure and its hierarchical organization. In Figure 6.2, the utilized symbols RH, LH, and MP mean right hand, left hand, and mental process.

Let us first consider motions and motor actions. A member of an algorithm O_1^ε depicts the fragment of the task when subjects move left and right hands simultaneously and grasp the pins. The first motion *R32B* is "Reach" with an average distance of 32 cm, when the position of an object varies slightly from cycle to cycle. *G1C1* is a motion of grasping a nearly cylindrical object with a diameter greater than 12 mm. Reach and Grasp are not just isolated motions. Together they make a goal-directed action, because these motions are integrated by a goal of an action. Let us consider some motor actions that are included in other members of the algorithm.

Letters above the line segments designate motions that are described using the MTM-1 system. These motions are integrated into actions. For example, the left part of the upper line for the member of the algorithm O_3^ε shows the following motor action: *M22B + mM10C*, with simultaneous turning of the pin 90° (*T90S* - the next line down). The right side of this line corresponds to the second action *P2SE + RL1* (installation of a pin into a hole and its release). All the above-described segments represent motions that together reflect performance of the single right-hand action (RH). So we use the MTM-1 system in a totally new way, by combining the concept of motion with the concept of motor actions.

M22B + mM10C means "move an object 22 cm to an approximate position and then, without interruption, move a hand 10 cm to an exact position (average combined distance is 32 cm). While moving a pin into the hole, the hand

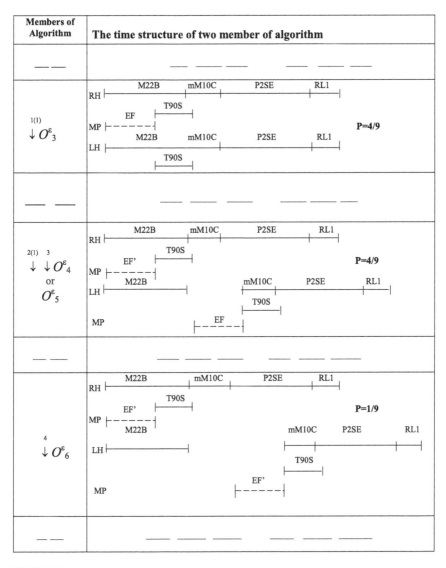

FIGURE 6.2
Graphical model of variable activity time structure during installation of pins (fragment).

simultaneously turns it 90° with a small effort. This movement is described by *T90S*. *P2S* means positioning a symmetrical object, of second class of fit (**class** of **fit** is variable depending on the care and precision needed), that is easy to handle. *RL1* describes a normal release or unfolding the fingers. *EF* means receiving information and making a Yes/No decision based on it. This is the only microelement in the MTM-1 system that describes a mental component of activity. It is applied in cases when a subject receives visual information and based on it performs the simplest "If-Then" or "Yes-No" decision. So elements that are designated by *EF* show *"look and makes a simple decision"* actions. Each *EF* on Figure 6.2 belongs to a member of the algorithm that describes cognitive components of activity. A cognitive element of activity that is an independent member of the algorithmic and is performed simultaneously with motor components, is designated by a dashed line. According to SSAT, this is the simplest cognitive action that includes a perceptual operation combined with a decision-making operation. These two mental operations require a high level of concentration of attention (the third level of attention concentration). SSAT offers rules that determine which elements of activity have to be performed in sequence and which ones can be performed simultaneously depending on the required level of attention concentration. These rules will be described in the following section.

EF as a simplest cognitive action, when decision is derived directly from the perceived information, can be used only in mass production when a worker performs the same production operation multiple times. As has been mentioned before, according to the MTM-1 system, it should take 03.sec. In cases when we study human-machine or human-computer interaction systems, the pace of performance is lower. Both the simultaneous perceptual action (receiving information) and the following simple decision-making action take 0.3 sec each. Thus, the performance time of these two simple cognitive actions takes 0.6 sec.

Member of the algorithm O_3^ε escribe the case when the pins without flutes are installed, and O_2^ε - O_5^ε shows the scenario when one pin has a flute and it's installed by a right or a left hand. A member of the algorithm O_6^ε when both pins have flutes, has the most complex time structure.

The time structure of activity shows a pause in the left-hand movement when an operator picks up one or two fluted pins. The pause can be noticed in the right-hand movement instead of the left-hand movement. These pauses coincide with the decision-making. Of course, the time structure for each trial can be slightly different. Figure 6.2 presents the most representative strategy around which other strategies vary.

The bold letters *P* at the right-hand side of Figure 6.2 reflect the probability of the appearance of the corresponding members of the algorithm. As has been covered in the previous chapters, we utilize a hierarchically organized system of units of analysis. The presented time structure consists of members of the algorithm that include cognitive and behavioral actions as well as cognitive and behavioral operations.

Let us describe the strategy of task performance more specifically. At the first step, a subject moves two hands simultaneously and grasps two pins. The left hand and the right hand perform the same type of action that can be described as $(R32A + G1A)$. These two actions are not shown in Figure 6.2. The time structure of three members of the algorithm O_3^ε, O_4^ε, O_5^ε, an O_6^ε that are performed simultaneously with cognitive actions *"EF"* (recognition and yes/no decision) are presented. The simplest member of the algorithm is O_3^ε, as it reflects the case when subjects have to put two pins into a hole in any position, because both pins are without flutes. O_4^ε and O_5^ε describe the strategy of performance when only one of two pins is fluted. One of them reflects the scenario when the fluted pin is grasped by the left hand, and in another situation, it is grasped by the right hand. The strategy of performance for these members of the algorithm is the same. O_6^ε represents the strategy of performance when both picked-up pins are fluted.

The strategy of performance of the first member of algorithm O_3^ε shows that there is only one cognitive action that starts at the same time as the motor actions by right and left hands, and there is no interruption in motor actions' performance. Analysis of O_4^ε and O_5^ε shows that there are two cognitive actions, and they are performed in sequence (when the first *EF* is completed, the second *EF* immediately follows). According to the SSAT rules, cognitive actions cannot be performed simultaneously, because they require a high level of attention concentration and this is supported by experiments (Bedny and Meister, 1997).

When subjects perform O_6^ε (both pins are fluted), subjects change their strategies of performance. This situation requires controlling how to turn pins by the left and the right hands in sequence. Subjects evaluate the position of the flute and decide how to turn the pin in one hand, and only after that evaluate the position of the flute in the other hand and decide how to turn it. As a result, the pause of the second hand movement is longer because the second decision can only be performed after the first hand has finished turning the pin. As it can be seen from Figure 6.2, the pause during performance of O_6^ε is longer than during performance of O_4^ε or O_5^ε. It is clear that the graphical method of the time-structure analysis makes it easier to decipher such complex strategies. Therefore, one should utilize both tabular and graphical form of time-structure description when performing a very complex time-structure analysis.

Based on the analysis of the time structure of the activity, it takes 3.32 seconds to grasp pins by two hands. Taking into consideration the fact that a subject grasps two pins and installs them into the holes 15 times, it takes 49.9 sec to perform this task. We stopped the experiment when subjects in the experimental group completed the task in 50.4 sec.

It is well known that existing methods of time study are adapted to the tasks or production operations that have the same sequence of task elements. Time-structure analysis presents a totally new approach to time study of task performance with a high level of variability, which is important not only for ergonomics but also for economics when studying the efficiency of performance and productivity. Time study is used for planning and scheduling work, for development of wage incentive plans, for evaluation of labor cost, and so on.

Analysis of the above-presented material shows that the cognition and activity in general is not only the process but also the structure that unfolds in time. Moreover, this process has a complicated logical organization.

One element of activity (member of a human algorithm) can follow the other in sequence or based on some logic, and transition from one element to another occurs with various probabilities. This process can be represented metaphorically as building a brick wall that has an unusual structure. Typically, brick walls are made of bricks of the same type. At the first stage, a mason puts down the first and lowermost row of bricks. At the next step he/she places the next row of bricks above the first row of bricks, and so on. The wall is constructed from the bottom up.

Cognition and activity in general also can be presented as a process and at the same time as a structure. It can also be seen as a special kind of a wall. However, construction of such a wall is different because it would consist of various cognitive and motor actions and their components. They would not be installed from the bottom up as normal bricks would be. The process of "constructing of a cognitive wall" can be represented as unfolding the time "paper roll" from left to right. Since the bricks would represent various elements of activity, they would have different colors and length and take up their positions automatically on the unfolded paper roll. Moreover, performance of various elements of activity requires different levels of attention concentration, and therefore some of them cannot be performed simultaneously. This means that not all bricks can unfold at the same time. For example, two bricks with red color designating thinking components cannot be under one another. These would be magic bricks and they would take their correct place very quickly. Thus, activity is a complex structure that unfolds in time as the considered hypothetical wall. In cognitive psychology, cognition and activity in general are considered as processes. In SSAT, activity is a structure that unfolds in time, which requires new methods of task analysis. Development of the time structure of activity during task performance is the main purpose of morphological analysis.

We have considered a fragment of a complex manual combinatory task because it included various logical rules of how to manipulate the pins. Its time structure is so complex that its development in tabular and in graphical form is very helpful. Without developing the graphical model of the time structure of the activity, it is very difficult to discover the most efficient strategies of task performance.

As can be seen, such presentation of the time structure is very useful for task complexity evaluation. The rules developed in SSAT determine the possibility of simultaneous performance of various cognitive and motor elements of activity. A detailed description of these rules can be found in Bedny and Meister (1997) and Bedny and Karwowski (2007). Combining elements of activity in time is defined by these rules and depends on strategies of task performance selected by a subject. The possibility of combining activity elements is also determined by equipment or software configuration. Changes in the interface design or equipment configuration leads to changes in the time structure of activity. If the time structure is excessively complex, this means that equipment or interface are not designed efficiently. Efficiency of design can be evaluated based on the analysis of the time structure of activity. The approach developed can be used not only as an experimental method but also as an analytical one, which is important for ergonomic design where analytical methods are erroneously replaced by experimental procedures. Activity time-structure models reflect an idealized description of activity during task performance. As a result of repeated executions, the data obtained during experimental studies only describe the idealized models of execution.

Time-structure analysis of variable tasks offers new ways of evaluating the reserve time of the system functioning. Time during which a man-machine system is transferred from an initial state to a required state is called the cycle of time regulation. A substantial part of the cycle of time regulation belongs to the time spent by an operator performing various tasks, including highly variable tasks. This time influences the system's reserve time, which can be defined as a surplus of time over the minimum that is required to detect and correct any deviation of the system parameters from permissible limits and bring the system to the required state (Kotick, 1974; Siegal and Wolf, 1969). According to data in SSAT, it is necessary to differentiate between objectively existing reserve time and an operator's subjective evaluation of that time. In some cases, they are not the same, which might lead to an inadequate evaluation of the situation in accidental conditions. The approach to time structure analysis and determining performance time of variable tasks in combination with analysis of reserve time that was developed in SSAT is based on application of queuing theory in human-error analysis (Bedny, 2015). The main purpose of this theory is coordination of call requests (signals that should be served per unit of time) and the ability of an operator to process these requests (the ability of an operator to respond to them) in a given period of time.

In conclusion, we want to underline the fact that traditional chronometrical studies that derive from observation of real tasks are not acceptable for design purposes. Verbal descriptions of tasks, produced by existing methods without observation of their real performance, are hard to understand. The method presented here of morphological analysis of human activity eliminates this problem.

6.5 Pace of Task Performance and the Process of Its Formation

In this chapter, we consider the concept of work pace. Activity time-structure development cannot be accomplished without taking into account the concept of pace. Without understanding the pace of task performance, it is impossible to determine the reserve time and its variation in emergency conditions. The pace of performance is an important factor in time study and task analysis, for the development of the time structure of activity, and therefore for the morphological analysis of an activity.

In ergonomics and engineering psychology, attention is paid to studying the psychological refractory period (PRP), the time between two stimuli or inter-stimulus interval (ISI), single-channel or multiple-channel processing of information, and so on. However, these methods are not useful for the study of task performance. Such an approach relies on the pace of task performance that depends on the inter-stimulus interval (force-pace) or response-stimulus interval (self-pace) stimulus rate with the critical factor in pacing being the inter-stimulus interval (Wickens and Hollands, 2000). However, a subject is not a channel of information processing. A subject develops strategies of task performance depending on conditions in which the task is performed. There are two basic situations. In one of them, the task is performed multiple times during the shift, and in the other, a subject performed various tasks that have different complexity and often emerge unexpectedly. Each task is a logically organized system of cognitive and behavioral actions, but not reactions. So a subject cannot be considered as a reactive system when a pace is defined by a number of unrelated responses initiated by various stimuli. The pace of the task performance is slower than the pace of various reactions performed in laboratory conditions. Even in forced-pace conditions, a subject can regulate his/her pace in a certain range. A pace is formed and maintained at a certain level due to the mechanisms of self-regulation and selected strategies of task performance rather than due to the principle of a human's reactivity. In SSAT, pace can be considered as the speed of performing various cognitive or behavioral actions that are structurally organized in time. Hence, the pace of performing can be defined as an operator's ability to sustain a specific speed (below maximum) of holistic activity that unfolds in time (Bedny, 2015). A subject should sustain such pace during the whole shift and subjectively evaluate it as an optimal pace. This discussion clearly demonstrates that not all laboratory studies are adequate for practical use.

The purpose of the time study is to determine a standard time for performance of a task. Usually this procedure is applied for repetitive tasks (production operation) that have the same sequence of elements. Each element has a starting and an end point. The performance time of each element is measured, and then the entire task performance time is determined by summation of time for each task element (Drury, 1995). Such a method is not applicable to studying complex contemporary tasks and particularly for computerized tasks with variable structure. Hicks's law and Fitts's law are popular theoretical basis in psychology, ergonomics, and engineering psychology for time study of manual components of activity. In our opinion, time study of manual work cannot be reduced to the application of these laws because a subject should not be considered as a reactive system that functions at the pace that corresponds to execution of discrete reactions with the maximum speed. Such methods consider a human being as a reactive system that performs a number of isolated independent reactions with the maximum speed (Bedny, 2015; Bedny et al., 2015).

It has already been demonstrated that a task is a logically organized system of cognitive and behavioral actions that influence each other and therefore the performance time of actions in the context of a task is not the same as the performance time of isolated reactions. A subject formulates a goal of the task, regulates actions, changes his/her strategy, and so on. All of the above influences the actions' performance time, which means that a subject cannot perform various actions with maximum speed in the context of a task. Moreover, a task can emerge unexpectedly, or it can be performed multiple times during the shift, and so on. These factors are also ignored in the studies of reaction time and the speed-accuracy trade-off in engineering psychology, where scientists concentrate on separate reactions in emergency conditions. Cognitive components of activity are structured as a logically organized system of cognitive actions that unfold in time.

Our study demonstrates that a task is not a sum of independent reactions but rather a system of logically organized actions integrated by the goal of this task. The pace of work activity depends on the strategies selected for the task performance.

There is no exact definition of work pace. For example, Barness (1980) defines work pace as the speed of an operator's motions. This definition is not accurate, because it ignores the cognitive components of work and the logical organization of behavioral and cognitive components of activity. We define pace as a specific speed of task performance that an operator is able to sustain during the shift. This pace is below maximum and therefore lower than time for various isolated reactions. This pace should be sustained during the work shift and be subjectively evaluated by an operator as an optimal pace. In emergency conditions, an operator might increase the speed of task performance. However, it still will be lower than the performance time of the isolated reactions.

There is a lot of difficulty in pace evaluation. One widely utilized method of pace evaluation in industry is based on subjective judgment, which is called rating. This method utilizes comparison the pace of a blue-collar worker's performance with the observer's own concept of normal or standard pace that can be maintained during a shift without excessive physical and mental effort. Such a pace assumes that a subject can deliver not only the required quantity but also the required quality of performance. In the past, motor components of activity dominated in the industry. So the speed of an average person walking on a level grade along a straight road has been selected as a standard pace. Physiological studies demonstrate that energy expenditure per unit of covered distance is minimal if the speed of walking is between 4 and 5 km/h (Lehmann, 1962; Frolov 1976).

We utilize the MTM-1 system for analysis of the motor components of activity. The pace of performance in an MTM-1 system corresponds to a level of walking 5.8 km/h. According to this system, highly skilled workers who perform the same or similar task during a shift can maintain this pace, which is considered to be an average or standardized pace. However, as demonstrated above, the optimal pace has the range of walking between 4 and 5 km/h. According to our study (Bedny and Karwowski, 2007) and studies of Gal'sev, (1973), the MTM-1 system pace exceeds the optimal one. In automated systems, an operator becomes a machine monitor rather than a performer of multiple repetitive tasks. In cases when an operator performs complex tasks that involve multiple decisions and have probabilistic structure, the pace of performance is also slower. In our studies (Bedny, 1981), when utilizing the MTM-1 system data, applying the coefficient of pace 1.1–1.2 to take into account the lower pace of performance was recommended. In special situations, this coefficient can be even higher. An operator has to perform some checking actions to control the correctness of the system's or her/his own performance. There is also a need to account for time required to correct the erroneous actions.

There are several different rating scales for evaluation of pace for physical work. For example, there is a scale where a standard or normal pace is designated by the number 100. If the actual pace of performance is less than normal, it receives a number less than 100 (see detailed description in Barness, 1980). The other method for evaluation of physical work is based on measurement of oxygen consumption in calories per minute and heart rate in beats per minute (Bedny and Karwowski, 2007). It is important to use the concept of optimal pace. There is evidence that the optimal pace can change during training. Training with gradual increase in the speed of task performance is known as above real-time training (Miller et al., 1997). It is a useful method for training operators to work in time-restricted conditions (Bedny, 1981; Bedny and Karwowski, 2008; Bedny et al., 2012). As a result, young workers quickly adapted to the requirements to qualitative and quantitative of task performance. According to the self-regulative concept of learning developed by Bedny (1981) and described in Bedny and Karwowski (2007) in a detailed manner, learning and training can be considered as a self-regulative process. In the process of learning and training, trainees switch from a well-known strategy of task performance to a new strategy or adapt the well-known strategy to a new one. The more difficult the task is, the more intermediate strategies should be used to learn how to do it. This dynamic of strategies implies that learning and training can be described as a sequence of stages. Learning or training activity is constructed through a sequence of these stages that can be described in an algorithmic or quasi algorithmic way and can be utilized for similar classes or types of tasks. They can also be described as written instructions that are adequate for specific tasks. The trainee uses different temporal strategies that are not included in the final stage of task performance. These strategies are not erroneous. They are required for an efficient acquisition of new strategies. Hence, the system of instruction should be changed during skill acquisition. Based on this, Bedny (1981) formulated the principle of dynamic orientation at different stages of skill acquisition. Thus, written instructions or algorithmic prescriptions and verbal instructions should be adequate to utilize strategies of task performance. If necessary, technical devices, simulators, and software should be adequately progressive adapted to match the stages of the training process. An adequate pace of task performance should be selected for each strategy of task performance.

Subjective judgment about the pace can also be useful. Chebisheva (1969) conducted the laboratory experiment for evaluation of the pace for manual work. Vocational school students had to sort wooden sticks with different colors matching the pace of strokes of a metronome. At the beginning of the experiment, the metronome strokes were set on a slow pace. The pace of the metronome strokes was gradually increased, and the students had to sort the sticks with the faster pace. The following levels of performance pace have been discovered:

1. Very slow pace that has been evaluated as uncomfortable
2. Optimal pace
3. Effortful or intense pace
4. Difficult to achieve pace
5. Unachievable pace

It has also been revealed that the transition from a very slow pace to an optimal one reduces the amount of errors. The optimal pace is conveyed by the most positive emotional state of the subjects. However, the further increase of the pace causes increase in the number of errors. The effortful or intense pace is evaluated as more difficult and causing emotional tension. This study, with some modification, can be performed with real operators. The required pace can be gradually increased from a slow to a higher one and vice versa. Thus, the subjective judgment about the pace in combination with the scale of the pace description can be a useful source of information about the task performance pace.

We recommend outlining three levels of pace in task analysis: *very high, high, and average* (Bedny, 1979; Bedny and Karwowski, 2007). The *very high pace* is slower than the operator's reaction time to various stimuli. This pace is possible only in cases when an operator reacts to isolated signals, using discrete actions in highly predictable situations. For example, an operator can have a high level of readiness to push a button when a particular signal appears.

The *high pace* is when an operator performs a sequence of logically organized mental and physical actions in response to the appearance of various signals. For motor activity, this is the pace similar to the one suggested by the MTM-1 system. Such pace should be used when an operator works in emergency conditions and performs not isolated reactions but various tasks.

The pace of performance of mental actions should be determined based on analysis of strategies of their performance in particular situations.

Chronometrical procedures for determining the time of their performance have some specifics, and the beginning and end of action should be clearly defined. Measurements of the performance time of cognitive actions should be conducted not only when a subject repeats the same isolated action multiple times but also when such actions are executed in the context of specially imitated tasks. This is critically important because it helps to discover the most preferable pace of cognitive actions.

The average pace is that in which an operator performs tasks at his or her own subjective time scale, and there are no time constraints for a task performance. The recommended pace can be corrected based on a subjective judgment and scaling procedures by using developed coefficients for correction of pace of task performance. In cases when errors or failures have serious consequences, there is a need for additional time to complete the task. In such situations the coefficient of pace can be higher.

Currently, there is no clear understanding about how to train blue-collar workers to perform production tasks within the required time frame. Usually instructors mostly pay attention to the quality of the trainees' performance, and only later focus on the quantitative aspects of their performance. For example, the instructors usually do not inform students about a required task's performance time at the first stage of the training process. Only at the final stage does he/she introduce requirements to perform the tasks according to prescribed time standard requirements. This is based on the assumption that introducing a time standard early in the training can decrease the quality of work because the students do not yet have the required knowledge and skills. However, problems arise when the students complete the training and start working and are unable to meet the quantity requirements in the specific time period. This reduces productivity, lowers the salary of young workers, and causes dissatisfaction and increases the turnover rate. Bedny (1981) conducted the study with blue-collar workers in various industries. Preliminary observation of students revealed that, in most cases, the students did not know how much time they should spend on a particular task. A stop-watch study discovered that students spent a great deal of time unproductively and did not properly organize their work in time. They could not perform tasks in the required period of time and regulate their pace of task performance. Performing tasks at a certain pace required reconstruction of skills, not just increasing the speed of task performance. After being trained without time limits, the students could not increase the pace of task performance in production conditions for a long period of time. Transition to faster pace of performance required reconstruction of the structure of the workers' skills. It has been demonstrated that introducing the goal of the task along with the required time standard that is adequate to the specific period of training, and teaching how to properly organize work in order to meet the time standard improved not just the productivity, but also the quality of work. Introducing the time-standard into the training process, and explaining its importance for the future salary reduced positively affected the training process, increased work motivation, improved discipline, and reduced fatigue. When students started working, they quickly adapted to production requirements. The obtained data is also important for the economic analysis of the training process. Companies evaluate training improvements mainly in terms of the financial effect that results. Thus, the important aspect of the time study is development the training process to perform tasks according to the time standard.

References

Bainbridge, L. and Sanderson, P. (1991). Verbal protocol. In Wilson, J. R. and Corlett E. N. (Eds.), *Evaluation of Human Performance. A Practical Ergonomics Methodology*, 2nd edn, London, UK: Taylor & Francis.

Barness, P. M. (1968/1980). *Motion and Time Study Design and Measurement of Work*. New York, NY: John Wiley and Sons.

Bedny, G. Z. (1979). *Psychophysiological Aspects of a Time Study*. Moscow, Russia: Economics Publishers.

Bedny, G. Z. (1981). *The Psychological Aspects of a Timed Study during Vocational Training*. Moscow, Russia: Higher Education Publisher.

Bedny, G. Z. (2015). *Application of Systemic-Structural Activity Theory to Design and Training*. Boca Raton, FL: CRC Press, Taylor & Francis Group.

Bedny, G. Z. and Karwowski, W. (2007). *A Systemic-Structural Theory of Activity. Application to Human Performance and Work Design*. Boca Raton, FL: CRC, Taylor & Francis.

Bedny, G. Z. and Karwowski, W. (2008). Time study during vocational training. In Chebykin O. Y., Bedny, G. Z. and Karwowski, W. (Eds), *Ergonomics and Psychology. Development in Theory and Practice* (pp. 41–70), London, UK, and New York, NY: Taylor & Francis

Bedny, G. Meister, D. (1997). *The Russian Theory of Activity: Current Application to Design and Learning*. Mahwah, NJ: Lawrence Erlbaum Associates.

Bedny, G., von Breven, H. and Synytsya, K. (2012). Learning and training: Activity approach. In Seel, N. M. (Ed.), *Encyclopedia of the Science of Learning*, 1st edn, (pp. 1800–1805). Boston, MA: Springer.

Bedny, G. Z., Karwowski, W., and Bedny, I. (2015). *Applying Systemic-Structural Activity Theory to Design of Human-Computer Interaction Systems*. CRC Press, Taylor & Francis Group.

Chebisheva, V. V. (1969). *The Psychology of Vocational Training*. Moscow, Russia: Education Publishers.

Drury, C. G. (1995). Method of direct observation of performance. In Wilson, J. R. and Corlett, E. N. (Eds.). *Evaluation of Human Work. A Practical Ergonomics Methodology* (pp. 45–68). London, UK: Taylor & Francis.

Fitts, P. M. (1954). The information capacity of the human motor system in controlling the amplitude of movement. *Journal of Experimental Psychology*, 47, 381–391.

Frolov, N. I. (1976). *Physiology of Movement*. Leningrad, Russia: Science Publishers.

Gal'sev, A. D. (1973). *Time Study and Scientific Management of Work in Manufacturing*. Moscow, Russia: Manufacturing Publishers.

Gilbreth, F. V. and Gilbreth, L. M. (1917). *Applied Motion Study*. New York, NY: Macmillan.

Kirwan, B. and Ainsworth, L. K. (Eds.) (1992). *A Guide to Task Analysis*. London, UK: Taylor & Francis.

Kotik, M. A. (1974). *Self-Regulation and Reliability of Operator*. Tallin, Estonia: Valgus

Landy, F. J., Conte, J. M. (2007). *Work in the 21st Century. An Introduction to Industrial and Organizational Psychology*. Hoboken, NJ: Blackwell Publishing.

Lehmann, G. (1962). *Practical Work Physiology*. Stuttgart, Germany: George Theme Verlag.

Lomov, B. F. (Ed.) (1982). *Handbook of Engineering Psychology*. Moscow, Russia: Manufacturing Publishers.

Maynard, H. B., Stegemerten, G. J., and Schawab, J. L. (1948). *Method-Time Measurement*. New York, NY: McGraw-Hill Book Co.

Miller, L., Staney, K. and Guckenberger, E. (1997). Above real-time training. *Ergonomics in Design*, Vol. 5, pp. 21–24.

Myasnikov, V. A. and Petrov, V. P. (Eds.) (1976). *Aircraft Digital Monitoring and Control Systems*. Leningrad, Russia: Manufacturing Publishers.

Ritchey, T. (1991). Analysis and synthesis. On scientific method-based on a study by Bernhard Riemann. *Systems Research*, 8 (4), 21–41.

Rodgers, S. H., Kenworth, D. A., and Eggleton, E. M. (1986) *Ergonomic Design for People at Work*, vol. 2, New York, NY: Van Nostrand Reinhold Company.

Schultz, D., Schultz, S. (1986). *Psychology and Industry Today. An Introduction to Industrial and Organizational Psychology*. New York, NY: Macmillan Publishing Company.

Siegal, A. I. and Wolf, J. J. (1969). *Man-Machine Simulation Models*. New York, NY: Wiley.

Simon, H. A. (1999). *The Sciences of the Artificial*. 3rd rev. ed. Cambridge, MA: MIT Press.

UK MTMA (2000). MTM-1. analyst manual, London. UK: The UK MTM Association.

Vygotsky, L. S. (1978). *Mind in Society. The Development of Higher Psychological Processes*. Cambridge, MA: Harvard University Press.

Wickens, C. D. and Hollands, J. G. (2000). *Engineering Psychology and Human Performance* (3rd edn), New York, NY: Harper-Collins.

Yarbus A. L. (1965). *The Role of Eye Movements in the Visual Process*. Moscow, Russia: Science Publishers.

Zarakovsky, G. M., Korolev, B. A., Medvedev, V. I., and Shlaen, P. Y. (1974). *Introduction to Ergonomics*. Moscow, Russia: Soviet Radio.

Zwicky, F. (1969). *Discovering, Invention, Research-Through the Morphological Approach*, Toronto, ON: The Macmillian Company.

7

Design of Computerized Tasks with Complex Logical and Probabilistic Structure

7.1 Qualitative Analysis of a Computerized Task

In a production environment, computerized tasks can be extremely complex. They include physical components, when an operator interacts with equipment and material tools. An operator also uses the software when the cognitive components of activity dominate. Information received by an operator from these two sources can create extremely variable situations that require performing the task in different ways. An operator has to omit or include different subtasks in order to achieve the goal of the task at hand. Moreover, each subtask also can be variable. Thus, the structure of the task can be presented via multiple versions of the same task which have a complex logical organization that includes numerous decision-makings that occur with different probability. Currently, there are no analytical methods that would include formalized and quantitative principles of study for tasks of such nature. In this section, we will present a new approach to the analysis and design of such tasks.

Here we present the totally new principle of analysis of extremely complex tasks with probabilistic structure. We have selected as the object of study an inventory receiving task which is one of the most important tasks in a warehouse production process. Currently, there is a variety of companies whose main business is receiving orders and distributing goods. The most well known are Amazon and eBay. However, there are also a lot of small- and medium-size distributors of various kinds. Such companies do not produce anything. They just receive goods and ship them to the consumers. For such companies, warehousing is one of the main production processes.

The companies that manufacture goods also have warehouses. They receive items that are used in the manufacturing process and put them in the warehouse before using them in their production process. They also warehouse their products before shipping them out. Warehousing includes receiving items, putting them away (shelving them), picking items up (taking them off the shelves), delivering them to the manufacturing departments, storing ready products, and shipping products out. We consider one of the main warehousing tasks, which is the *inventory receiving task* that utilizes a computer system. One of the specific features of the considered task is the extreme variability of human activity during its performance. Furthermore, this task has a complicated logical structure.

It consists of a Setup task (subtask 0) and the main task that includes three versions of a task that occur with different probabilities. Thus, the task as a whole has a complex logical and probabilistic structure. The Setup task involves choosing the right menu option, opening the box with items, taking the purchase order out of the box, and keying or scanning the PO number. The first version of the main part of the task included getting the item out of the box, matching the item number on the physical item with the item number on the list of items on the computer screen, activating the screen with the information for the item at hand, and comparing received quantity with order quantity.

The considered task includes three versions that emerged in the structure of the task with different probabilities. We designate the setup task and versions of the main task as follows: *Subtask 0* is performed in the beginning of the shift and every time the new box with received items is open; *Version 1* of the task is performed when the item quantity is not equal to the ordered quantity and the difference according to the rules cannot be accepted (the item is rejected due to the unacceptable quantity); *Version 2* of the task occurs when the item quantity can be accepted but the item price cannot (the item is rejected due to the price difference as per given rules); *Version 3* of the task is performed when both item quantity and price are acceptable and the item is received. All the above-described versions of the task occur with various probabilities during the shift.

For better understanding of such complex tasks, we present the relationship between subtask 0 and item processing versions of the task in Figure 7.1.

Task analysis of these types of tasks and analytical descriptions of the tasks is extremely difficult and has not yet been developed in the field of

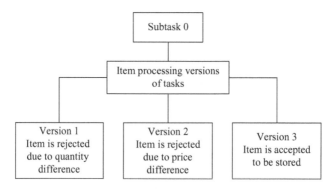

FIGURE 7.1
Relationship between subtask 0 and item processing versions of task.

task analysis. Here, it will be demonstrated for the first time how to analyze and describe such tasks with complex probabilistic structure using new analytical methods. We start our study with an objectively logical analysis that includes numerous steps. The purpose of these steps is to identify the content of the task and its relation to other tasks. Observation, discussion with supervisors and operators, review of documentation, and so on were utilized. This stage enables the researcher to obtain a general understanding of the technological process and work methods. It required careful analysis because computer-based tasks are frequently variable and vague. It has been discovered that the production process consists of three subsystems: (a) *stocking*, (b) *record keeping*, and (c) *work-in-process* (WIP). WIP refers to the products that are partially completed. Stocking occurs when material or items are received from different vendors, and the material or items are put into a specific place in the warehouse. Stock can be increased either by purchasing or by returning items from manufacturing (WIP) to stock. Stock is decreased by sale of products, by sending WIP products for further manufacturing, or by scrapping.

Any physical movement of items is registered in the computer system (*record keeping process*). Such movement includes moving ready products in the warehouse for stocking, moving of items and products out of stock for manufacturing (WIP), sale of ready products, or returning the items to the vendor. The record keeping process uses a complicated computerized system that has to track all physical movements of purchases, different parts, intermediate products, and so on.

The first task is called *inventory receiving*. Operators responsible for registration of all purchases and movement of intermediate and final products perform this task. The task includes two parts. One part includes reception of parts from different vendors to restock the warehouse and fulfill special and emergency orders. The second part includes receiving intermediate or finished products for WIP. The inventory receiving task is completed when the receiver places a tote filled with items on the special stock belt which delivers the tote with items into the warehouse or on the belt that goes into the workshop floor.

There are the following four tasks which can be performed after performance of the first task (receiving) before improvement: (1) Receiving → (2) Putting Away→ (3) Pick Up → (4) Delivery to Workshop → (5) Production Process Stage. The first task is the inventory receiving task. The second task (2) is performed by the put-away worker. He/she takes items from the tote that have been delivered to the warehouse and places them on the corresponding shelves. The next task (3) is pick up. The "pick up" operator takes the parts that have been ordered from the shelves and places them in the tote that is delivered to the workshop to be used in production. The "pick up" operator also can place ready-for-sale products into the other tote. Thus, this task has two outputs: delivery of items into the workshop and delivery of ready products for sale. Production Process Stage (5) is not a separate task. It includes all production tasks that are involved in the manufacturing process. The intermediate products of the production process should be returned to the first task, called "receiving."

The object of our study is the inventory receiving task. Parts arrive to the plant in special boxes that are delivered to the reception area. Figure 7.2 depicts a view from above of the workplace for the receiving task. The arrows in Figure 7.2 show the movements of the receiver around the workplace during performance of the inventory receiving task.

Figure 7.2 shows that the workplace consists of various equipment components. This task performance requires noticeable walking and moving separate parts and boxes with items. It also includes mental components of work that involve the computer system. Existing equipment is designated by solid lines. Equipment that has been introduced later based on our improvement recommendations is designated by dashed lines. That equipment did not exist before improvement. Base units 6 and 7 have metal balls that are installed at the surfaces near the belt 8 or belt 9. Balls have springs under them. As a result, when a receiver has to push totes with items to belts 8 or 9, sliding friction has been replaced by rolling friction. This reduces the physical effort when a receiver pushes totes with items to the corresponding belt.

Number 1 depicts a receiver who opens the boxes placed on the base unit (5). For this purpose, a receiver utilizes a special knife. After opening the box, a receiver removes a packing slip and reads it. Then a receiver uses

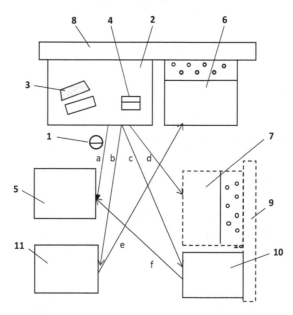

FIGURE 7.2
A view of the receiving task workplace from above. 1-receiving operator; 2-work table; 3-computer; 4-tag printer; 5-base unit for unpacking; 6-base unit for stocking process; 7-base unit for work-in-process (WIP); 8-belt for stocking; 9-belt for WIP; 10-put aside area; 11-totes' station.

a computer-based warehouse management system (3). A receiver enters a purchase order (PO) number listed on a packing slip and presses the F3 key to check what is still open on the PO. The receiver returns to the base unit for unpacking and takes parts out of the box and compares the order quantity with the received quantity. A receiver chooses the item number from the list on the screen, gets the item information, changes or confirms the quantity and the price, and assigns allocation for the part, if necessary. If allocation is already reserved for the item, the system will select it automatically. All required information is shown on the screen; later this information is printed on the label.

There are two kinds of subtasks: the first one is the setup task, and the second one is the main task. The setup task includes login, menu selection, key in PO number, and so forth. The main task begins when an item is taken out of the box and ends when it is put in the tote or rejected.

The arrows a-f in Figure 7.2. depict the movements of the receiver around the workplace during performance of the inventory receiving task.

Line *a* depicts how a receiver goes to a base unit for unpacking and opens a box, gets a packing slip out of the box (subtask 0), or gets the next item to be received (versions 1, 2, and 3);

b - A receiver goes to the totes' station to put received parts in the corresponding tote (Version 3);

c - A receiver goes to the put-aside area to set aside disputed arrivals (Versions 1 and 2);

d - A receiver goes to the base unit for work-in-process and puts the item into the corresponding department tote (Version 3);

e - When the tote is filled up, a receiver brings it to the base unit for the stocking process and pushes it into the belt;

f - A receiver goes back to the base unit for unpacking and get the next item (Versions 1, 2, or 3) or open another box (subtask 0).

7.2 Morphological Analysis of Computerized Task with Complex Probabilistic Structure

7.2.1 Morphological Analysis of Subtask 0

As we have described above in Section 7.1, the studied task consists of several subtasks that appear with various probabilities in the task structure. The subtask 0 is performed every time the receiver has to open a new box. This subtask reflects an operator's activity when processing a new purchase order (PO). As it is shown further in Table 7.1, subtask 0 starts with O_1^α

TABLE 7.1

Algorithmic Description of Activity and its Time Structure during Inventory Receiving Task Performance (Setup Task or Subtask 0 before Improvement)

Members of Algorithm (Psychological Units of Analysis Described in Consolidated Manner)	Description of Elements of Task (Technological Units of Analysis)	Description of Elements of Activity (Psychological Units of Analysis Described in Detailed Manner)	Time (sec)
O_1^α	Check for presence of inventory receiving screen menu before PO number has been entered	Simultaneous perceptual action (ET + EF)	$0.42 + 0.3 = 0.72$
$\overset{1}{\downarrow}O_2^\varepsilon$	Type 1 and then press ENTER to choose ADD INVENTORY RECEIVING screen	(R50B + AP1) + (R30B + AP1)	$1.68 \times 1.2 = 2.01$
O_3^α	Check to see if you are at the ADD TRANSACTION screen	Simultaneous perceptual action (ET + EF)	$0.42 + 0.3 = 0.72$
$\overset{1}{l_1}\uparrow$	If you are at the right screen, go to O_4^ε. Otherwise, press F3 for exit and go back to O_4^ε	Decision-making action from two alternatives	0.3
$\begin{array}{c}17(1)\,9(1)\,7(1)\\ \downarrow\ \downarrow\ \downarrow\ \downarrow\end{array}\,O_4^\varepsilon$	Go to base unit 5 and take a packing slip by the left hand from the box placed on based unit 5 (see Figure 7.2) and go to working area with computer	(1) Leg movements: "performance four steps" TBC1 + WP + WP + TBC2 (2) Hand action (R50B + G1B + M40B) (3) Leg movements: TBC1 + TBC2 + WP + WP	$7.5 \times 1.2 = 9$
$O_5^{\alpha\mu}$	Find PO (purchase order) number on the packing slip, select the first 3 or 4 digits as perceptual units of information and keep them in working memory until keying them in (requires moving eyes from PO number on the packing slip in the worker's left hand to the screen)	Three simultaneous perceptual actions [(ET + EF) + (ET + EF) + (ET + EF)] combined with mnemonic operations — "keeping information in working memory" One double check perceptual action	2.46

(Continued)

TABLE 7.1 (CONTINUED)

Algorithmic Description of Activity and its Time Structure during Inventory Receiving Task Performance (Setup Task or Subtask 0 before Improvement)

Members of Algorithm (Psychological Units of Analysis Described in Consolidated Manner)	Description of Elements of Task (Technological Units of Analysis)	Description of Elements of Activity (Psychological Units of Analysis Described in Detailed Manner)	Time (sec)
$O_6^{\alpha\mu}$	Find PO field on the screen (see Figure 7.3)	Simultaneous perceptual action (ET + EF) combined with mnemonic operations —"keeping information in working memory"	$0.42 + 0.3 = 0.72$
$O_7^{\epsilon\mu}$	Key in 3 or 4 digits selected from PO number as perceptual units of information ($O_7^{\epsilon\mu}$ requires keeping selected digits in memory until keying is completed)	Motor actions: (R26B+AP1) +(R6B+AP1) + (R6B+AP1) + (R6B+AP1)	$2 \times 1.2 = 2.4$
$O_8^{\alpha\mu}$	Find PO (purchase order) number on the packing slip, select the second 3 or 4 digits as perceptual units of information and then move eyes to the PO field on the screen (required keeping digits in working memory until key in is completed (EF – 0.3 sec and EF – 0.6 sec)	Three simultaneous perceptual actions [(ET + EF) + (ET + EF) + (ET + EF) combined with mnemonic operations — "keeping information in working memory." One double check perceptual action	2.46
$O_9^{\epsilon\mu}$	Key in selected as perceptual units of information last 3 or 4 digits from PO number (requires keeping selected digits in memory until keying is completed)	Practically the same as in $O_7^{\epsilon\mu}$	2.7
O_{10}^{ϵ}	Press ENTER when keying PO is completed and put the packing slip held in the left hand on the work table (Figure 7.2)	Motor actions: (R30B+AP1) + (R50B+RL1); Where: (R30B+AP1) is right-hand action and (R50B+RL1) is left-hand action	$1.47 \times 1.2 = 1.76$
O_{11}^{α}	Look at the screen message	Simultaneous perceptual action (ET + 2EF)	$0.42 + 0.6 = 1.02$

(Continued)

TABLE 7.1 (CONTINUED)

Algorithmic Description of Activity and its Time Structure during Inventory Receiving Task Performance (Setup Task or Subtask 0 before Improvement)

Members of Algorithm (Psychological Units of Analysis Described in Consolidated Manner)	Description of Elements of Task (Technological Units of Analysis)	Description of Elements of Activity (Psychological Units of Analysis Described in Detailed Manner)	Time (sec)
$\overset{2}{l_2}\uparrow$	If the screen displays an error message, INVALID PO NUMBER, then go to O_{12}^{α} ($P=0.05$). If PO number is correct ($P=0.95$), go to O_{15}^{ε}	Decision-making action from two alternatives	0.4
O_{12}^{α}	Compare PO number on the screen with the number on the packing slip and include one double check perceptual action ($P=0.05$)	Five simultaneous perceptual actions (ET+EF)	$(0.42+0.3)=0.72\times5$ $=3.6\times0.05=0.18$
O_{13}^{ε}	Change incorrect digits (suppose an incorrect digit is in the middle). $P=0.05$	Motor actions: (1) press 3 times left arrow (AP1+AP1+AP1); (2) press backspace key (R11B+AP1); (3) key in correct digit (R15B+AP1)	$(1.95\times1.2)\times0.05=0.12$
$\overset{3}{l_3}\uparrow$	If the PO number is correct and error message persists (system can't find purchase order), go to O_{14}^{ε} ($P=0.01$). If PO number is correct, go to $O_{15}^{\varepsilon\mu}$ ($P=0.99$)	Decision-making action with two alternatives and different probabilities (this member of algorithm is not consider due to low probability)	—— ——
O_{14}^{ε}	Call manager	Verbal and motor actions (this member of algorithm does not considered)	—— ——
$\overset{2}{\downarrow}O_{15}^{\varepsilon}$	Press ENTER and the system populates current date	Motor action (R14B+AP3)	$0.57\times1.2=0.68$
O_{16}^{ε}	Press F8 to look up items on the purchase order and return to base unit 5 for unpacking (see Figure 7.4)	Motor action with right hand (R17B+AP1) and four steps to base unit # 5 (TBC1+TBC2+WP+WP)	$(0.61+3.1)\times1.2=4.45$
Total Performance Time	Beginning of Individual Item Processing versions of the task		32.1

```
ADD INVENTORY RECEIVING                 P.O.#:                RECV-DATE:

LINE ITEM-NO.      QTY-RCV UNIT-COST TY CMP ASGN-QTY  BIN-LOCATE  CONFIRM
-------------------------------------------------------------------------




ESC=Abort,  F3=Exit,  F8=Look
```

FIGURE 7.3
Add inventory receiving screen before PO number is enter.

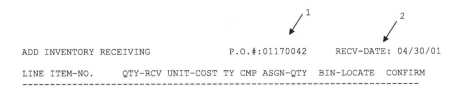

```
ADD INVENTORY RECEIVING            P.O.#:01170042      RECV-DATE: 04/30/01

LINE ITEM-NO.      QTY-RCV UNIT-COST TY CMP ASGN-QTY  BIN-LOCATE  CONFIRM
-------------------------------------------------------------------------




ESC=Abort, F3=Exit, F8=Look
```

FIGURE 7.4
Add inventory receiving screen. 1-purchase order number; 2-received date.

and ends with O_{16}^{ε}. Subtask 0 is performed with probability $P=1$ when an operator opens a new box with various parts.

The following is the description of receiver's activity when performing subtask 0.

The receiver approaches the computer monitor and checks for the presence of inventory receiving screen menu, then types 1 and presses Enter in order to choose Add Inventory Receiving screen. After that, she/he checks if it appears in caps on the screen (see Figure 7.3).

If the Add Inventory Receiving screen does not appear, the receiver goes back by pressing F3, types 1, and presses Enter again. If the right screen (Figure 7.3) does appear, she/he goes to a base unit 5 (see Figure 7.2), opens the box, takes a packing slip out and goes back to the working area with the computer. The receiver finds a purchase order number on the packing slip, selects the first 3 or 4 digits, and keeps them in working memory. He/she then finds the PO field on the screen and keys in the memorized numbers. The last steps (memorizing and keying in the second part of PO number) are repeated to complete entering the PO number (see Figure 7.4).

Then the receiver presses Enter. If there are no error messages on the screen and the date is populated, the receiver presses F8 to produce a list of items (see Figure 7.5).

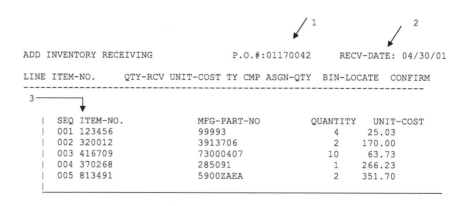

```
ADD INVENTORY RECEIVING                P.O.#:01170042    RECV-DATE: 04/30/01

LINE ITEM-NO.    QTY-RCV UNIT-COST TY CMP ASGN-QTY  BIN-LOCATE  CONFIRM
------------------------------------------------------------------------
  3
      |  SEQ ITEM-NO.          MFG-PART-NO        QUANTITY    UNIT-COST
      |  001 123456            99993                 4         25.03
      |  002 320012            3913706               2        170.00
      |  003 416709            73000407             10         63.73
      |  004 370268            285091                1        266.23
      |  005 813491            5900ZAEA              2        351.70
      |
```

ESC=Abort, F3=Exit, F8=Look

FIGURE 7.5
Add inventory receiving screen. 1-purchase order number; 2-received date; 3-item number.

If an error message saying that the PO number was not found appears, he/she checks if the right number has been keyed in or not. The receiver compares the PO number on the screen with the PO number on the packing slip and makes corrections. If the PO is still not found, she/he contacts the manager. A qualitative description of the task was the basis for its algorithmic description. An operator performs perceptual, executive, and mnemonic actions. A qualitative analysis of the task also shows that it includes decisions related to the situations (1) when an operator decides if he/she is at the right screen or not; (2) if the screen displays an error message or such a message is absent; (3) if the PO number is correct and error message persists or error message does not appear on the screen. Depending on the information on the screen, an operator decides what course of action should be taken. An algorithmic description of the task based on the method developed in SSAT is given in Table 7.1.

Below we consider principles of describing the individual members of the algorithm, the actions that are included in their content, and we explain the method of evaluation of their performance time. Motor actions are described using the symbolic system that is utilized in the MTM-1 system. Simultaneous perceptual actions are described in a similar way. In the MTM-1 system, there are the eye travel time ET that determines time of eye movement depending on the distance of such movement, and eye focus time EF (0.27 sec). The latter includes recognition and "yes/no"-type decisions. According to various engineering psychology handbooks, such perceptual actions or the simplest decisions require 0.3–0.4 sec. The member of the algorithm O_1^{α} includes only one simultaneous perceptual action. So it contains two psychological operation ET (eye movement $t = 0.42$ sec) and EF (simultaneous recognition $t = 0.3$ sec). The performance time of O_3^{α} and $O_6^{\alpha\mu}$ is determined similarly. We use the same principle for evaluating the performance time of $O_5^{\alpha\mu}$. The distance of eye movement here is different, and eye focuses for 0.4 sec. Analysis of strategies that are utilized during performance of O_{11}^{α} demonstrates that an operator performs the simultaneous perceptual action that includes $(ET + 2EF)$ because the presented stimulus is more complex. Decision-making or logical conditions are of the simplest type in the considered subtask 0. They usually require the same performance time as the simultaneous perceptual actions. Thus, performance of l_1 takes 0.3 sec, and 0.4 sec has been assigned for performance of l_2. Decision-making (logical conditions) l_3 has a very low probability ($P = 0.01$) and is considered. It should be pointed out that only simple cognitive actions can be described using the MTM-1 system. More complex cognitive actions should be described by utilizing standardized descriptions of cognitive actions or psychological units of analysis developed in SSAT. If we determine the performance time of individual cognitive actions that are included in the considered member of algorithm, then the performance time of all the cognitive members of the algorithm can be defined.

Let us consider motor actions that are included in the motions or motor operations. The concept of motor action does not exist in the MTM-1 system. The description of motor action as a system of standardized motions that are integrated by a goal of a motor action has been introduced in SSAT. Thus, for a description of motor actions, we utilize the MTM-1 system, using the SSAT approach.

As the first example, we consider the member of the algorithm $O_7^{\varepsilon\mu}$. The action it depicts is performed based on visual information and requires keeping this information in working memory until the motor actions are completed (symbols $\varepsilon\mu$ are used to reflect this fact). During keying in the PO number, the following actions are performed (distance of hand movement varies and is selected with some approximation): after returning to the computer area with the packing slip in the left hand, an operator moves the right hand to the corresponding key (R265B) and presses it. R26B means "reach a single object in the location that may vary slightly from cycle to cycle" when distance of movement is 26sm; AP-1 means "press a corresponding key". This is how the first number is keyed in; then he/she moves a hand to key in the other three numbers. This requires moving fingers three times to an average distance of 6sm (R6B). At the end of each move, the second, third, and fourth corresponding keys are pressed. The motor actions can be described as follows: (1) $(R26B + AP1)$; (2) $(R6B + AP1)$; (3) $R6B + AP1)$; (4) $(R6B + AP1)$. Each action includes two motions that are integrated by the corresponding goal of motor actions. The performance time is calculated as follows:

Key in the first four digits
Action 1: Key in the first digit - $(R26B + AP1) = 0.43\ sec + 0.38\ sec = 0.81\ sec$;
Action 2: Key in the second digit - $(R6B + AP1) = 0.17\ sec + 0.38\ sec = 0.55\ sec$;

Action 3: Key in the third digit - (R6B + AP1) = 0.17 sec + 0.38 sec = 0.55 sec;

Action 4: Key in the fourth digit - (R6B + AP1) = 0.17 sec + 0.38 sec = 0.55 sec;

Keying all four digits - 0.81 sec + 0.55 sec + 0.55 sec + 0.55 sec = 2.07 sec ≈ 2 sec

The pace used in the MTM-1 system is too high for performance of tasks in the automatized or computerized systems. Thus, here it is necessary to use coefficient 1.2 for the slower pace of computerized task performance. Such a coefficient is useful because the considered type of work method is not rigid and an operator can make mistakes.

According to the algorithmic description (see Table 7.1), an operator has to key in the second four numbers when he/she performed $O_9^{\varepsilon\mu}$. This requires a little more time (the difference is 0.3 sec) because when he/she turns his/her hand and eyes to the packing slip in the left hand, his/her right hand involuntarily moves back and this requires 0.3 second to return the right hand to the starting position in order to key in the last four digits of the PO number.

Let us briefly consider the member of the algorithm O_{14}^{ε}. It includes two motor actions: (R50B + AP1) and (R30B + AP1). R50B is a standardized motion, "reach a single object in a location that may vary slightly from cycle to cycle." The distance of motion is 50 cm. The second motion is AP1 (apply pressure). The second action includes the same type of motions, but the distance of the motion is 30 cm. Based on this description, we can define the performance time of these two actions. As we have mentioned in this chapter, the pace of motions in MTM-1 cannot be considered as the optimal one for the considered type of tasks. So in SSAT we usually utilize a coefficient of pace 1.1–1.2 or even 1.3. According to MTM-1, the performance time of the above-considered two actions is 1.68 sec. If we apply the coefficient 1.2, the performance time would be equal to 2.01 sec. The performance time of other motor actions and therefore the performance time of the members of the algorithm is calculated similarly.

Let us now consider how the performance time of the member of the algorithm O_{13}^{ε} is calculated. In column two of Table 7.1, the probability of appearance this member of the algorithm in the structure of the task performance is listed as $P = 0.05$ (the same as for O_{12}^{α}). Taking into consideration this probability, the performance time of O_{13}^{ε} in the structure of the task is defined as:

$$T = (1.95 \times 1.2) \times 0.05 = 0.12, \tag{7.1}$$

where:

1.95	is the performance time according to the pace in the MTM-1 system
1.2	is the coefficient of the pace
$P = 0.05$	is the probability of appearance of O_{13}^{ε} in the task structure

The symbolic description of motor and cognitive components clearly demonstrates that we utilize psychological units of analysis to develop the time structure of the task.

In order to define the decision-making outputs for subtask 0, the subject matter experts (SME) utilized the method of defining different scale diapasons developed in SSAT. In subtask 0, the decision-making that is depicted as the logical condition l_2 has two outputs with different probabilities. Analysis of this logical condition allows demonstration of the principle of determining probabilities of decision-making outputs for logical conditions in general. The logical condition l_2 occurs when the receiver makes a decision based on visual information. If the warning "invalid PO number" is presented, then the receiver performs O_{12}^α. If the message does not appear, the receiver should do the next step, O_{15}^α. Subject matter experts (SMEs) evaluate the probability of these two outputs. The warning "invalid PO number" is a rare event because in most cases the receiver does not make a mistake when keying in the PO number. Thus, the SMEs have to determine the probability of the receiving operator performing O_{12}^α or O_{15}^α. According to this approach, the probability can be considered as a degree of certainty or degrees of belief of the SME, who can rationally estimate the probability of events. An experienced expert assigns $P = 0.05$ to the event "invalid PO number" and $P = 0.95$ to the opposite event with confidence. As can be seen, an expert placed these two events into two diapasons: (1) extremely high scale diapason $P = 0.9$–1.00 and (2) extremely low scale diapason $P = 0.1$–0.01 (see Chapter 5, Figure 5.1).

If the message "invalid PO number" has probability $P = 0.05$, then the following members of the algorithm O_{12}^α and O_{13}^ε are also performed with this probability. The members of the algorithm l_3 and O_{14}^ε are not considered due to their extremely rare occurrence and involvement of the manager in such situations. Thus, after performance of l_2, the receiver should immediately perform either O_{15}^ε or O_{12}^α. After performance of O_{13}^ε the receiver also goes to the O_{15}^ε with probability $P = 0.05$. If the PO number is keyed in correctly, then O_{15}^ε is performed with probability 0.95. Hence, the probability of O_{15}^ε can be calculated as $P = 0.05 + 0.95 = 1$. Below Figure 7.6 shows the probability of performance of the following members of the algorithm that depend on decision-making outputs of logical condition l_2.

There are more complex situations when it is required to determine the probability of the outputs of various decisions that are depicted by logical conditions in the algorithmic description of rule-based tasks with a complex probabilistic structure. We will describe them in Chapter 8.

The analysis of the algorithmic description of the subtask 0 before improvement (see Table 7.1) indicates that a substantial part of it is related to keying in the PO number. For this purpose, an operator uses the following preferred strategy of task performance. She/he finds the PO number on the packing slip, selects the first three or four digits as perceptual units of information and keeps them in working memory until keying them in. This also requires moving her/his eyes from PO number on the packing slip in the left hand to the screen. This part of the task is repeated two times, because the PO number usually includes eight digits. This part of the task is represented

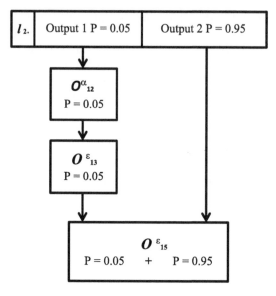

FIGURE 7.6
Probability of performance of the members of the algorithm that depend on decision-making outputs of logical condition l_2.

by members of the algorithm $O_5^{\alpha\mu}$ through O_{10}^{ε}. Using the bar scanner can eliminate this part of the task. The new, improved version of the task performance is described in Table 7.2. The advantage of performing the improved version of the subtask 0 can be easily discovered if we compare the algorithmic description before and after improvement. Similar cognitive and behavioral actions are performed from O_1^{α} to O_4^{ε} before and after improvements. After innovation, members of the algorithm from $O_5^{\alpha\mu}$ up to $O_{10}^{\alpha\mu}$ are eliminated. O_{11}^{α} before improvement is similar to the member of the algorithm O_7^{α} after improvement. Starting from this member of the algorithm, an operator performs the subtask the same way before and after improvement. As a result, the mental workload and quantity of motor and cognitive actions were significantly reduced after innovation.

These data are important because they demonstrate that we have here applied a non-traditional method of analysis. In ergonomics, specialists utilize experiments as the main method of comparison of different methods of task performance. Here we use analytical models that should be used as the main methods of the design process. Experiments should be applied as an additional method of study during the design.

It should be noted that in our further studies of the inventory receiving task, we used a variety of methods, including observation, chronometrical analysis, and the MTM-1 system. Some data about temporal characteristics of cognitive and behavioral components of human performance were obtained in applied activity theory (Beregovoy et al., 1978; Dobrolensky et al., 1975;

TABLE 7.2

Algorithmic Description of Activity and its Time Structure during Inventory Receiving Task Performance (Setup Task or Subtask 0 – after Improvement)

Member of Algorithm (Psychological Units of Analysis Described in Consolidated Manner)	Description of Elements of Tasks (Technological Units of Analysis)	Description of Elements of Activity (Psychological Units of Analysis Described in Detailed Manner)	Time (sec)
O_1^α	Check for presence of inventory receiving screen	Simultaneous perceptual action (ET + EF)	$0.42 + 0.3 = 0.72$
$\overset{1}{\downarrow} O_2^\varepsilon$	Type 1 and then press ENTER to choose ADD INVENTORY RECEIVING screen	(R50B + AP1) + (R30B + AP1)	$1.68 \times 1.2 = 2.01$
O_3^α	Check to see if you are at the ADD TRANSACTION screen	Simultaneous perceptual action (ET + EF)	$0.42 + 0.3 = 0.72$
$l_1^1 \uparrow$	If you are at the right screen, go to O_4^ε. If the screen is wrong, press F3 for exit and go back to O_2^ε	Decision-making action from two alternatives	0.3
$\overset{8(1)5(1)}{\downarrow} \to O_4^\varepsilon$	Go to base unit 5 and take a packing slip by the left hand from the box placed on the base unit 5 (see Figure 7.3) and go to working area with computer	(1) Leg movements: "performance four steps" TBC1 + WP + WP + TBC2 (2) Hand action (R50B + G1B + M40B) (3) Leg movements: TBC1 + TBC2 + WP + WP	$7.5 \times 1.2 = 9$
O_5^α	Find PO (purchase order) number on the packing slip	Simultaneous perceptual action (ET + EF)	$0.42 + 0.3 = 0.72$
O_6^ε	Take barcode scanner and scan the PO number (right hand moves and grasps scanner and moves it in exact position for scanning, then scans PO number and puts the scanner back in initial position)	(R50B + G1A) + (M40B + mM10C + AP1) + (M50B + RL1 + R30E)	$3.4 \times 1.2 = 4.08$
O_7^α	Look at the screen to check if the PO number has been populated	Simultaneous perceptual action (ET + EF)	$0.42 + 0.3 = 0.72$

(Continued)

TABLE 7.2

Algorithmic Description of Activity and its Time Structure during Inventory Receiving Task Performance (Setup Task or Subtask 0 - after Improvement)

Member of Algorithm (Psychological Units of Analysis Described in Consolidated Manner)	Description of Elements of Tasks (Technological Units of Analysis)	Description of Elements of Activity (Psychological Units of Analysis Described in Detailed Manner)	Time (sec)
$l_2^2 \uparrow$	If the screen displays an error message ($P=0.01$), INVALID PO NUMBER, go to O_8^ε. If PO number is correct (0.99), go to O_9^ε	Decision-making action from two alternatives	0.4
O_8^ε	Key in PO number manually or call manager	Verbal or motor actions (This member of algorithm is not considered due to low probability)	----
$\downarrow O_9^{R4}$	Press ENTER and the system populates current date	Motor action (R14B+AP1)	$0.57 \times 1.2 = 0.68$
O_{10}^ε	Press F8 to look up items on the purchase order (Figure 7.4) and return to base unit 5	Motor action with right hand (R17B+AP1) and four steps to base unit #5 (TBC1 + TBC2 + WP + WP)	$(0.6+3.1) \times 1.2 = 4.44$
Total Performance Time	Beginning of Individual Item processing Versions of Task		23.79

Zarakovsky and Pavlov, 1987; Petrov and Myasnikov, 1976) The data obtained in cognitive psychology have also been utilized (Card et al., 1983). For example, when we were working on determining the temporal characteristics of human performance of computerized tasks, we utilized the following temporal data suggested by Card and Morgan: 0.28 sec for a keystroke; 0.1 sec for a mouse button press or release; 1.1 sec (average) for a mouse move; 0.4 sec to hold the hand at a keyboard or mouse.

For determining possible strategies of task performance and obtaining temporal characteristics of activity, we also utilized abbreviated experimental procedures. Such procedures can be performed in the laboratory or real operating conditions. Usually elements of tasks that should be reproduced in laboratory conditions are for one or several members of the human algorithm. Specialists can find some elements of activity in already existing systems that have similarities with the designed element of activity for a new system. With some modification, such elements can be used as prototypes for the designed component of activity. Such a method is especially effective for the analysis of the temporal characteristics of the individual elements of the designed activity.

In this section, we have presented the morphological analysis of subtask 0. After performing the subtask, an operator performs three possible versions of the task with different probabilities. The first version involves scanning an item number, checking its quantity, and rejecting the item due to unacceptable quantity; the second version is performed when the quantity of the items can be accepted and an operator checks the item price and rejects the item by this parameter. The third version (the main version of the task) is performed only if such criteria as quantity and price are acceptable (see Figure 7.7).

This slightly different version of Figure 7.1 facilitates a clear understanding of the further analysis of this complex task. Versions 1, 2, and 3 of the task

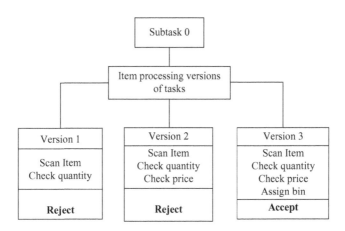

FIGURE 7.7
Relationship between subtask 0 and item processing versions of task.

are followed by the subtask 0 only if the new box with items is opened. After that, these three versions follow each other with certain probability until the box is empty and the receiver starts the next box.

7.2.2 Morphological Analysis of Version 1 of Inventory Receiving Task Performance

In the previous chapter, we described subtask 0 that is performed only when a new box is opened. This subtask is not included in the main part of the inventory receiving task, which is involved in receiving and registering individual items. Performance of the main task involves three different versions that occur with various probabilities and in diverse order. The purpose of this chapter is to demonstrate for the first time the principles of utilizing analytical procedures for analysis, description, and further improvement of such complex tasks.

Version 1 is involved in evaluation of the quantity of the items. It is completed when an item is rejected due to its unacceptable quantity. For example, the received quantity of items is greater than the ordered quantity and an operator rejects items based on the quantity criterion.

Version 2 occurs when quantity of an item is acceptable but an item is still rejected due to unacceptable price.

Version 3 describes the scenario when an item is accepted according to all criteria and put into a corresponding tote for further movement into the warehouse or workshop.

All three versions have different probabilities of occurrence. Version 3 has the highest probability. The version number does not determine its sequence of appearance. All three versions not only have different probabilities but can also be performed in a variety of orders. The creation of analytical models of a task with such a complex probabilistic structure is an important theoretical and practical undertaking.

This means that specialists can use analytical models (and utilize simplified experimental procedures as a supplemental method) for analysis of such complex task or for analysis of separate members of the algorithm. This is a true design method where not experimental, but analytical, methods play a leading role in task analysis.

In this chapter, we concentrate on considering morphological analysis of Version 1.

Version 1 describes the task execution when the item is not received all the way through but rejected and put aside due to the unacceptable quantity of the received item. The goal of Version 1 is assessment of the quantity of the received items and rejection of them if their quantity does not match the ordered quantity and is not a partial delivery. There are other reasons the items should be rejected, also, according to the acceptability rules for the item quantity.

Version 1 starts with processing the individual item. In order to do that, the receiver takes the item from the box and returns to his/her computer.

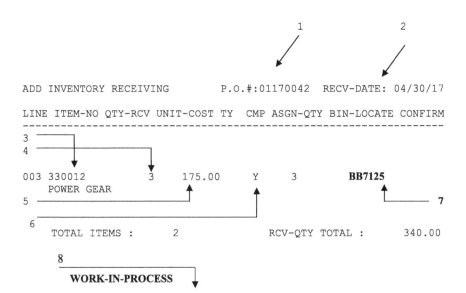

FIGURE 7.8
Add receiving screen with detailed item information. 1-purchase order number; 2-received date; 3-item number; 4-receved quantity; 5-unit cost; 6-completion flag; 7-bin location; 8-work-in-process option.

Then he/she reads the item number on the physical item and browses the list of item numbers on the screen in order to find the matching number (see Figure 7.5, subtask 0).

When the item number on the screen is located, the receiver moves the arrow key to the found item number and presses enter. Figure 7.8 depicts the screen with detailed item information that appears at this point.

The receiver then compares the received quantity of the physical item with the ordered quantity that is displayed on the screen. If the ordered and received quantity are the same, then the item is processed further (Versions 2 and 3). If the quantities are not the same, the receiver should apply the following rules:

> *If the received quantity is the same as the quantity in the PO on the screen, the receiver presses ENTER and proceeds to the next field.*
>
> *If the received quantity is different from the quantity in the PO on the screen, the receiver changes the quantity on the screen and the following question is presented to the receiver: "The received quantity is different from the ordered quantity. Do you want to accept?" and the following default appears:*
>
> 1. *If the quantity is less, the system defaults to Y.*
> 2. *If the quantity is greater, the system defaults to N.*

If the receiver accepts the default, she/he presses ENTER and

1. *If the default was N, she/he returns to the prior screen.*
2. *If the default was Y, the receiver proceeds to the next field.*

The following are the considerations for the decision to accept or change the system default:

1. *If the received quantity is less, then in most cases the default would not be changed because the rest of the order might come at a later time.*
2. *If the received quantity is greater: a) if it's much greater or/and the item is expensive, do not accept; b) if the quantity is not much greater and the item is not expensive, then accept.*

If the item is not rejected, this scenario belongs to Versions 2 or 3. If the item is rejected, the receiver puts it in the Put-Aside area (see Table 7.3; member of algorithm O_{26}^{ε}) and then checks if there are other items in the box to receive.

Table 7.3 reflects the morphological analysis of Version 1 before its improvement. It includes the algorithmic and time structure of the considered task. Let's selectively consider some members of the algorithm. Each member of the algorithm is as a quasi-system that consists of cognitive or behavioral actions that are described using technological and psychological units of analysis. This version of the task can be performed after subtask 0 is completed. Hence, the first member of this algorithm is O_{17}^{ε} because it's the continuation of the algorithm of subtask 0 before improvement. The detailed morphological analysis presented in this section has allowed us to develop a precise algorithmic description of the first version of the computerized task with the complex variable probabilistic structure. The most representative strategy of performance of O_{17}^{ε} (the psychological unit of analysis) can be described as follows. An operator takes an item out of the box that is located on the base unit 5 for unpacking using both hands (see Figure 7.2, an view from above of the workplace). Then he/she returns to the computer area. While performing these steps, his/her right hand releases the item in order to hold the item in the left hand only. This common language with some modification describes this fragment of task performance by using technological units of analysis (see column 2 of Table 7.3). In the third column, each motor action is described not only by common language but also using psychological units of analysis. For example, the first motor action is "simultaneously reach and grasp an item with two hands". This is a technological unit of analysis for the first motor action. In terms of psychological units of analysis each hand performs an action (R60B + G1G). This motor action includes two standardized motions: 1) R60B - reach an object in a location that may slightly vary; and G1B - grasp a small object or an object lying on a flat surface. The second

TABLE 7.3

Algorithmic Description of Version 1 of Inventory Receiving Task Performance (Version 1 before Improvement – Item Is Put Aside due to Unacceptable Quantity; $P=0.07$)[a]

Members of Algorithm (Psychological Units of Analysis)	Description of Elements of Task (Technological Units of Analysis) Beginning of Individual Item Processing Version 1 of the Task	Description of Elements of Activity (Psychological Units of Analysis)	Time (sec)
$\overset{17(2)\,11(2)\,7(2)}{\downarrow\downarrow\downarrow}\;O_{17}^{\varepsilon}$	Take an item out of the box, and return to computer area again (while a worker takes several steps, his/her right hand releases the item and worker holds the item only by left hand)	Motor actions - simultaneously by two-hand reach and grasp item (R60B+G1B); (3) – take out item from box – (M40B+RL1); (4) perform again four steps to computer area #3 (TBC1+TBC2+WP+WP)	$4.74\times1.2=5.7$
O_{18}^{ou}	Look at item number on the physical item and compare it with item numbers on the screen (Figure 7.5, field 3)	At least five simultaneous perceptual actions (ET+EF); combined with simple mnemonic and scanning operations	Average time ≈ 6
$\overset{4}{I_4\uparrow}$	If item number is on the first page, go to O_{20}^{ε}. If item number is not on the first page, go to O_{19}^{ε}	Decision-making action at sensory-perceptual level	0.3
O_{19}^{ε}	Press arrow key (repeat if required)	Motor action (R10B+AP1)	$0.52\times1.2=0.62$
$\overset{4}{\downarrow}O_{20}^{\varepsilon}$	Put cursor on the selected line (Figure 7.5) and press ENTER to go to the screen with detailed item information (Figure 7.8)	Three motor actions: (R30B+G1A)+(M5B+AP1+RL1)+(R30B+AP1+R30E)	$2.2\times1.2=2.64$
O_{21}^{ou}	**Compare received quantity with PO (purchase order) quantity (Figure 6, field 4)**	Combination of two simultaneous perceptual actions $-2\times(ET+EF)$ with simultaneously performed mnemonic operations (MO)	$(0.42+0.4)\times2=1.64$

(Continued)

TABLE 7.3 (CONTINUED)

Algorithmic Description of Version 1 of Inventory Receiving Task Performance (Version 1 before Improvement – Item Is Put Aside due to Unacceptable Quantity; $P = 0.07$)[a]

Members of Algorithm (Psychological Units of Analysis)	Description of Elements of Task (Technological Units of Analysis) Beginning of Individual Item Processing Version 1 of the Task	Description of Elements of Activity (Psychological Units of Analysis)	Time (sec)
$I_5 \uparrow^{5}$	If received quantity and ordered quantity are the same, go to O_{29}^{ε} (This output is performed in other subtasks). If received quantity is greater or less than ordered quantity, go to O_{22}^{ε}. (In this version of the task, this output is performed always with $P = 1$)	Decision-making actions that are performed based on visual information	0.4
O_{22}^{ε}	Type the received quantity and press ENTER to get a question at the bottom of the screen	Motor action (R20B + AP1) + (R12B + AP1) (example with two-digit number)	$(1.39 \times 1.2) = 1.7$
O_{23}^{α}	Read the statement: THE RECEIVED QUANTITY AND ORDERED QUANTITY DO NOT MATCH. DO YOU ACCEPT? (YES/NO). Scan and read ≈ four words	Successive perceptual action ET + 4 × EF	$(0.42 + 4 \times 0.18) = 1.14$
** $O_{24}^{\mu th}$	Recall instructions and perform required calculation and estimation (relationship between quantity and price)	Combination of mnemonic action (retrieve simple information from memory) and logical thinking action	≈ (1.2 + 3) = 4.2
$I_6 \uparrow^{6}$	If quantity is not accepted (computer defaults to 'N'), go to O_{25}^{ε}. (In this version of the task, considered output is performed always with $P = 1$) Otherwise, go to O_{28}^{ε} (This output is performed in other version of task)	Complex decision-making actions that include mnemonic and thinking operations	1.5

(Continued)

TABLE 7.3 (CONTINUED)

Algorithmic Description of Version 1 of Inventory Receiving Task Performance (Version 1 before Improvement – Item Is Put Aside due to Unacceptable Quantity; $P = 0.07$)[a]

Members of Algorithm (Psychological Units of Analysis)	Description of Elements of Task (Technological Units of Analysis) Beginning of Individual Item Processing Version 1 of the Task	Description of Elements of Activity (Psychological Units of Analysis)	Time (sec)
O_{25}^{ε}	Press ENTER (default is conformed)	Motor action (R26B + AP1)	$0.7 \times 1.2 = 0.84$
O_{26}^{ε}	Put rejected item in the Put-Aside Area, Figure 7.2, area 10. Return to the base unit 5. O_{26}^{ε} includes: (1) the left hand grasps the item; (2) worker turns body and takes approximately six steps to the put-aside area; moves the left hand and releases item; (3) worker turns body and takes approximately four steps to the base unit 5 for unpacking	Motor actions by hand and leg movements: (R6B + G1B) + (TBC1 + 6WP + TBC2) + (R40B + RL1) + (TBC1 + 4WP + TBC2)	$10.4 \times 1.2 = 12.5$
Total performance time for the first version of the task	**Task of receiving an item ends here**	$T_1 = T_{cog} = T_{ex}$	**39.18**
O_{27}^{α}	Check if there are other items in the box to receive	Simultaneous perceptual action (ET + EF)	$(0.42 + 0.3) = 0.72$
$l_7^{(1-2)}$↑	If there are no more items in the box, go to O_4^{ε}, (subtask 0) otherwise, go to O_{17}^{ε}	Simple decision-making	0.3
Performance time of O_{27}^{α} and l_7			1.03

[a] Version 1 of the task is performed with probability 0.07 during the shift (see event tree Figure 7.9). Each member of algorithm inside of this task has probability 1 inside this Version.

motor action is "take an item out of the box" (technological units of analysis). When we use psychological units of analysis, this action can be described as (M40B + RL1), where M40B means "move an object 40sm to an approximate or indefinite location" and RL1 means normal release performed by opening fingers as an independent motion. RL1 is crossed by an inclined line, which means that motion "release" is overlapped by other motor motions and therefore the time of its performance should not be taken into consideration during calculation of the general performance time of O_{17}^{ε}. The last motor action contains motor motions of the receiver's legs. TBC1 and TBC2 are steps with simultaneous turning of the body up to 45°. TBC1 is completed when a leading leg contacts the floor. TBC2 is completed when a leading leg contacts the floor. WP means that an operator performs unobstructed steps. As mentioned above, the coefficient 1.2 is used for reducing the pace of motor action performance. This example demonstrates that such a detailed analysis is effortful and time-consuming. However, such tasks can be performed the same way for many years. In our further analysis, we do not discuss any other motor members of the algorithm.

When an operator performs $O_{18}^{\alpha\mu}$ he/she looks at an item number and then compares it with an item number on the screen. Analysis of this element of cognitive activity shows that he/she has to perform on average five simultaneous perceptual actions and keep information in working memory when comparing these numbers (ET means move eyes and EF involves receiving information). Symbol $\alpha\mu$ means performing perceptual actions in combination with a mnemonic operation.

The most important part of the considered task involves comparison of the received quantity with the PO quantity (Figure 7.8, field 4). In order to depict Version 1 algorithmically, it is necessary to consider the rules described below and the decisions associated with them involved in the task performance. These rules are the following.

An operator has to evaluate two things—whether: (1) the ordered quantity and received quantity are the same; (2) the ordered quantity and received quantity are not the same. If the quantity is the same, an operator performs the next operation (evaluates the price of the items). This scenario does not belong to Version 1. If the ordered quantity is not the same, an operator types the received quantity and presses ENTER to get a question at the bottom of the screen. Thus, an operator has to make a decision about the following actions. This decision is designated as logical condition l_5. The system can default to yes to accept or not accept an item. Based on this information, an operator decides to accept or reject the item (YES or NO response). This requires a new decision after which different actions would follow. This decision is depicted as logical condition l_6. Each considered logical condition has two outputs with different probabilities. Moreover, outputs of l_5 influence outputs of l_6. We have to determine the probability of these outputs because they influence the probability of the following members of the algorithm. For this purpose, we utilize the event tree technique that has been described in Chapter 5. The method developed in SSAT helps to transfer qualitative data about outputs into the quantitative data when the probability is determined by subject matter experts (SME) utilizing the scale of the subjective probability of events (see Figures 5.1 and 5.2 in Chapter 5). In Figure 7.9 we present the event tree that depicts the subjective probability of the considered events (see Figure 7.9).

The first decision l_5 has two outcomes. The first outcome reflects the situation when the received quantity and the ordered quantity are the same. The extremely high scale diapason covers very frequent events with P from 0.9 to 0.99. SMEs related the event when ordered and received quantity of the item are equal to the low border of the high diapason ($P=0.9$–0.93) according to Figure 5.3 (see Chapter 5). Therefore, SMEs determined $P=0.9$ as the probability of this event, which falls into the extremely high scale diapason.

The second outcome of l_5 depicts the scenario when the ordered and received quantity are not the same. Accordingly, probability of the opposite outcome is 0.1. This latter probability can be related to the low scale diapason (see Figure 5.2). The extremely low scale diapason covers very infrequent events with P from 0.01 to 0.1.

Let us now consider the second decision or logical condition l_6. The probability of logical condition l_6 is $P=0.1$. Therefore, in order to determine outcomes of this logical condition, SMEs utilized the extremely low scale diapason depicted on Figure 5.2. SMEs evaluated that an item is rejected according to quantity not matching the order quantity as a highly probable event inside this diapason, assigning to it the probability $P=0.07$. The opposite outcome has the probability $P=0.03$. The event tree in Figure 7.9 demonstrates that the probability of an event when items are accepted based on the quantity criterion is $P=0.93$ and are rejected with $P=0.07$.

After determining the probability of outputs of decision-making actions (l_5 and l_6) we can determine the probability of other members of the algorithm.

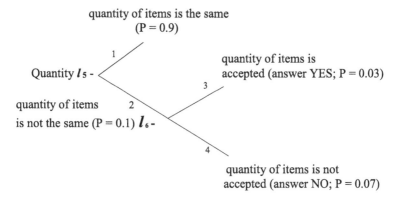

FIGURE 7.9
Event tree of decisions l_5 and l_6 when item quantity is evaluated.

Table 7.3 depicts the algorithmic and time-structure description of Version 1 that only takes place when the received quantity is not acceptable and the item is rejected. For this version this output is performed with probability $P=1$ (see Table 7.3 logical condition l_5). Hence, members of the algorithm O_{22}^{ε}; O_{23}^{ε}, $O_{24}^{\mu th}$, and l_6 are performed with the same probability $P=1$. Version 1 can have only one output of logical condition l_6 "if quantity of items is not accepted then go to O_{25}^{ε}."

These members of the algorithm will otherwise be combined with the next stage of task performance that involves price evaluation (go to O_{28}^{ε}). The combination of quantity evaluation with price evaluation is performed in Version 2, which is going to be described in the next chapter.

The sequence of the performance of subtask 0 and Version 1 is presented in the Figure 7.10.

This figure shows that Version 1 is completed only when the quantity is rejected and an operator puts an item in the Put-Aside Area (see O_{26}^{ε}).

The following discussion is dedicated to Version 1 of the task performance after improvement and its comparison with the before-improvement version.

The sequence of performance of subtask 0 and Version 1 after improvement is presented in Figure 7.11, which shows that Version 1 after improvement starts with O_{11}^{ε}.

Let us now consider the usefulness of introducing the barcode scanner in the first version of the task. For this purpose, we compare described

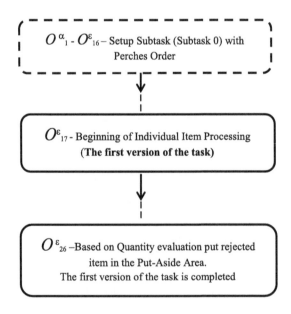

FIGURE 7.10
Setup task with the new Purchase Order and the first version of the task before improvement with probability $P=0.07$.

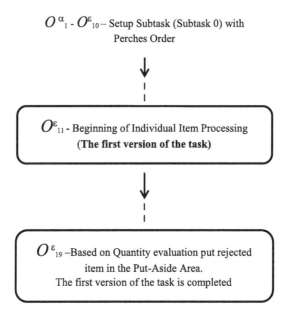

$O^{\alpha}{}_1$ - $O^{\varepsilon}{}_{10}$ – Setup Subtask (Subtask 0) with Perches Order

$O^{\varepsilon}{}_{11}$ - Beginning of Individual Item Processing
(The first version of the task)

$O^{\varepsilon}{}_{19}$ –Based on Quantity evaluation put rejected item in the Put-Aside Area.
The first version of the task is completed

FIGURE 7.11
Setup task with the new Purchase Order and the first version of the task after improvement performed with probability $P = 0.07$.

members of algorithm from $O_{18}^{\alpha\mu}$ to O_{20}^{ε} before improvement with members of the algorithm O_{12}^{ε} and O_{13}^{ε} after innovation when an operator uses the barcode scanner. An analysis of the algorithmic and time-structure description of Version 1 performance before improvement demonstrates that an operator has to look at the item number on the physical item, move his/her eyes to the computer screen and then compare these numbers (perform $O_{18}^{\alpha\mu}$). This requires performance of perceptual actions combined with mnemonic operation. After that, an operator makes a decision (logic condition l_5). The performance time of these two members of the algorithm that described cognitive components of activity will be 6 sec + 0.3 sec = 6.3 sec. Cognitive components are usually more complex elements of activity than motor components. Really, as we show later, motor activity in Version 1 before improvement and after improvement belongs to the first or second category of complexity according to the level of concentration of attention. Cognitive components belong to IV or III category of complexity according to this criterion. This means that elimination of these components of the task that take 6.3 sec is a more important factor than a slight increase in duration of motor activity. In general, reduction of cognitive components more often is more useful than reduction of motor components of activity.

The performance time of motor activity before improvement was 3.26 sec (see O_{19}^{ε} and O_{20}^{ε}). The performance time of motor activity after improvement was 3.8 sec (O_{12}^{ε} and O_{13}^{ε}). This means that after improvement, time for motor activity increased by 0.54 sec (3.8 to 3.26 sec) and in both situations this motor activity belongs to the second category of complexity. However, an insignificant increase of performance time for relatively simple motor components of activity after improvement is accompanied by much more noticeable reduction of the performance time of more complex cognitive components of activity (reduction of time for performance of $O_{18}^{\alpha\mu}$ and l_5). The performance time of the whole Version 1 before improvement was 39.18 sec and the performance time of the same task after improvement was 32.68 sec. Thus, introduction of a barcode scanner is a reasonable innovation. This innovation is not a complicated one and has been chosen to demonstrate the method of justifying any innovation. Such a detailed step-by-step analysis of each member of the algorithm of the task performance allows finding the critical points in the task performance and predicting the need for the innovation. Some innovations are costly or hard to implement. So such analysis allows evaluation to determine if the implementation of the innovation is justified.

Table 7.4 shows the algorithmic description of Version 1 of the Inventory Receiving Task Performance after improvement.

7.2.3 Morphological Analysis of Version 2 of Task Performance

As has been mentioned before, the inventory receiving task has preparatory subtask 0 and three deferent versions of its performance that appear with different probabilities during the shift depending on the quantity and price of the received items. Version 2 is performed when the quantity of items is accepted but the item is still rejected due to an unacceptable price. Hence, Version 2 is completed when a receiver puts a rejected item in the put-aside area based on the *price* criterion. Version 2 can be performed in any order in relation to other versions. For example, Version 2 can be performed before Version 1 and after Version 3 or after the setup task (subtask 0) is completed.

Members of the algorithm O_1^α - O_{16}^ε in the top box on Figure 7.12 belong to Subtask 0, which takes place only when all the items in the box are processed, the box is empty, and the receiver has to get a new box of items, put it on base unit 5 (Figure 7.2), open it, and enter the new PO number. This is why this box is indicated by a dotted line showing that Subtask 0 precedes Version 2 in the algorithm. The stages of Version 2 are presented in Figure 7.12.

Version 2 starts with the same actions as Version 1 but unlike in Version 1, the item quantity is accepted, and the receiver moves to evaluation of the item price. If we return to the event tree (Figure 7.9) for Version 1 when the item quantity is evaluated, we can see that there are four branches with three different probabilities of occurrence. The first branch shows that the received item quantity is the same as the PO quantity, with probability 0.9, and the item is processed further in Versions 2 or 3. The third branch depicts the scenario when the item quantity is different, but it's still acceptable ($P = 0.03$), and the

TABLE 7.4

Algorithmic Description of Version 1 of Inventory Receiving Task Performance (Version 1 after Improvement – Item Is Put Aside due to Unacceptable Quantity; $P = 0.07$)

Member of Algorithm (Psychological Units of Analysis Described in Consolidated Manner)	Description of Elements of Tasks (Technological Units of Analysis) Beginning of Individual Item Processing Version 1 of the Task	Description of Elements of Activity (Psychological Units of Analysis Described in Detailed Manner)	Time (sec)
$^{8\,(2)\,5\,(2)}_{\downarrow\ \downarrow}\ O_{11}^{\varepsilon}$	Take an item out of the box, and return to computer area again (while a worker takes several steps, his/her right hand releases the item and worker holds the item only by left hand)	Motor actions – performed simultaneously by two hands: reach and grasp item (1) (R60B + G1B); (2) – take item out of the box – (M40B + RL1); (3) take four steps to computer area 3 (TBC1 + TBC2 + WP + WP)	$4.74 \times 1.2 = 5.7$
O_{12}^{ε}	Take the barcode scanner and scan the item number. (The matching item is highlighted)	(R50B + G1A) + (M50B + AP1) + (M50B + RL1)	$2.44 \times 1.2 = 2.9$
O_{13}^{ε}	Press ENTER to go to the screen with detail item information (Figure 7.8)	Motor action (R26B + AP1)	$0.76 \times 1.2 = 0.9$
O_{14}^{qu}	**Compare received quantity with PO (purchase order) quantity (Figure 7.8, field 4)**	Combination of two simultaneous perceptual actions - $2 \times (ET + EF)$ with simultaneously performed mnemonic operations (MO)	$(0.42 + 0.4) \times 2 = 1.64$
$l_3^{\,3} \uparrow$	If received quantity and ordered quantity are the same, go to O_{22}^{ε} ($P = 0.9$). If received quantity is greater or less than ordered quantity, go to O_{15}^{ε} ($P = 0.1$)	Decision-making actions that are performed based on visual information	*0.4*
O_{15}^{ε}	Type the received quantity and press ENTER to get a question at the bottom of the screen ($P = 0.1$)	Motor action (R20B + AP1) + (R12B + AP1) (example with two-digit number)	$1.39 \times 1.2 = 1.7$

(Continued)

TABLE 7.4 (CONTINUED)

Algorithmic Description of Version 1 of Inventory Receiving Task Performance (Version 1 after Improvement – Item Is Put Aside due to Unacceptable Quantity; $P = 0.07$)

Member of Algorithm (Psychological Units of Analysis Described in Consolidated Manner)	Description of Elements of Tasks (Technological Units of Analysis) Beginning of Individual Item Processing Version 1 of the Task	Description of Elements of Activity (Psychological Units of Analysis Described in Detailed Manner)	Time (sec)
O_{16}^{α}	Read the statement: THE RECEIVED QUANTITY AND ORDERED QUANTITY DO NOT MATCH. DO YOU ACCEPT? (YES/NO). ($P = 0.1$). Scan and read $\approx$ four words	Successive perceptual action. $ET + 4 \times EF$	$0.42 + 4 \times 0.18 = 1.14$
* $O_{17}^{\mu th}$	Recall instructions and perform required calculation and estimation (relationship between quantity and price)	Combination of mnemonic action (retrieve simple information from memory) and logical thinking action	$\approx (1.2 + 3) = 4.2$
$l_4^4 \uparrow$	If quantity is not accepted (computer defaults to 'N'), go to O_{18}^{ε}. ($P = 0.07$). Otherwise, go to O_{21}^{ε} ($P = 0.03$)	Complex decision-making actions that includes mnemonic and thinking operations	1.5
O_{18}^{ε}	Press ENTER (default is conformed and after performing O_{19}^{ε} start to work with new item). $P = 0.07$	Motor action (R26B + AP1)	$0.7 \times 1.2 = 0.84$

(*Continued*)

TABLE 7.4 (CONTINUED)

Algorithmic Description of Version 1 of Inventory Receiving Task Performance (Version 1 after Improvement – Item Is Put Aside due to Unacceptable Quantity; $P = 0.07$)

Member of Algorithm (Psychological Units of Analysis Described in Consolidated Manner)	Description of Elements of Tasks (Technological Units of Analysis) Beginning of Individual Item Processing Version 1 of the Task	Description of Elements of Activity (Psychological Units of Analysis Described in Detailed Manner)	Time (sec)
O_{19}^{ε}	Put rejected item in the Put-Aside Area, Figure 7.2, area 10. Return to the base unit 5. ($P = 0.07$). O_{19}^{ε} includes: (1) the left hand grasps the item; (2) worker turns body and takes approximately six steps to the put-aside area; moves the left hand and releases item; (3) worker turns body and take approximately four steps to the base unit 5	One motor actions by hand and leg movements: (R6B + G1B) + (TBC1 + 6WP + TBC2) + (R40B + RL1) + (TBC1 + 4WP + TBC2)	$10.4 \times 1.2 = 12.5$
Total performance time of the first version of the task	**Task of receiving an item ends here**	$T_1 = T_{\text{cog}} = T_{\text{ex}}$	33.42
O_{20}^{α}	Check if there are other items in the box to receive	Simultaneous perceptual action ET + EF	$0.42 + 0.3 = 0.72$
$I_5^{5(1-2)}\uparrow$	If there are no more items in the box, go to O_4^{ε}, otherwise, press F3 to return to previous screen (Figure 7.5) and go to O_{11}^{ε}	Simple decision-making	0.3
Performance time of O_{20}^{α} and I_5.			1.02

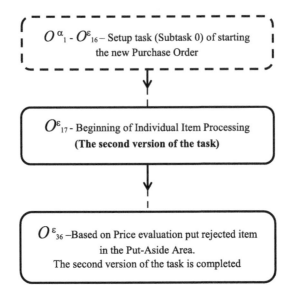

FIGURE 7.12

Setup task and Version 2 of the task before improvement that is performed with probability $P = 0.025$.

item is also processed in Versions 2 or 3. So the process moves to Version 2, with $P = 0.93$. Let us consider the algorithmic description of Version 2 before improvement (see Table 7.5).

Analysis of the algorithm of version 2 shows three stages of task performance. The first stage, from the beginning and including O_{20}^{ε} is the same as in Version 1. The second stage involves evaluation of the item quantity. It describes the situation when the received quantity of items is the same as the received quantity or is not the same but is still acceptable. The third stage of this version describes the evaluation of the price. Version 2 is completed when a receiver puts a rejected item in the put-aside area due to an unacceptable price.

If we compare logical condition l_5 in Version 1 and Version 2, they are similar but not identical. In Version 1, l_5 has only one output of a decision that belongs to this version. In Version 2, this decision has two outputs, with probabilities $P = 0.9$ and $P = 0.1$. Let us consider an example of determining the probability of some members of the algorithm that are associated with logical conditions l_5. The accumulated probability of O_{29}^{ε} also depends on this logical condition.

In Version 2, logical condition l_5 has two outputs. The first output is when the received quantity and the ordered quantity are the same, go to O_{29}^{ε} ($P = 0.9$). The second output reflects the situation when these quantities are not the same but the receiver decided to process the item further anyway ($P = 0.1$). This allows determining the probability of O_{29}^{ε}. Member of algorithm O_{28}^{ε} is

TABLE 7.5

Algorithmic Description of the Time Structure of Performance of Inventory Receiving Task (Version 2 before Improvement – Item Is Put Aside due to Price Rejection; $P = 0.03$)[a]

Members of Algorithm (Psychological Units of Analysis)	Description of Elements of Task (Technological Units of Analysis) Beginning of Individual Item Processing Version 2 of the task	Description of Elements of Activity (Psychological Units of Analysis)	Time (sec)
**	Take an item out of the box, and return to computer area again (while a worker takes several steps, his/her right hand releases the item and worker holds the item only by left hand)	Motor actions - simultaneously by two hand reach and grasp item (R60B+G1B); (3) – take out item from box – (M40B+RL1); (4) perform again four steps to computer area #3 (TBC1+TBC2+WP+WP)	$4.74 \times 1.2 = 5.7$
$\underset{\rightarrow}{17\,(2)} \underset{\rightarrow}{11\,(2)} \underset{\rightarrow}{7\,(2)} O^{\varepsilon}_{17}$			
$O^{\alpha\mu}_{18}$	Look at item number on the physical item and compare it with item numbers on the screen (Figure 7.5, field 3)	At least five simultaneous perceptual actions (ET+EF); combined with simple mnemonic and scanning operations	Average time ≈ 6
$l_4 \uparrow^4$	If item number is on the first page, go to O^{ε}_{20}. If item number is not on the first page, go to O^{ε}_{19}	Decision-making action at sensory-perceptual level	0.3
O^{ε}_{19}	Press arrow key (repeat if required)	Motor action (R10B+AP1)	$0.52 \times 1.2 = 0.62$
$\rightarrow O^{\varepsilon}_{20}$	Put cursor on the selected line (Figure 7.5) and press ENTER to go to the screen with detail item information (Figure 7.8)	Three motor actions: (R30B+G1A)+(M5B+AP1+RL1) +(R30B+AP1+R30E)	$2.2 \times 1.2 = 2.64$
$O^{\alpha\mu}_{21}$	**Compare received quantity with PO (purchase order) quantity (Figure 7.8, field 4**	Combination of two simultaneous perceptual actions - $2 \times (ET + EF)$ with simultaneously performed mnemonic operations (MO)	$(0.42 + 0.4) \times 2 = 1.64$

(Continued)

TABLE 7.5　(CONTINUED)

Algorithmic Description of the Time Structure of Performance of Inventory Receiving Task (Version 2 before Improvement – Item Is Put Aside due to Price Rejection; $P = 0.03$)[a]

Members of Algorithm (Psychological Units of Analysis)	Description of Elements of Task (Technological Units of Analysis) Beginning of Individual Item Processing Version 2 of the task	Description of Elements of Activity (Psychological Units of Analysis)	Time (sec)
$\overset{5}{I_5}\uparrow$	If received quantity and ordered quantity are the same, go to O_{29}^{ε} ($P=0.9$). If received quantity is greater or less than ordered quantity, go to O_{22}^{ε}. ($P=0.1$)	Decision-making actions that are performed based on visual information	0.4
O_{22}^{ε}	Type the received quantity and press ENTER to get a question at the bottom of the screen ($P=0.1$)	Motor action (R20B + AP1) + (R12B + AP1) (example with two-digit number)	$(1.39 \times 1.2) \times 0.1 = b\ 0.17$
O_{23}^{α}	Read the statement: THE RECEIVED QUANTITY AND ORDERED QUANTITY DO NOT MATCH. DO YOU ACCEPT? (YES/NO). ($P=0.1$). Scan and read $\approx$ four words	Successive perceptual action $ET + 4 \times EF$	$(0.42 + 4 \times 0.18) = 1.14 \times 0.1 = 0.11$
$**O_{24}^{\mu th}$	Recall instructions and perform required calculation and estimation (relationship between quantity and price). $P=0.1$	Combination of mnemonic action (retrieve simple information from memory) and logical thinking action	$\approx (1.2 + 3) = 4.2 \times 0.1 = 0.42$

(Continued)

TABLE 7.5 (CONTINUED)

Algorithmic Description of the Time Structure of Performance of Inventory Receiving Task (Version 2 before Improvement – Item Is Put Aside due to Price Rejection; $P = 0.03$)[a]

Members of Algorithm (Psychological Units of Analysis)	Description of Elements of Task (Technological Units of Analysis) Beginning of Individual Item Processing Version 2 of the task	Description of Elements of Activity (Psychological Units of Analysis)	Time (sec)
$I_6 \uparrow^{6}$	If quantity is not accepted (computer defaults to 'N') go to O^ε_{25}. (This output is performed in other version of the task). Otherwise, go to O^ε_{28}. In this version of the task considered, output is always performed with ($P=0.1$)	Decision-making action that includes simple syllogistic conclusion	$1.5 \times 0.1 = 0.15$
$O^\varepsilon_{25}; O^\varepsilon_{26}; O^\varepsilon_{27}$ and $I_7;$	These members of algorithm are performed in Version 1 only	—	—
$\overset{6}{\rightarrow} O^\varepsilon_{28}$	Change "N" to "Y" (quantity is accepted) and go to O^ε_{29} (O^ε_{28} is performed with probability $P=0.1$, see event tree in Figure 7.9)	One motor action: (R60B+AP1)+)	$(1.05 \times 1.2) \times$ $0.1 = 1.261 \times 0.1 = 0.126$
$\overset{5}{\rightarrow} O^\alpha_{29}$	Press ENTER and go to O^α_{30}. The member of the algorithm O^α_{29} is performed with $P=0.1$ after O^ε_{28} and with probability $P=0.9$ after I_7 (the accumulated probability of O^α_{29} is $P=1$)	One motor action (R40 B+AP1)	$(0.84 \times 1.2) = 1.01$

(Continued)

TABLE 7.5 (CONTINUED)

Algorithmic Description of the Time Structure of Performance of Inventory Receiving Task (Version 2 before Improvement – Item Is Put Aside due to Price Rejection; $P = 0.03$)[a]

Members of Algorithm (Psychological Units of Analysis)	Description of Elements of Task (Technological Units of Analysis) Beginning of Individual Item Processing Version 2 of the task	Description of Elements of Activity (Psychological Units of Analysis)	Time (sec)
*** O_{30}^{α}	Compare price of the item on the packing slip with price on the screen. (O_{26}^{α} is performed with probability 1)	Combination of three simultaneous perceptual actions $-2 \times (ET + EF) + (ET + EF)$ (includes one double check perceptual action)	$2 \times (0.42 + 0.4) +$ $0.42 + 0.3) = 2.36$
$l_8 \uparrow^8$	If price is the same, go to O_{38}^{ε}. (This output is performed in Version 3 of the task). If price is different go to O_{31}^{thu}. In this version of the task, considered output is performed with probability $P = 1$	Simple decision-making action	0.4
O_{31}^{thu}	Mentally calculate the *price difference* and store it in working memory ($P = 1$)	Simple arithmetic calculation and mnemonic operations	$\approx (0.6 + 3) = 3.6$
$O_{32}^{\varepsilon\mu}$	Key in new price and press ENTER and continue to store *price difference* in working memory ($P = 1$)	Four motor actions and mnemonic operations $(R20B + AP1) + 2 \times (R6B + AP1) + (R12B + AP1)$	$(2.48 \times 1.2) = 2.976$

(Continued)

TABLE 7.5 (CONTINUED)

Algorithmic Description of the Time Structure of Performance of Inventory Receiving Task (Version 2 before Improvement – Item 1s Put Aside due to Price Rejection; $P = 0.03$)[a]

Members of Algorithm (Psychological Units of Analysis)	Description of Elements of Task (Technological Units of Analysis) — Beginning of Individual Item Processing Version 2 of the task	Description of Elements of Activity (Psychological Units of Analysis)	Time (sec)
$O_{33}^{\alpha\mu}$	Read question "Price is not the same. Do you accept the price?" and continue to store *price difference* in working memory ($P=1$)	Successive perceptual action. $ET + 4 \times EF$ and mnemonic operation of keeping price difference in working memory	$\approx (0.42 + 4 \times 0.18) = 1.14$
$\overset{\mu\ 9}{l_9}\uparrow$	If price is less or if the price is greater but the difference is $\leq 10\%$, go to O_{34}^{ε}. This output is performed in Version 3 of the task. If price is greater and the difference is $>10\%$, go to O_{35}^{ε} ($P=1$)	Complex decision – making actions from three alternatives that requires actualization of information in memory about rules of decision and utilizing information in memory about price difference (thinking operations)	$\approx (2 + 2) = 4$
O_{34}^{ε}	Type "Y". This member of algorithm is performed in Version 3 of the task ($P=1$)	——	——
$\overset{9}{\to} O_{35}^{\varepsilon}$	Type "N" ($P=1$)	Motor action: (R30B + AP1)	$0.74 \times 1.2 = 0.9$

(Continued)

TABLE 7.5 (CONTINUED)

Algorithmic Description of the Time Structure of Performance of Inventory Receiving Task (Version 2 before Improvement – Item Is Put Aside due to Price Rejection; $P = 0.03$)[a]

Members of Algorithm (Psychological Units of Analysis)	Description of Elements of Task (Technological Units of Analysis) Beginning of Individual Item Processing Version 2 of the task	Description of Elements of Activity (Psychological Units of Analysis)	Time (sec)
O_{36}^ε	Put rejected item in the Put-Aside Area, Figure 7.2, area 10. Return to the base unit 5. O_{36}^ε includes: (1) the left-hand grasps the item; (2) worker turns body and takes approximately six steps to the put-aside area; moves the left hand and releases item; (3) worker turns body and takes approximately four steps to the base unit 5 for unpacking. ($P = 1$)	Motor actions, including waking: (R6B + G1B) + (TBC1 + 6WP + TBC2) + (R40B + RL1) + (TBC1 + 4WP + TBC2)	$10.4 \times 1.2 = 12.5$
Total performance time of the second version of the task	**Task of receiving an item ends here**	$T_2 = T_{cog} + T_{ex}$	$20.52 + 26.64 = 47.16$
O_{37}^α	Check if there are other items in the box to receive ($P = 1$)	Simultaneous perceptual action	0.3
l_{10} ↑ 10 (1–2)	If there are no more items in the box, go to O_4^ε, otherwise, go to O_{17}^ε ($P = 1$)	Simple decision-making	0.3
Performance time O_{37}^α and l_{10}	Time is not included in task performance		0.6

[a] The second version of the task is performed with probability 0.03 (see event tree in Figure 7.13).

performed with probability $P=0.1$. It is followed by O_{29}^{ϵ} and there are no logical conditions that can change this probability between them. Hence, member of algorithm O_{29}^{ϵ} is performed with the same probability, $P=0.1$. The first output from l_5 is when quantity and ordered quantity are the same, go to O_{29}^{ϵ} $(P=0.9)$. This member of the algorithm O_{29}^{α} has probability $P=1$, but it is an accumulated probability because it is performed either when received and ordered item quantity are the same $(P=0.9)$ or when they are not the same but the receiver decided to process the item further anyway $(P=0.1)$. Thus, when it is necessary to evaluate the performance time of the members of algorithm depicting this process, it is necessary to multiply the calculated time by the probability $P=0.1$. The third stage of task performance is totally new and includes the evaluation of the item price, which requires additional time for its performance.

Let us perform the qualitative analysis of task performance that involves price evaluation.

The price is different relatively seldom and has the following outputs:

The price is less than the ordered price;

The price is greater, but the difference is less or equals 10%.

The price is greater, and the difference is greater than 10%.

The receiver should apply the rule:

If the price is the same or less, or greater but the difference is $\leq 10\%$, an item can be processed.

If the price is greater and the difference is greater than 10% ($>10\%$), an item should be rejected.

For example, if an item's ordered price is $10, the following outputs are possible:

The price of the item is exactly $10 and it is processed.

The price of the item is $9, which also results in an item being processed.

The price of the item is $10.50, and although the price is now greater, the difference is $<10\%$, and an item should be processed.

The price of the item is now $11, and this means that the price has been raised 10%, and according to the rule, the item can still be processed.

The price rose to $12, and this is unacceptable according to the rules, and the item should be rejected.

This qualitative analysis is used in our further discussion of the price evaluation. We need to determine the probability of Version 2, when a receiver

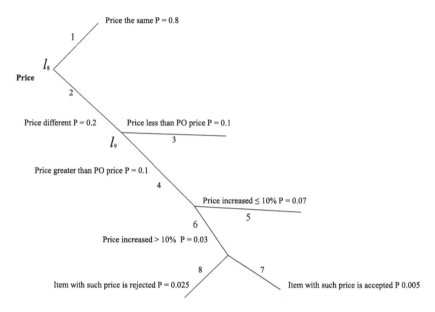

FIGURE 7.13
Event tree of the items price estimation.

puts the rejected item in the put-aside area based on the *price* criterion. Only after this event is the second version of the task completed. The probability of the considered events correlates with the probability of the second version of the task. After its completion, a receiver should take another item and start the whole task again.

Based on qualitative analysis of the price evaluation, its algorithmic description is presented in Table 7.5. This is the final stage of Version 2 (see Table 7.5). For determining probabilistic characteristics of this stage, the event tree has been developed (see Figure 7.13). It has multiple branches. If the item is accepted based on the price criterion (Branches 1, 3, 5, and 7 in Figure 7.13), the receiver moves to assigning the bin location. This scenario belongs to Version 3, which will be considered in the next chapter. Branches 2, 4, 6, and 8 show the scenario when the item is rejected, and Version 2 is completed.

Let us consider in greater detail this event tree model. It has two logical conditions, l_8 and l_9. Branches 1, 3, 5, and 7 show the scenarios when price can be the same, less, or greater when the difference is $\leq 10\%$ or in rare cases as an exception accepted even when the difference is $>10\%$. In all such situations, an item is accepted. The probability of various branches depends on decisions or logical conditions' outcomes. The probabilities of these branches have been determined by the subject matter experts (SME). They assign probabilities for all decision-making outputs by utilizing the scale of subjective probability evaluation of events (see Section 8.3). It should be noted that the assessment of subjective probability has several stages. At the first stage,

SMEs select the diapason of the subjective scale of probability in which the considered event can fall, and at the second stage they define if the probability of the considered event is close to the top, middle, or the bottom of this diapason.

Let us consider the output of l_8 when the received and ordered price are the same. In most cases, the price should be the same because an item has been ordered from the vendor with a specifically defined price. According to the SMEs, output when price is the same is the most frequent situation. Hence, at the first step the SME decides that this event should fall into the diapason or interval "rather frequent event with probability $P=0.7–0.89$." At the second stage, experts defined the probability of the considered event to be in the middle of this interval. Therefore, the probability of the event when price is the same is $P=0.8$. Presented in the Figure 7.13, the event tree demonstrates this output of logical condition l_8 (Branch 1).

Let us consider a scenario when the received price is different from the ordered price. The price can be different in two ways. It can be greater than the PO price or less than the PO price. Thus, at the second stage, SMEs define the probability of outputs when the price is not the same. The situation when the price is different is a relatively infrequent event, with $P=0.2$ (Branch 2).

At the next step, SMEs have to determine the probability of outputs of the decision l_9 when the price is different. This decision has two outputs: (1) price less or greater than ordered price but the difference is $\leq 10\%$, and item is processed; (2) price is increased and the difference is $>10\%$.

Branches 3 and 4 derived according to SMEs have the same probabilities, $P=0.1$. Branch 4 also has two branches, 5 and 6, derived from it that have the combined probability $P=0.1$. The diapason with probabilities $P=0.01–0.1$ is called "very seldom event diapason". According to SMEs, considered events don't have similar probabilities. The occasion when the price is close to the PO price has greater probability than the branch where the price is significantly greater than the PO price. The considered diapason has three subareas, $P_1=0.07–0.1$; $P_2=0.04–0.06$; and $P_3=0.01–0.03$. The SMEs selected the lower probability of the first subarea with $P_1=0.07$ (Branch 5).

Branch 6 with probability $P=0.03$ has two outputs. SMEs stated that for events when the price increase is $>10\%$, one branch with some approximation should have $P=0.025$ and another $P=0.005$. Branch 7 is not reflected in the algorithm due to the extremely low probability of this event. This last event is not relevant for Version 2. The obtained probabilities allow determining the probability of various members of the algorithm, the time of their performance, and the time of performance of Version 2 and the probability of its occurrence. It should be stressed that in Version 2 the price of the received item is never the same as the ordered price. So in Version 2, members of the algorithm related to Branch 2 are performed with probability $P=1$.

The event tree for the price evaluation includes logical conditions l_8 and l_9, which are performed in both Version 2 and Version 3 with different probabilities.

From Table 7.5 it can be seen that the evaluation of price starts from the member of algorithm O^{α}_{30}, after which a receiver makes a decision which is presented as logical condition l_8. In Version 2, this logical condition has only one output when the price is different. The probability of this output is 1. All members of the algorithm that follow this logical condition have the same probability, $P = 1$. This probability can be changed only after logical condition l_9. However, this logical condition in Version 2 also has only one output with probability $P = 1$. Hence all members of the algorithm after this logical conditions also have probability $P = 1$. Therefore, probability 0.1 is not utilized, when the performance time of separate members of the algorithm that depict evaluation of an item price is calculated for Version 1.

The performance of members of algorithm $O^{thμ}_{31}$, $O^{εμ}_{32}$, $O^{αμ}_{33}$, and $l^{μ}_9$ requires mental calculation of the price difference and keeping this information in working memory, which presents additional mental workload at the final stage of Version 2. Let's consider members of the algorithm associated with the price evaluation (see Figure 7.13).

The logical condition l_8 depicts the decision-making at the first stage of price evaluation. If the price of the received item matches the PO price, which happens with probability $P = 0.8$, the next member of the algorithm to perform is $O^{ε}_{38}$ (performed in Version 3). $O^{thμ}_{31}$ is performed otherwise, which means that the operator keys in the new price and presses ENTER (see Table 7.5). The logical condition l_9 describes how the price difference is evaluated. If the price has decreased or the increase is less or equal to 10% ($P = 0.17$), then the item is still accepted (see $O^{ε}_{34}$). Even if the price increase is over 10%, the items might still be accepted in rare cases for special business reasons ($P = 0.005$). Such low-probability events are not considered in our algorithm. All members of the algorithm that described the scenario when the item price is accepted are performed only in Version 3. The item is rejected based on the price criteria with probability $P = 0.025$, but in Version 2, $O^{ε}_{35}$ and $O^{ε}_{36}$ are performed with probability $P = 1$ because in Version 2 the item is always rejected due to the price being too high.

Let us now consider the second version of the task after innovation. The first member of the algorithm after improvement is $O^{ε}_{11}$. Its content is the same as that of $O^{ε}_{17}$ in Table 7.5 before improvement because the numbering scheme depends on the subtask 0. After improvement, the scanning device is used, which does not just reduce the number of members of the algorithm but also makes performance subtask 0 easier.

In Version 2 after improvement, the first stage is involved in matching the physical item number with the number on the screen. Version 2 has the same enhancement for this stage as Version 1, which is to scan the item number instead of keying it in. In Version 2 before improvement, a receiver performs $O^{αμ}_{18}$ (see Table 7.5). After improvement, a receiver uses a scanning device and performs $O^{ε}_{12}$ which involves simple motor actions. Complex cognitive actions of $O^{αμ}_{18}$ are now avoided at this stage of performance. Before improvement, after $O_{18}^{\alpha\mu}$ a receiver performed l_4; O_{19}^{α} and O_{20}^{α}. Moreover, O_{20}^{α} (before improvement) requires more motor actions than is needed for O_{13}^{α} after improvement in spite of the fact that O_{13}^{ε} has the same purpose in task performance. Thus, the first stage of Version 2 of the task after improvement has not only fewer members of the algorithm but they are simpler and the cognitive components of activity are significantly reduced.

The following stage of the task before and after improvement is involved in comparison of the received quantity with the PO (purchase order) quantity. It starts with $O_{21}^{\alpha\mu}$ (before improvement) and $O_{14}^{\alpha\mu}$ (after improvement). After that, a receiver performs l_5 (before improvement) and l_4 after improvement. A comparison of members of the algorithm $O_{14}^{\alpha\mu}$ - O_{22}^{ε} after improvement and $O_{21}^{\alpha\mu}$ - O_{29}^{ε} before improvement shows that they are identical.

In the left column of Table 7.5 there are O_{18}^{ε} - O_{20}^{α}; and l_5 after l_4. In the second column for these members of the algorithm, there is the statement "these members of the algorithm are performed in Version 1 of the task." All these members are presented here to keep the same numbering scheme for all three versions of the algorithm. The next member of the algorithm that is performed in Version 2 is O_{21}^{ε} and the next logical condition is l_6.

At the next stage of this version a receiver compares the price of the item with the order price. This part of the task after improvement starts with O_{23}^{α} and with O_{30}^{α} before improvement. Before improvement, the price difference is calculated by the receiver. This requires performing mental calculations and keeping information in working memory (see members of algorithm $O_{31}^{th\mu}$, $O_{32}^{\varepsilon\mu}$, $O_{33}^{\alpha\mu}$ and l_9^{μ}). In the after improvement version, the price difference is calculated automatically by the computer system, and if the difference is >10%, the warning "Price increase is >10%. Do you want to proceed?" is produced on the screen. And the cursor is placed at the "N" default. The logical condition l_9^{μ} is also more complex than l_7 after improvement because the first one requires not only actualization of information about rules of price evaluation but also keeping information about the price difference in working memory and performing deductive reasoning. As a result, the performance time of l_9^{μ} before improvement is 4 sec, and the performance time of l_7 after improvement is 3.5 sec. The last two members of the algorithm before and after improvement are similar. Comparison of the algorithms of Version 2 of the task performance before and after improvement shows that the later version significantly reduces cognitive effort and the task takes less time to perform.

The total performance time of the second version of the task after improvement is $T_2 = 37.25$ sec (see Table 7.6).

7.2.4 Morphological Analysis of Version 3 of Task Performance

Version 3 of task performance is the most probable one. In the previously described first two versions, the item has been rejected due to either inadequate quantity or price. In this version, the item is accepted and processed

TABLE 7.6

Algorithmic Description of Version 2 of the Inventory Receiving Task after Improvement (Item is Put Aside Due to Price Rejection; $P = 0.03$)

Member of Algorithm (Psychological Units of Analysis Described in Consolidated Manner)	Description of Elements of Tasks (Technological Units of Analysis)	Description of Elements of Activity (Psychological Units of Analysis Described in Detailed Manner)	Time (sec)
$\begin{smallmatrix}8\,(2)\,5\,(2)\\ \downarrow\downarrow\end{smallmatrix}\ O_{11}^{\varepsilon}$	Take an item out of the box, and return to computer area again (while a worker takes several steps his/her right hand releases the item and worker holds the item only by left hand)	Motor actions – simultaneously by two hand reach and grasp item (1) (R60B + G1B); (2) – take out item from box – (M40B + RL1); (3) perform again four steps to computer area #3 (TBC1 + TBC2 + WP + WP)	$4.74 \times 1.2 = 5.7$
O_{12}^{ε}	Take the barcode scanner and scan the item number. (The matching item is highlighted)	(R50B + G1A) + (M40B + mM10C + AP1) + (M50B + RL1)	$2.44 \times 1.2 = 2.9$
O_{13}^{ε}	Press ENTER to go to the screen with detail item information (Figure 7.8)	Motor action (R26B + AP1)	$0.76 \times 1.2 = 0.9$
$O_{14}^{\alpha\mu}$	**Compare received quantity with PO (purchase order) quantity (Figure 7.8, field 4)**	Combination of two simultaneous perceptual actions $-2 \times (ET + EF)$ with simultaneously performed mnemonic operations (MO)	$(0.42 + 0.4) \times 2 = 1.64$
$\begin{smallmatrix}3\\ \uparrow\end{smallmatrix}\ l_3$	If received quantity and ordered quantity are the same, go to O_{22}^{ε} ($P = 0.9$). Otherwise, go to O_{15}^{ε} ($P = 0.1$)	Decision-making actions that are performed based on visual information	0.4
O_{15}^{ε}	Type the received quantity and press ENTER to get a question at the bottom of the screen ($P = 0.1$)	Motor action (R20B + AP1) + (R12B + AP1) (example with two-digit number)	$1.39 \times 1.2 \times 0.1 = 0.17$

(Continued)

TABLE 7.6 (CONTINUED)

Algorithmic Description of Version 2 of the Inventory Receiving Task after Improvement (Item is Put Aside Due to Price Rejection; $P = 0.03$)

Member of Algorithm (Psychological Units of Analysis Described in Consolidated Manner)	Description of Elements of Tasks (Technological Units of Analysis)	Description of Elements of Activity (Psychological Units of Analysis Described in Detailed Manner)	Time (sec)
O_{16}^{α}	Read the statement: THE RECEIVED QUANTITY AND ORDERED QUANTITY DO NOT MATCH. DO YOU ACCEPT? (YES/NO). ($P=0.1$). Scan and read $\approx$ four words	Successive perceptual action. $ET + 4 \times EF$	$(0.42 + 4 \times 0.18) = 1.14 \times 0.1 = 0.11$
* $O_{17}^{\mu th}$	Recall instructions and perform required calculation and estimation (relationship between quantity and price). $P = 0.1$	Combination of mnemonic action (retrieve simple information from memory) and logical thinking action	$\approx (1.2 + 3) = 4.2 \times 0.1 = 0.42$
$l_4 \uparrow$	If quantity is not accepted (computer defaults to 'N'), go to $O_{18}^{\varepsilon} \cdot$ ($P=0$). Otherwise, go to O_{21}^{ε} ($P=0.1$)	Decision-making action that includes simple syllogistic conclusion	$1.5 \times 0.1 = 0.15$
$O_{18}^{\varepsilon}; O_{19}^{\varepsilon};$ $O_{20}^{\alpha}; l_5$	These members of algorithm are performed in Version 1 of the task	————	————
$\overset{4}{\longrightarrow} O_{21}^{\varepsilon}$	Change "N" to "Y", quantity is accepted. ($P=0.1$)	One motor action: (R60B + AP1)	$(1.05 \times 1.2) \times 0.1 = 0.126$
$\overset{3}{\longrightarrow} O_{22}^{\varepsilon}$	Press ENTER ($P=1$)	One motor action (R40 B + AP1)	$0.84 \times 1.2 = 1.01$
** O_{23}^{α}	**Compare price of the item on the packing slip with price on the screen.** (Starting from O_{23}^{α} till l_8 all members of algorithm have probability $P=1$)	Combination of two simultaneous perceptual actions including one double check perceptual action– $2 \times (ET + EF) + (ET + EF)$ (includes one double check perceptual action)	$2 \times (0.42 + 0.4) + (0.42 + 0.3) = 2.36$

(Continued)

TABLE 7.6 (CONTINUED)

Algorithmic Description of Version 2 of the Inventory Receiving Task after Improvement (Item is Put Aside Due to Price Rejection; $P = 0.03$)

Member of Algorithm (Psychological Units of Analysis Described in Consolidated Manner)	Description of Elements of Tasks (Technological Units of Analysis)	Description of Elements of Activity (Psychological Units of Analysis Described in Detailed Manner)	Time (sec)
$l_6 \uparrow$	If the price on the screen and packing slip are different, go to O_{24}^ε ($P = 0$). If price is the same go to O_{30}^ε ($P = 1$)	Simple decision-making action	0.4
O_{24}^ε	Key in the new price and press ENTER ($P = 1$)	Four motor actions and mnemonic operations $(R20B + AP1) + 2 \times (R6B + AP1) + (R12B + AP1)$	$(2.48 \times 1.2) = 2.98$
O_{25}^α	Look at information on the screen (Cursor moves to the next field if price is accepted. Otherwise, there is a message on the screen "Price increase is >10%". Do you wish to proceed?) $P = 1$. Scan and read ≈ four words	Successive perceptual action. $ET + 4 \times EF$	$(0.42 + 4 \times 0.18) = 1.14$
$l_7 \uparrow$ $7(1-3)$	If difference >10% go to O_{26}^ε ($P = 1$). However if there is a special reason to accept go to O_{29}^ε. If there were no message and cursor moved to the next field go to O_{30}^ε (see the third version of task)	Decision-making action from three alternatives that includes simple syllogistic conclusion and requires actualization of information in memory.	$\approx (2 + 1.5) = 3.5$
$7(1)$ $\downarrow O_{26}^\varepsilon$	Press ENTER to accept system default ("N"). $P = 1$	Motor action $(R26B + AP1)$	$(0.7 \times 1.2) = 0.84$

(Continued)

TABLE 7.6 (CONTINUED)

Algorithmic Description of Version 2 of the Inventory Receiving Task after Improvement (Item is Put Aside Due to Price Rejection; $P = 0.03$)

Member of Algorithm (Psychological Units of Analysis Described in Consolidated Manner)	Description of Elements of Tasks (Technological Units of Analysis)	Description of Elements of Activity (Psychological Units of Analysis Described in Detailed Manner)	Time (sec)
O_{27}^{ε}	Put rejected item in the Put-Aside Area, Figure 7.2, area 10. Return to the base unit 5. O_{27}^{ε} includes: (1) the left hand grasp the item; (2) worker turn body and make approximately 6 steps to the put-aside area; move the left hand and release item; (3) worker turn body and make approximately 4 steps to the base unit 5 ($P = 0.025$)	One motor action by hand and leg movements: (R6B + G1B) + (TBC1 + 6WP + TBC2) + (R40B + RL1) + (TBC1 + 4WP + TBC2)	$(10.4 \times 1.2) = 12.5$
Total performance time of the second version of the task	**Task of receiving an item ends here**	$T_2 = T_{cog} + T_{ex}$	$10.12 + 27.13 = 37.25$
*** O_{28}^{α}	Check to see if there are other items in the box to receive	Simultaneous perceptual action	0.3
l_8 (1–2) ↑	If there are no more items in the box, go to O_4^{ε}, otherwise, press F3 to return to previous screen (Figure 7.5) and go to O_{11}^{ε}	Simple decision-making	0.3
Performance time O_{28}^{α} and l_8	This time is not included in task performance	Not included in content of task	

Total performance time of the second version of task after improvement is $T_2 = 37.25$ sec.

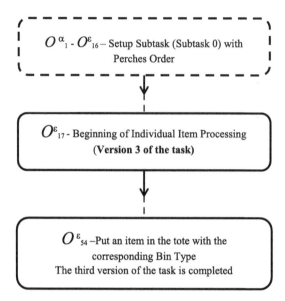

FIGURE 7.14
Setup task and the third version of the task performed with probability $P = 0.905$.

all the way through to be put in the corresponding tote either to be put away in the warehouse or to be delivered to the department that has put it on order. The latest (delivery to the workshop) is available only in the improved version of the task. The stages of Version 3 of the task performance are presented in Figure 7.14. Members of the algorithm O_1^{α} - O_{16}^{ε} in the first box in Figure 7.14 belong to Subtask 0 and take place only when all the items in the box are processed and a receiver has to get a new box and put it on base unit 5 (Figure 7.2). After that, a receiver opens the box, takes the packing slip out, returns to the computer and keys in the PO number from the packing slip. After that, the third version of the task can be performed. The first box is indicated by a dashed line because subtask 0 is performed not always, but only in a particular situation.

Figure 7.14 shows that the third version of the task starts with the same motor actions as Versions 1 and 2. They are depicted by O_{17}^{ε}. Version 3 is completed after O_{54}^{ε} is performed.

The algorithmic description of this version before improvement (see Table 7.7) shows that members of the algorithm from O_{17}^{ε} to l_9 are the same as in Version 2, but in Version 2 the item price is unacceptable and the item is rejected, and in this version the price is accepted and the receiver continues to process the item.

Version 3 in the beginning has the same stages as Version 2 of the task. A receiver takes an item out of the box and returns to the computer (see $O_{18}^{\alpha\mu}$). This stage of the task is completed when a receiver activates the screen with detailed item information (see O_{20}^{ε}). The second stage of this version involves

TABLE 7.7

Algorithmic Description of Activity and Its Time Structure (Morphological Analysis) during Computerized Task Performance (Version 3 of the Task before Improvement – Item is Accepted and Put in the Tote; $P = 0.9$)

Members of Algorithm (Psychological Units of Analysis)	Description of Elements of task (Technological Units of Analysis) Beginning of the Individual Item processing Version 3 of the task.	Description of Elements of Activity (Psychological Units of Analysis)	Time (sec)
$\xrightarrow{17\,(2)\,11\,(2)\,7\,(2)}\rightarrow\downarrow O^{\varepsilon}_{17}$	Take an item out of the box and return to computer area again (while a worker takes several steps, his/her right hand releases the item and worker holds the item only by left hand)	Motor actions - simultaneously by two hand reach and grasp item (R60B + G1B); (3) – take out item from box – (M40B + RL1); (4) perform again four steps to computer area #3 (TBC1 + TBC2 + WP + WP)	$4.74 \times 1.2 = 5.7$
$O^{\alpha\mu}_{18}$	Look at item number on the physical item and compare it with item numbers on the screen (Figure 7.5, field 3)	At least five simultaneous perceptual actions (ET + EF); combined with simple mnemonic and scanning operations	Average time ≈ 6
$\overset{4}{l_4}\uparrow$	If item number is on the first page, go to O^{ε}_{20}. If item number is not on the first page, go to O^{ε}_{19}	Decision-making action at sensory-perceptual level	0.3
O^{ε}_{19}	Press arrow key (repeat if required)	Motor action (R10B + AP1)	$0.52 \times 1.2 = 0.62$
$\overset{4}{\downarrow} O^{\varepsilon}_{20}$	Put cursor on the selected line (Figure 7.5) and press ENTER to go to the screen with detail item information (Figure 7.8)	Three motor actions: (R30B + G1A) + (M5B + AP1 + RL1) + (R30B + AP1 + R30E)	$2.2 \times 1.2 = 2.64$
$O^{\alpha\mu}_{21}$	**Compare received quantity with PO (purchase order) quantity (Figure 7.8, field 4)**	Combination of two simultaneous perceptual actions – $2 \times (ET + EF)$ with simultaneously performed mnemonic operations (MO)	$(0.42 + 0.4) \times 2 = 1.64$

(Continued)

TABLE 7.7 (CONTINUED)

Algorithmic Description of Activity and Its Time Structure (Morphological Analysis) during Computerized Task Performance (Version 3 of the Task before Improvement – Item is Accepted and Put in the Tote; $P = 0.9$)

Members of Algorithm (Psychological Units of Analysis)	Description of Elements of task (Technological Units of Analysis) Beginning of the Individual Item processing Version 3 of the task.	Description of Elements of Activity (Psychological Units of Analysis)	Time (sec)
$I_5^5 \uparrow$	If received quantity and ordered quantity are the same, go to O_{29}^ε ($P = 0.9$). If received quantity is greater or less than ordered quantity, go to O_{22}^ε. ($P = 0.1$)	Decision-making actions that are performed based on visual information	0.4
O_{22}^ε	Type the received quantity and press ENTER to get a question at the bottom of the screen ($P = 0.1$)	Motor action (R20B + AP1) + (R12B + AP1) (example with two-digit number)	$(1.39 \times 1.2) \times 0.1 = 0.17$
O_{23}^α	Read the statement: THE RECEIVED QUANTITY AND ORDERED QUANTITY DO NOT MATCH. DO YOU ACCEPT? (YES/NO). ($P = 0.1$). Scan and read $\approx$ four words	Successive perceptual action $ET + 4 \times EF$	$(0.42 + 4 \times 0.18) = 1.14 \times 0.1 = 0.11$
** $O_{24}^{\mu th}$	Recall instructions and perform required calculation and estimation (relationship between quantity and price). $P = 0.1$	Combination of mnemonic action (retrieve simple information from memory) and logical thinking action	$\approx (1.2 + 3) \times 0.1 = 4.2 \times 0.1 = 0.42$
$I_6^6 \uparrow$	If quantity is not accepted (computer defaults to 'N'), go to O_{25}^ε. (This output is performed in Version 1 of the task). Otherwise, go to O_{28}^ε In this version of the task, considered output is always performed with $P = 0.1$)	Complex decision-making actions that include mnemonic and thinking operations	$1.5 \times 0.1 = 0.15$

(Continued)

TABLE 7.7 (CONTINUED)

Algorithmic Description of Activity and Its Time Structure (Morphological Analysis) during Computerized Task Performance (Version 3 of the Task before Improvement – Item is Accepted and Put in the Tote; $P = 0.9$)

Members of Algorithm (Psychological Units of Analysis)	Description of Elements of task (Technological Units of Analysis) Beginning of the Individual Item processing Version 3 of the task.	Description of Elements of Activity (Psychological Units of Analysis)	Time (sec)
O_{25}^{ε}; O_{26}^{ε}; O_{27}^{ε}; l_{7};	These members of algorithm are performed in Version 1 of the task	—	—
$\overset{6}{\underset{\downarrow}{}} O_{28}^{\varepsilon}$	Change "N" to "Y" (quantity is accepted) and go to O_{29}^{ε} (O_{28}^{ε} is performed with probability $P = 0.1$, see event tree in Figure 7.9)	One motor action: (R60B+AP1)+)	$(1.05 \times 1.2) \times$ $0.1 = 1.26 \times 0.1 = 0.126$
$\overset{5}{\underset{\downarrow}{}} O_{29}^{\varepsilon}$	Press ENTER and go to O_{30}^{α}. After O_{28}^{ε} member of algorithm O_{29}^{ε} is performed with $P = 0.1$ and after l_{7} with probability $P = 0.9$ (the accumulated probability of O_{29}^{α} is $P = 1$)	One motor action (R40 B+AP1)	$0.84 \times 1.2 = 1.01$
*** O_{30}^{α}	**Compare price of the item on the packing slip with price on the screen**	Combination of three simultaneous perceptual actions – $2 \times (ET + EF) + (ET + EF)$ (includes one double check perceptual action)	$2 \times (0.42 + 0.4) + (0.42 + 0.3)$ $= 2.36$
$\overset{8}{\underset{l_8}{\uparrow}}$	If price is the same, go to go to O_{38}^{ε}; ($P = 0.8$) If price is different, go to O_{31}^{thu} ; ($P = 0.2$)	Simple decision-making action	0.4
O_{31}^{thu}	Mentally calculate the *price difference* and store it in working memory ($P = 0.2$)	Simple arithmetic calculation and mnemonic operations	$\approx (0.6 + 3) \times 0.2 = 3.6 \times 0.2 = 0.72$
$O_{32}^{\varepsilon\mu}$	Key in new price and press ENTER and continue to store *price difference* in working memory ($P = 0.2$)	Four motor actions and mnemonic operations (R20B+AP1) + $2 \times$ (R6B+AP1) + (R12B+AP1)	$(2.48 \times 1.2) \times 0.2 = 0.59$

(Continued)

TABLE 7.7 (CONTINUED)

Algorithmic Description of Activity and Its Time Structure (Morphological Analysis) during Computerized Task Performance (Version 3 of the Task before Improvement – Item is Accepted and Put in the Tote; $P = 0.9$)

Members of Algorithm (Psychological Units of Analysis)	Description of Elements of task (Technological Units of Analysis) Beginning of the Individual Item processing Version 3 of the task.	Description of Elements of Activity (Psychological Units of Analysis)	Time (sec)
$O_{33}^{\alpha\mu}$	Read question, "Price is not the same. Do you accept the price?" and continue to store *price difference* in working memory. ($P=0.2$)	Successive perceptual action. ET $+ 4 \times$ EF and mnemonic operation of keeping price difference in working memory	$\approx (0.42 + 4 \times 0.18) \times 0.2$ $= 1.14 \times 0.2 = 0.23$
$^{9} l_9^{\mu}\uparrow$	If price is less or if the price is greater but the difference is $\leq 10\%$), go to O_{34}^{ε} ($P=0.2$). If price is greater and the difference is $>10\%$, go to O_{35}^{ε} (performed in version 2)	Complex decision-making actions from three alternatives that requires actualization of information in memory and includes thinking operations	$\approx (2+2) \times 0.2 = 4 \times 0.2 = 0.8$
O_{34}^{ε}	Type "Y" ($P=0.2$)	Motor action: (R40B + AP1)	$(0.84 \times 1.2) \times 0.2 = 1.01 \times 0.2 = 0.2$
O_{35}^{ε}; O_{36}^{ε}; $_a O_{37}^{\varepsilon}$; l_{1v}	These members of algorithm are performed in Version 2 of the task	————	— —
$^{8} \to O_{38}^{\varepsilon}$	Press ENTER to go to Completion Flag field (Figure 7.8, field 6)	Motor action: (R30B + AP1). *Hand movement from home position to the ENTER key is 0.8 and accepted as a distance of hand movement*	$0.7 \times 1.2 = 0.89$
$*** O_{39}^{\alpha}$	Check system default "Y" or "N" flag on the screen	Simultaneous perceptual action (ET + EF)	$0.42 + 0.3 = 0.72$
$^{11} l_{11}^{\mu}\uparrow$	If system default is accepted, go to O_{41r} ($P=0.95$); otherwise, go to O_{40}^{ε} ($P=0.05$). In most cases the system default is accepted)	Decision-making actions at a verbal thinking level that requires actualization of information from memory	*0.4*

(Continued)

TABLE 7.7 (CONTINUED)

Algorithmic Description of Activity and Its Time Structure (Morphological Analysis) during Computerized Task Performance (Version 3 of the Task before Improvement – Item is Accepted and Put in the Tote; $P=0.9$)

Members of Algorithm (Psychological Units of Analysis)	Description of Elements of task (Technological Units of Analysis) / Beginning of the Individual Item processing Version 3 of the task.	Description of Elements of Activity (Psychological Units of Analysis)	Time (sec)
O_{40}^{ε}	Change system default ('Y' to 'N' or 'N' to 'Y'), go O_{41}^{ε} (Probability O_{40}^{ε} $P=0.05$)	Motor action (R50B+AP1) or (R60B+AP1) Average time calculation is required.	$(0.94+1.05):2\times1.2\times0.05$ $=1.19\times0.05=0.06$
$\overset{11}{\rightarrow} O_{41}^{\varepsilon}$	Press ENTER and go O_{42}^{α}	Motor action: (R18B+AP1)	$0.72\times1.2=0.86$
O_{42}^{α}	**Check to see if the bin number is assign (see field 7, Figure 7.8)** Bin number includes bin type plus shelf's number.	Simultaneous perceptual action	0.3
$l_{12}^{12}\uparrow$	If bin number is assigned (bin type plus shelf number), go to O_{48}^{ε} ($P=0.9$). If only the bin type is assigned, go to $O_{43}^{\alpha\mu}$ ($P=0.1$)	Simple decision-making from two alternatives	0.3
$O_{43}^{\alpha\mu}$	Based on recalled bin type category (from BB0 to BB9), check the bin type ($P=0.1$)	Perceptual action combined with mnemonic action	$(1.2+0.6)\times0.1=1.8\times0.1=0.18$
$l_{13}^{\mu}\uparrow$	If you accept bin type, go to O_{48}^{ε} ($P=0.08$). If you do not accept bin type, go to $O_{44}^{\varepsilon\mu}$. ($P=0.02$)	Decision-making action at verbal thinking level with mnemonic operation	$0.6\times0.1=0.06$
$O_{44}^{\varepsilon\mu}$	Change the bin type from BB0 to BB9 by keying in its number. ($P=0.02$).	Motor action with mnemonic operations (R37B+AP1).	$0.85\times1.2\times0.02=1.02\times0.02$ $=0.02$
$\omega_2\uparrow$	Always false logical conditions	—	—

(Continued)

TABLE 7.7 (CONTINUED)

Algorithmic Description of Activity and Its Time Structure (Morphological Analysis) during Computerized Task Performance (Version 3 of the Task before Improvement – Item is Accepted and Put in the Tote; $P = 0.9$)

Members of Algorithm (Psychological Units of Analysis)	Description of Elements of task (Technological Units of Analysis) Beginning of the Individual Item processing Version 3 of the task.	Description of Elements of Activity (Psychological Units of Analysis)	Time (sec)
O_{45}^{ε}	In case of doubts about the Bin Type refer to Bin Type Chart displayed next to the monitor by taking one step to the left. ($P = 0.02$)	One side step –SSC2	$(1.23 \times 1.2) \times 0.02 = 1.48 \times 0.02 = 0.03$
O_{46}^{α}	Visually scan the Bin Type Chart and select adequate bin type ($P = 0.02$)	Complex perceptual action that includes scanning and recognition of bin type ($ET + Scanning + EF$)	$(0.42 + 0.9 + 4) \times 0.02 = 1.72 \times 0.02 = 0.03$
O_{47}^{ε}	Take a step to the right to get back to the computer. Change the bin type from BB0 to BB9 by keying in its number. ($P = 0.02$)	One side step and hand movements: SSC2 + (R30B + AP1 + AP1) + {[(R8B + R15B)/2]+AP1}	$(1.23 + 1.02 + 0.53) \times 1.2 \times 0.02 = 2.78 \times 1.2 \times 0.02 = 0.067$
$13 \; 12 \; \omega 2$ $\uparrow \downarrow \uparrow \rightarrow$ O_{48}^{ε}	**Press ENTER to print the label for corresponding bin (see place for tote 11, Figure 7.2).**	One motor actions: (R15B + AP1)	$0.58 \times 1.2 = 0.7$
O_{49}^{ε}	Peel the label off the printer and put it on the item. O_{48}^{ε} - includes: (1) side step to the right; (2) grasp and peel label with the right hand; (3) move the item with the left hand to the right and put it on the worker table; (4) move the left hand and grasp the label; (5) bring the label by two hands to the item's surface and stick it on the item; (6) smooth the label over with the right hand fist.	11887201809750(1) SSC2 + (2) (R50B + G1A + M15B) + (3) (M60B + RL1) + (4) (R20B + G1A) + (5) (M10B + AP1 + RL1) + (6) (R12B + M12B)	$5 \times 1.2 = 6$

(Continued)

TABLE 7.7 (CONTINUED)

Algorithmic Description of Activity and Its Time Structure (Morphological Analysis) during Computerized Task Performance (Version 3 of the Task before Improvement – Item is Accepted and Put in the Tote; $P = 0.9$)

Members of Algorithm (Psychological Units of Analysis)	Description of Elements of task (Technological Units of Analysis) Beginning of the Individual Item processing Version 3 of the task.	Description of Elements of Activity (Psychological Units of Analysis)	Time (sec)
O_{50}^{ε}	Take the item and go with it to the place for tote – number 11 (see Figure 7.2); (1) the left hand grasps the item; (2) worker turns his/her body and take approximately six steps to the tote area and turn body in working position	(R6B + G1B) + (TBC1 + 6WP + TBC2)	$5.58 \times 1.2 = 6.7$
O_{51}^{α}	Compare Bin type on the label with the Bin Type on the Tote (place for tote 11, see Figure 7.2)	Approximately two simultaneous perceptual actions: $(ET_1 + EF_1) + (ET_{2+} EF_2)$, $ET_1 \approx 20sm$; $ET_2 \approx 40sm$; $EF_1 = 0.3$ sec; $EF_2 = 0.6$ sec	$0.57 + 1.14 = 1.71$
$l_{14}^{14}\uparrow$	If Bin Type on the label and the Bin Type on the Tote do not match, go to O_{52}^{ε}. If Bin Type on the label and the Bin Type on the Tote match, go to O_{54}^{ε}. Combined $P = 1$	Simple decision-making	0.4
O_{52}^{ε}	Take two corrective steps in average ($P = 0.1$ for each tote out of ten totes, $P = 1$ to get to any tote)	Walk $\approx$ two steps (WP + WP)	$1.1 \times 1.2 = 1.32$
O_{53}^{α}	Compare bin type on the label with the bin type on the tote again ($P = 0.1$ for each tote out of ten, $P = 1$ for any tote)	The same as O_{51}^{α}	1.71
O_{54}^{ε}	Put an item in the tote with corresponding Bin Type (tote BB0; tote BB1; tote BB2…)	One motor action (R60B + RL1 + R60 E)	$(1.54 \times 1.2) = 1.8$

(Continued)

TABLE 7.7 (CONTINUED)

Algorithmic Description of Activity and Its Time Structure (Morphological Analysis) during Computerized Task Performance (Version 3 of the Task before Improvement – Item is Accepted and Put in the Tote; $P = 0.9$)

Members of Algorithm (Psychological Units of Analysis)	Description of Elements of task (Technological Units of Analysis) Beginning of the Individual Item processing Version 3 of the task.	Description of Elements of Activity (Psychological Units of Analysis)	Time (sec)
Total performance time of the third version of the task	Task of receiving an item ends here	$T_3 = T_{cog} + T_{ex}$	$19.34 + 29.5 = 48.84$
$\omega 1$ $\downarrow O_{55}^{\alpha}$	Check if there are other items in the box to receive	Simultaneous perceptual action	0.3
$l_{15}^{15\,(1-2)} \uparrow$	If there are no more items in the box, decide go to O_4^{ε}, otherwise go to O_{17}^{ε}	Simple decision-making	0.3
Performance time O_{55}^{α} and l_{15}	This time is not included in task performance	Do not included in content of task	0.6

comparison of the received quantity and the PO (purchase order) quantity. This stage is completed when the quantity is accepted. The third stage of Version 3 is associated with comparing the price of the item on the packing slip with the price on the screen. Unlike in the second version of the task, an item is not rejected based on this criterion but is accepted. In Version 3, an item is accepted with the probability $P=1$. So a receiver can proceed to the final stage of the task performance. This stage includes determining the bin number for the item and placement of it in the corresponding tote associated with the assigned Bin Type.

The last stage of task performance occurs only in Version 3 of the task performance. O_{42}^α shows the part of the process when a receiver checks the screen to see if the bin number is assigned for the item at hand. The assigned bin number includes a bin type and a shelf number, for example, BB7125 (see Figure 7.15, field 7). If just two letters and one number are assigned, this means that only the Bin Type (BB3) is assigned, because it is a new item that does not have an assigned shelf number yet. A bin number is determined and appears on the screen for every item that has been received before and has an assigned shelf in the warehouse. The shelf will be created for the new items during the put-away process. When the new item is ordered by the Procurement Department, the Bin Type is specified. During the receiving process, the Bin Type can be corrected if needed. If the receiver does not

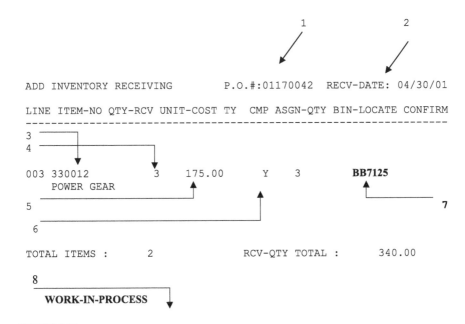

FIGURE 7.15
Add receiving screen with detailed item information. 1-purchase order number; 2-received date; 3-receved quantity; 4-receved quantity; 5-unit cost; 6-completion flag; 7-bin location; 8-work-in-process option.

accept the Bin Type assigned by the system, she/he can change it, print the label, place it on the item, and put the item in a corresponding box.

The completion flag on the screen is used to identify if the ordered quantity is received in full or more items for this order are coming. For example, if ten items have been ordered and only five have been received, then the completion flag is set to N. So the next time this PO is processed, the line with this item number is still on the list and the remaining quantity of the ordered item can be received. If the quantity of ten has been ordered and the same quantity has been received, the completion flag is set to Y. So the next time this PO is processed, the line for this item is no longer on the list.

There are four scenarios in the performance of this stage of the considered task before improvement:

1. A received item has a determined bin location and the system defines it;
2. An item is received for the first time and does not have an assigned bin;
3. The system suggests a bin type (area in the warehouse) and a receiver accepts it;
4. The system suggests a bin type and a receiver changes it.

Considered events can have different probabilities. Hence, for algorithmic description of the final stage of Version 3, it is necessary to develop an event tree and based on it to determine the probability of the above-described events. The task is completed when a receiver places the label on the item, goes to the boxes on the base unit 11, finds the corresponding tote, and puts the item in the tote (O_{54}^{ε}).

In order to determine probabilistic characteristics of these events, we use data from qualitative and algorithmic analyses of this task in combination with the probability assessment of events by subject matter experts (SME). SMEs utilized the scale of the subjective probability of events for this purpose. These procedures are discussed in the prior chapters. Figure 7.16 shows the event tree that has four branches that correspond to the above-listed events.

Bin location identification is the major distinctive stage of Version 3 of task performance. Hence, let us now consider members of the algorithm that show the bin assignment process, which is the last stage of Version 3. It starts with the member of the algorithm O_{42}^{α} presented in Table 7.7. This table gives the algorithmic description of activity and its time structure during performance of the third version of the task before improvement.

The total performance time of the third version of the task before improvement is $T_3 = 48.84$.

From member of algorithm O_{17}^{ε} up to O_{34}^{ε}, the receiver performs the part of task that is similar to the second version of the task before improvement. We start our selective analysis of the data in the considered table from the

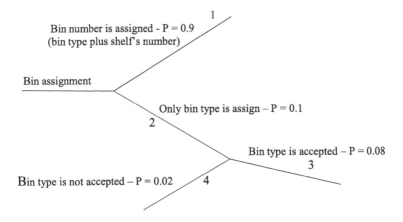

FIGURE 7.16
Event tree for bin location identification for Version 3 of the task performance before improvement.

member of the algorithm O_{42}^α. The member of the algorithm O_{42}^α includes only one simultaneous perceptual action with a performance time of 0.3 sec. Based on obtained visual information, a receiver makes a simple decision l_{12} out of two alternatives, as described in Table 7.7. Probabilities of considered outputs are taken from the event tree (see Figure 7.16, Branch 1 and Branch 2). $O_{43}^{\alpha\mu}$ has the same probability as the second output of logical condition l_{12}. This is taken into consideration when the performance time is calculated for this member of the algorithm, multiplying 1.8 sec by probability 0.1. Outputs of l_{13}^μ are shown by Branch 3 and Branch 4 in Figure 7.16. As can be seen, performance of $O_{43}^{\alpha\mu}$, l_{13}^μ, and $O_{44}^{\epsilon\mu}$, in this version of the task, requires keeping information in working memory. Here, the MTM-1 system is utilized to calculate the performance time for the motor components of activity, but the time is multiplied by the coefficient of 1.2 because the MTM-1 pace is too high for performing this type of task (Bedny and Karwowski, 2007).

Let us consider an example that demonstrates how the activity time structure that involves interaction with the computer is described using *analytical* procedures developed in SSAT. The member of the algorithm O_{47}^ϵ shows motor actions of a receiver after she/he decides to key in the bin type code utilizing the most preferable strategy of task performance. The bin type category consists of two capital letter ("BB") and a number from 0 to 9. Altogether there are ten bin types from BB0 to BB9. Prior to performing this element of the task, a receiver is positioned slightly to the left of the computer. She/he has to make one side step to the computer in order to be in front of it. Then a hand moves from home position to the key with letter B and presses this key two times. Then he/she moves a hand to one of the ten number keys. Thus, this step involves one leg movement and several hand motor actions. First, we have determined the distance of the hand movements. According to the MTM-1 system, the measured distance of the hand movement from the start position to the corresponding key is multiplied by a coefficient of 1.3, because the movement is not strictly linear. There is approximately 25 sm from home position to the letter B. Then we have to determine an average distance of the hand movement from the letter B to any of the ten numbers. The distance from the key with the letter B to numbers 7 or 8 is 6 sm and to the key with number 1 is 12 sm. The movement from home position to the letter B is depicted by R30B and the movement from letter B to the nearest number 8 or 7 is R8B, and to the most remote numbers 1 or 0 it is R15B. Letter R means Reach, B is the type of movement, and the number specifies the distance of the movement. The performance time of R8B is 0.2 sec, and for R15B it is 0.3 sec. The average performance time of this movement is determined as $(0.2 + 0.3)/2 = 0.25$ sec. The performance time of R30B is 0.46 sec. A receiver has to press a key. This type of movement in the MTM-1 system is known as AP (Press). There are several types of such motions in this system. For the considered situation, the most applicable one is AP1. This type of motion requires 0.38 sec. However, for the considered task we have utilized the data from the GOMS system, according to which it takes 0.28 sec to press the key. The side step can be described as SSC1 with an approximate distance of 30 sm. Thus, leg movement and hand actions during performance of this member of the algorithm can be described as $\text{SSC2} + (\text{R30B} + \text{AP1} + \text{AP1}) + \{[(\text{R8B} + \text{R15B})/2] + \text{AP1}\} = 2.78$ sec. There are two hand motor actions: $(\text{R30B} + \text{AP1} + \text{AP1})$ and $\{[(\text{R8B} + \text{R15B})/2] + \text{AP1}\}$. This is the symbolic description of the considered member of the algorithm. For this task, we utilize the pace coefficient 1.2 for motor components of activity. Therefore, the performance time of this member of the algorithm is 3.33 sec. In Table 7.6 this time is multiplied by 0.02, because this member of the algorithm has probability 0.02. According to SSAT, the standardized or symbolic description of activity is always required, even when one can easily determine the performance time experimentally. This allows understanding the content of motor actions that are involved in performance of the considered member of the algorithm.

Above we have considered a completely new part that is performed only in Version 3 of this task. Now we will briefly review the beginning parts of the third version of the task that are also performed in Version 2 but have a different outcome here.

When the item price is evaluated in Version 2, it is always higher than the order price, and the increase is always >10%. So the item is rejected due to this reason with the probability $P = 1$.

In Version 3, the price evaluation is depicted by logical conditions l_8 and l_9, and in this case the price of the received item can be the same as the price on the order, it can be decreased or increased by <10%, but in all these cases, the item is accepted and received.

At this stage of task performance, the receiver checks if the bin number is assigned for the item. Items that have been received before have an assigned shelf in the warehouse that has a corresponding bin number. If the item is received for the first time and does not have the bin number, the bin type needs to be determined. The bin type determines the area in the warehouse where this type of item is stored. When the put-away operator brings the item to the assigned bin area, she/he creates the shelf for the item, puts the label on the shelf with the item number, and scans it, and the system now has a bin number for this item.

Figure 7.17 depicts the totes with examples of the labels that show the bin type or workshop number they should be delivered to.

The final stage of the considered fragment of the task is completed when an operator changes the bin type if needed. An operator prints the label for the corresponding bin (O_{48}^{ε}) and puts the item into the corresponding tote. This part of the activity has a low probability of occurrence, which reduces the complexity of the task performance during the shift. However, this situation happens unexpectedly, and an operator has to perform multiple cognitive and motor actions. If the considered stage of task performance occurs, the complexity of this task increases sufficiently. This stage of task performance includes two decisions, and one is performed based on extraction of information from memory. The member of the algorithm for executive activity $O_{44}^{\varepsilon\mu}$ is associated with the memory workload. It also should be taken into consideration that an operator performs repetitive work during the shift. We have concluded that this stage of task performance can be simplified, and the innovation can be simple and not require a lot of effort for its realization. It has been suggested that when the bin type assigned by the system is rejected by an operator, she/he should use the drop-down menu with all bin types and their descriptions on the screen. We will consider this innovation when discussing Version 3 after improvement. This innovation is not a complicated one, but it would take a lot of time and effort for its testing and implementation if it has to be done as a separate project. Our analysis allows predicting the need for this improvement during the software design.

Version 3 is the most complex version of the task that includes problem-solving components (members of algorithm $O_{41}^{th\mu}$ and $O_{31}^{th\mu}$). However, in production processes repetitive tasks are predominantly rule-based tasks. It

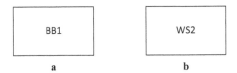

a   b

FIGURE 7.17
The totes with the label that shows the bin type or the workshop number it should be delivered to: (a) base unit 6 for stocking process in the warehouse; (b) base unit 7 for work-in-process for delivering to corresponding workshops (base units 6 and 7 are shown in Figure 7.2).

should be noted that even in complex computerized tasks problem-solving components should be reduced or simplified, especially in emergency conditions. Nine logical conditions that are included in the considered task made it variable in terms of content, time performance, memory workload, and so on. This is not the quantitative but the qualitative analysis that demonstrates that even though this task is repetitive, it is also a complex one.

We will now consider Version 3 after improvement. The algorithmic description of Version 3 after improvement is presented in Table 7.8. Like Version 2 of the task after improvement, this version also starts with O_{11}^ε. The item quantity evaluation in Versions 2 and 3 after improvement are the same (see Tables 7.6 and 7.8). The difference starts with the evaluation of the price, because in Version 2 the price is always unacceptable and in Version 3 it's always accepted. Similar comparative analysis has been done for Versions 2 and 3 before improvement, and we are not going to repeat it.

Let us compare the performance of Version 3 before and after improvement. This version after improvement has the same innovations as Version 2: the item number is scanned instead of being keyed in, and the system calculates the price difference so the receiver does not need to calculate it and keep it in her/his working memory and recall the price evaluation rules. These enhancements have been described above. The new enhancement here is in the last version of the algorithm that depicts the bin assignment starting with O_{34}^α. So we will concentrate our attention on this part of the algorithm.

We remind the reader that the operation starts when a receiver takes an item out of the box that is placed on the base unit 5 (Figure 7.2). Then a receiver takes parts out of the box and compares the order quantity and price with the received quantity and price. Then he/she changes or confirms the quantity and price, and assigns the bin location in the warehouse if necessary. If the shelf is already reserved for the item in the warehouse, the system will select it automatically. After that, an item is placed into a tote with the corresponding label. For this purpose, totes with different labels are put on the table dedicated for totes (see Figure 7.2). When a tote is filled up, a receiver takes it and brings to the base unit 6 for stock processing, then pushes the tote into belt 8 for stocking. We do not cover this part of the work process in our algorithmic description, limiting it to the description of processing individual items.

The belt transports the totes into the warehouse. The next operation after the items arrive into the warehouse is called "putting-away." The put-away operator takes items from the tote and puts them on the corresponding shelves. The next task is performed by the "pick up" operator. This operator takes the items that have been ordered from the shelves and places them into the totes. The totes are delivered either into the workshop according to their request or if it's has the final product, the items are shipped to the customers. Thus, before improvement, all items received from the vendors are placed into the storage kept in the warehouse shelves. Analysis of the production process demonstrates that some items on a special order should be delivered

TABLE 7.8

Algorithmic Description of Activity and Its Time Structure (Morphological Analysis) for Version 3 of the Inventory Task after Improvement ($P = 0.9$)

Member of Algorithm (Psychological Units of Analysis)	Description of Elements of Tasks (Technological Units of Analysis)	Description of Elements of Activity (Psychological Units of Analysis)	Time (sec)
$\overset{8\,(2)\,5\,(2)}{\downarrow} \downarrow O^{\varepsilon}_{11}$	Take an item out of the box, and return to computer area again (while a worker takes several steps his/her right hand releases the item and worker holds the item only by left hand)	Motor actions - simultaneously by two-hand reach and grasp item (1) (R60B + G1B); (2) – take out item from box – (M40B + RL1); (3) take four steps to the computer area 3 (TBC1 + TBC2 + WP + WP)	$4.74 \times 1.2 = 5.7$
O^{ε}_{12}	Take the barcode scanner and scan the item number. (The matching item is highlighted)	(R50B + G1A) + (M40B + mM10C + AP1) + (M50B + RL1)	$2.44 \times 1.2 = 2.9$
O^{ε}_{13}	Press ENTER to go to the screen with detail item information (Figure 7.8)	Motor action (R30B + AP1)	$0.74 \times 1.2 = 0.9$
$O^{\alpha\mu}_{14}$	**Compare received quantity with PO (purchase order) quantity (Figure 7.8, field 4)**	Combination of two simultaneous perceptual actions - $2 \times (ET + EF)$ with simultaneously performed mnemonic operations (MO)	$(0.42 + 0.4) \times 2 = 1.64$
$\overset{3}{I_3} \uparrow$	If received quantity and ordered quantity are the same, go to O^{ε}_{22} ($P = 0.9$). If received quantity is greater or less than ordered quantity, go to O^{ε}_{15} ($P = 0.1$)	Decision-making actions that are performed based on visual information	*0.4*
O^{ε}_{15}	Type the received quantity and press ENTER to get a question at the bottom of the screen ($P = 0.1$)	Motor action (R20B + AP1) + (R12B + AP1) (example with two-digit number)	$1.39 \times 1.2 \times 0.1 = 0.17$

(Continued)

TABLE 7.8 (CONTINUED)

Algorithmic Description of Activity and Its Time Structure (Morphological Analysis) for Version 3 of the Inventory Task after Improvement ($P = 0.9$)

Member of Algorithm (Psychological Units of Analysis)	Description of Elements of Tasks (Technological Units of Analysis)	Description of Elements of Activity (Psychological Units of Analysis)	Time (sec)
O_{16}^{α}	Read the statement: THE RECEIVED QUANTITY AND ORDERED QUANTITY DO NOT MATCH. DO YOU ACCEPT? (YES/NO). ($P = 0.1$). Scan and read ≈ four words	Successive perceptual action. $ET + 4 \times EF$	$(0.42 + 4 \times 0.18) = 1.14 \times 0.1 = 0.11$
* $O_{17}^{\mu th}$	Recall instructions and perform required calculation and estimation (relationship between quantity and price). $P = 0.1$	Combination of mnemonic action (retrieve simple information from memory) and logical thinking action	$\approx (1.2 + 3) = 4.2 \times 0.1 = 0.42$
$\overset{4}{l_4} \uparrow$	If quantity is not accepted (computer defaults to 'N'), go to O_{18}^{ε} ($P = 0$). Otherwise, go to O_{21}^{ε} ($P = 0.1$)	Complex decision-making actions that include mnemonic and thinking information	$1.5 \times 0.1 = 0.15$
$O_{18}^{\varepsilon};\ O_{19}^{\varepsilon};\ O_{20}^{\alpha};\ l_5;$	These members of algorithm are performed in the Version 1	———	———
$\overset{4}{\rightarrow} O_{21}^{\varepsilon}$	Change "N" to "Y", quantity is accepted. ($P = 0.1$)	One motor action: (R60B + AP1)	$(1.05 \times 1.2) \times 0.1 = 0.126$
$\overset{3}{\rightarrow} O_{22}^{\varepsilon}$	Press ENTER	One motor action (R40 B + AP1)	$0.84 \times 1.2 = 1.01$
** O_{23}^{α}	Compare price of the item on the packing slip with price on the screen. (Starting from O_{23}^{α} till O_{29}^{α})	Combination of two simultaneous perceptual actions including one double check perceptual action- $2 \times (ET + EF) + (ET + EF)$ (includes one double check perceptual action)	$2 \times (0.42 + 0.4) + (0.42 + 0.3) = 2.36$

(Continued)

TABLE 7.8 (CONTINUED)

Algorithmic Description of Activity and Its Time Structure (Morphological Analysis) for Version 3 of the Inventory Task after Improvement ($P = 0.9$)

Member of Algorithm (Psychological Units of Analysis)	Description of Elements of Tasks (Technological Units of Analysis)	Description of Elements of Activity (Psychological Units of Analysis)	Time (sec)
$I_6^6 \uparrow$	If the price on the screen and packing slip are different, go to O_{24}^ε ($P = 0.2$). If price is the same go to O_{30}^ε ($P = 0.8$)	Simple decision-making action	*0.4*
O_{24}^ε	Key in the new price and press ENTER ($P = 0.2$)	Four motor actions and mnemonic operations (R20B + AP1) + 2 × (R6B + AP1) + (R12B + AP1)	(2.48 × 1.2) × 0.2 = 0.59
O_{25}^α	Look at information on the screen (Cursor moves to the next field if price is accepted. Otherwise there is a message on the screen "Price increase is >10%". Do you wish to proceed?) $P = 0.2$. Scan and read ≈ four words	Successive perceptual action. ET + 4 × EF	(0.42 + 4 × 0.18) = 1.14 × 0.2 = 0.23
$I_7^{7(1-3)} \uparrow$	If difference >10%, go to O_{26}^ε. However, if there is a special reason to accept, go to O_{29}^ε ($P = 0.2$). (O_{26}^ε and O_{29}^ε are performed in version 2). If there was no message and cursor moved to the next field, go to O_{30}^ε. Probability of O_{30}^ε is 0.8	Complex decision from three alternatives that requires actualization of information in memory and includes simple thinking operations.	≈ (2 + 1.5) = 3.5 × 0.2 = 0.7
$O_{26}^\varepsilon; O_{27}^\varepsilon; O_{28}^\alpha; I_8,$	These members of algorithm are performed in the version 2 of the task	—	—

(Continued)

TABLE 7.8 (CONTINUED)

Algorithmic Description of Activity and Its Time Structure (Morphological Analysis) for Version 3 of the Inventory Task after Improvement ($P = 0.9$)

Member of Algorithm (Psychological Units of Analysis)	Description of Elements of Tasks (Technological Units of Analysis)	Description of Elements of Activity (Psychological Units of Analysis)	Time (sec)
$7(2)$ $\rightarrow O_{29}^{\varepsilon}$	Type "Y". $P = 0.2$. Starting with O_{29}^{ε} till the end of the algorithm, all members of the algorithm have blanket $P = 0.905$. In further calculation, we accept this as $P = 1$.	Motor action: (R30B + AP1)	$0.74 \times 1.2 \times 0.2 = 0.178$
$7(3)\ 6$ $\rightarrow O_{30}^{\varepsilon}$	Press ENTER to go to the Completion Flag (Figure 7.8, field 6). $P = 0.8 + 0.2 = 1$	One motor action: (R25B + AP1)	$0.69 \times 1.2 = 0.83$
*** O_{31}^{α}	Check system default "Y" or "N" flag on the screen	Simultaneous perceptual action (ET + EF)	$0.42 + 0.3 = 0.72$
$l_9^\mu \uparrow$ 9	If you accept the system default, go to O_{33}. ($P_1 = 0.95$). Otherwise O_{32}. ($P_2 = 0.05$). (In most cases the system default is accepted). General probability is $P_1 + P_2 = 1$	Simple decision-making action at verbal thinking level with mnemonic operation	0.4
O_{32}^{ε}	Change system default ('Y' to 'N' or 'N' to 'Y'). ($P = 0.05$)	Motor action (R50B + AP3) or (R60B + AP1). Average time calculation is presented.	$0.995 \times 1.2 \times 0.05 = 0.06$
9 $\rightarrow O_{33}^{\varepsilon}$	Press ENTER to go the next field	Motor action: (R18B + AP1)	$0.72 \times 1.2 = 0.86$

(Continued)

TABLE 7.8 (CONTINUED)

Algorithmic Description of Activity and Its Time Structure (Morphological Analysis) for Version 3 of the Inventory Task after Improvement ($P = 0.9$)

Member of Algorithm (Psychological Units of Analysis)	Description of Elements of Tasks (Technological Units of Analysis)	Description of Elements of Activity (Psychological Units of Analysis)	Time (sec)
O_{34}^{α}	Look at the screen to check if the message **"WORK-IN-PROCESS"** has been displayed. Read two words (Figure 7.16, $P_1 = 0.2$) or read information about a bin location for the item (see Figure 7.8, field 7) with probability $P_2 = 0.8$, Figure 7.16).	Combination of successive and simultaneous perceptual actions. $[(ET + 2 \times EF) \times 0.2] + (ET + EF) \times 0.8$	$[(0.42 + 0.36) \times 0.2] + (0.42 + 0.3) \times 0.8 = 0.74$
$\overset{10(1-3)}{l_{10}} \uparrow$	If screen displays a message "WORK-IN-PROCESS" ($P_1 = 0.2$) go to O_{35}^{ε}. If bin number is assigned, go to O_{44}^{ε} ($P_2 = 0.72$). If only the bin type is assigned by the system, go to $O_{40}^{\alpha u}$ to check correctness of assignment ($P_3 = 0.08$)	Simple decision-making from three alternatives	$0.5 \times 0.2 = 0.1$
$\overset{10(1)}{\downarrow} O_{35}^{\varepsilon}$	Press ENTER to print the label for WORK-IN-PROCESS (labels WS1; WS2; or WS3). $P = 0.2$	One motor actions: (R15B + AP1)	$(0.58 \times 1.2) \times 0.2 = 0.82 \times 0.2 = 0.14$

(Continued)

TABLE 7.8 (CONTINUED)

Algorithmic Description of Activity and Its Time Structure (Morphological Analysis) for Version 3 of the Inventory Task after Improvement ($P = 0.9$)

Member of Algorithm (Psychological Units of Analysis)	Description of Elements of Tasks (Technological Units of Analysis)	Description of Elements of Activity (Psychological Units of Analysis)	Time (sec)
O_{36}^{ε}	Peel the label off the printer and put it on the item. O_{36}^{ε} - includes: (1) side step to the right; (2) grasp and peel label with the right hand; (3) move the item with the left hand to the right and put it on the worker table; (4) move the left hand and grasp the label; (5) bring the label with two hands to the item's surface and stick it on the item; (6) smooth the label over with the right hand fist. $P = 0.2$	One side step and motor actions by hand: (1) SSC2 + (2) (R50B + G1A + M15B) + (3) (M60B + RL1) + (4) (R20B + G1A) + (5) (M10B + AP1 + RL1) + (6) (R12B + M12B)	$(4.96 \times 1.2) \times 0.2 = 5.95 \times 0.2 = 1.19$
O_{37}^{ε}	Take item and go with it to base unit 7 (see Figure 7.2); (1) the left hand grasp the item; (2) worker turn his body and make approximately 4 steps to the tote area and turn body in working position $P = 0.2$	(R6B + G1B) + (TBC1 + 4WP + TBC2)	$(4.48 \times 1.2) \times 0.2 = 5.4 \times 0.2 = 1.1$
O_{38}^{α}	Compare department number on the label with the department number on the tote on the base unit for WIP (tote WS1; tote WS2; tote WS3) and perform O_{39}^{ε}. $P = 0.2$	Approximately two simultaneous perceptual actions: $(ET_1 + EF_1) + (ET_{2+} EF_2)$. $ET_1 \approx 20$sm; $ET_2 \approx 40$sm; $EF_1 = 0.3$ sec; $EF_2 = 0.6$ sec	$(0.57 + 1.14) \times 0.2 = 1.71 \times 0.2 = 0.34$

(Continued)

TABLE 7.8 (CONTINUED)

Algorithmic Description of Activity and Its Time Structure (Morphological Analysis) for Version 3 of the Inventory Task after Improvement ($P = 0.9$)

Member of Algorithm (Psychological Units of Analysis)	Description of Elements of Tasks (Technological Units of Analysis)	Description of Elements of Activity (Psychological Units of Analysis)	Time (sec)
O_{39}^{ε}	Put item in the tote with corresponding department number (tote WS1; tote WS2; tote WS3). $P = 0.2$	One motor action (R60B + RL1 + R60E)	$(1.54 \times 1.2) \times 0.2 = 0.37$
$\omega_1 \uparrow$	Always false logical condition (go to O_{49}^{α})	—	—
10 (3) $\downarrow O_{40}^{\alpha\mu}$	**Based on recalled bin type category (from BB0 to BB9) check correctness the bin type assign by the system ($P = 0.08$)** Bin number includes bin type plus shelf's number	Perceptual action combined with mnemonic action	$(1.2 + 0.6) \times 0.08 = 1.8 \times 0.08$ $= 0.15$
11 $l_{11}^{\mu} \uparrow$	If you accept bin type go to O_{44}^{ε} ($P_1 = 0.07$). If you do not accept bin type go to O_{41}^{ε} ($P_2 = 0.01$) $\Sigma P_1 = P_1 + P_2 = 0.08$	Decision-making action at verbal thinking level with mnemonic operation	$0.6 \times 0.08 = 0.048$
O_{41}^{ε}	Click arrow to get a drop down menu with Bin Types and their descriptions. ($P = 0.01$). Includes two motor actions: (1) move hand and grasp the mouse; (2) move pointer and point and click	Motor actions (R20B + G1A) + (M2B + AP1).	$0.78 \times 1.2 \times 0.01 = 0.936 \times 0.01$ $= 0.00936$

(Continued)

TABLE 7.8 (CONTINUED)

Algorithmic Description of Activity and Its Time Structure (Morphological Analysis) for Version 3 of the Inventory Task after Improvement ($P = 0.9$)

Member of Algorithm (Psychological Units of Analysis)	Description of Elements of Tasks (Technological Units of Analysis)	Description of Elements of Activity (Psychological Units of Analysis)	Time (sec)
O^α_{42}	**Visually scan drop-down menu and select required item (menu contains list for ten types of Bins and each line contains one–two words).** $P = 0.01$	Complex perceptual action that includes scanning and recognition of bin type (ET + Scanning + EF)	$(0.42 + 0.9 + 0.3) = 1.62 \times 0.01$ = 0.016
O^ε_{43}	**Move the cursor to pick up the chosen Bin type and left-click to select the item.** $P = 0.01$	One motor action (M2B + AP1 + RL1)	$0.0422 \times 1.2 \times 0.01 = 0.0506 \times 0.01$ = 0.0005
$^{11\ 10(2)}_{} \to O^\varepsilon_{44} \to$	**Press ENTER to print the label for corresponding bin.** $P = 0.72 + 0.07 + 0.01 = 0.8$	One motor action: (R15B + AP1)	$0.58 \times 1.2 \times 0.8 = 0.7 \times 0.8 = 0.56$
O^ε_{45}	Peel the label off the printer and put it on the item. O^ε_{45} includes: (1) side step to the right; (2) grasp and peel label by the right hand; (3) move the item by the left hand to the right and put it on the worker table; (4) move the left hand and grasp the label; (5) bring the label by two hands to the item's surface and stick it on the item; (6) smooth the label over with the right hand fist	One side step and motor actions by hand: (1) SSC2 + (2) (R50B + G1A + M15B) + (3) (M60B + RL1) + (4) (R20B + G1A) + (5) (M10B + AP1 + RL1) + (6) (R12B + M12B)	$(5 \times 1.2) \times 0.8 = 6 \times 0.8 = 4.8$

(Continued)

TABLE 7.8 (CONTINUED)

Algorithmic Description of Activity and Its Time Structure (Morphological Analysis) for Version 3 of the Inventory Task after Improvement ($P = 0.9$)

Member of Algorithm (Psychological Units of Analysis)	Description of Elements of Tasks (Technological Units of Analysis)	Description of Elements of Activity (Psychological Units of Analysis)	Time (sec)
O_{46}^{ε}	Take an item and go to base unit 11 (Figure 2); (1) the left hand grasps the item; (2) worker turns his/her body and takes approximately four steps to the tote area and turns body to working position	One motor action by hand and leg movements: (R6B + G1B) + (TBC1 + 4WP + TBC2)	$(4.48 \times 1.2) \times 0.8 = 5.4 \times 0.8 = 4.32$
O_{47}^{α}	Compare Bin Type on the label with the Bin Type on the tote on the base unit 11 (tote BB1; tote BB2; tote BB3,...) and perform O_{48}^{ε}	Approximately two simultaneous perceptual actions: $(ET_1 + EF_1) + (ET_2 + EF_2)$. $ET_1 \approx 20$sm; $ET_2 \approx 40$sm; $EF_1 = 0.3$ sec; $EF_2 = 0.6$ sec	$(0.57 + 1.14) \times 0.8 = 1.71 \times 0.8 = 1.34$
O_{48}^{ε}	Put item in the tote with corresponding Bin Type (tote BB0; tote BB1; tote BB2,...). $P = 0.8$	One motor action (R60B + RL1 + R60E)	$(1.54 \times 1.2) \times 0.8 = 1.85 \times 0.8 = 1.48$
Total performance time of the third version of the task	**Task of receiving an item ends here**	$T_3 = T_{cog} + T_{ex}$	$10.7 + 27.31 = 37.56$
$\omega 1 \downarrow O_{49}^{\alpha}$	Check if there are other items in the box to receive	Simultaneous perceptual action	0.3
l_{12}	If there are no more items inbox, go to O_4^{ε}, otherwise go to O_{11}^{ε}	Simple decision-making	0.3
Performance time O_{49}^{α} and l_{12}	This time is not included in task performance	Do not included in content of task	

into the workshops immediately, bypassing the warehouse. It was impossible to do this under the existing conditions. The presented analysis demonstrates that two routes of delivery are needed: one proceeds to the warehouse, and the other should be sent directly to the workshops. The last situation is called "work-in-process" (WIP). It was empirically discovered that about 20% of the items should be immediately delivered through the WIP process. Such innovation would improve economic efficiency. Figure 7.15 shows that after improvement, if the item is on order from any of the workshops, a receiver gets a Work-in-Process message at the button of the screen (field 8). This option has not been available before innovation.

In Version 3 after improvement, when the receiver presses ENTER after setting the completion flag, she/he might get the message in the left bottom corner of the screen "WORK IN PROCESS" which means that the system checked the orders created by the workshops and found an order for this item. If the item is on order, it should not go into the warehouse but should be delivered directly to the workshop that ordered it. So if this new message appears and the receiver presses ENTER again, the WS1 (or WS2, WS3, or WS4) appears under the BIN-LOCATE on the screen indicating that the item should be put in the tote marked with the corresponding workshop number. If the message "WORK IN PROCESS" does not appear, it means that there is no order from the workshops for this item and it should be shelved. Instead the bin number or the bin type is produced in the BIN-LOCATE column and the item is processed the same way as in the previous version.

Below we will explain what the branches of the above event tree represent.

Branch 1 shows the scenario when the item is on order from the workshop because it is needed for the production process (if the produced label has WS2 on it, it means that the item should be delivered to Workshop 2).

Branch 2 describes the case when the item needs to be stored in the warehouse.

Branch 3 shows the situation when a received item has a defined bin number. In other words, this item already has a dedicated shelf in the warehouse.

Branch 4 is associated with the case when an item is received for the first time and does not have an assigned bin.

Branch 5 is related to the situation when the system defaults a bin type (area in the warehouse) and a receiver accepts it.

Branch 6 shows the case when the system defaults a bin type and a receiver overwrites it.

The probabilistic characteristics of these events have been determined based on qualitative and algorithmic analysis of the task and the assessment of the subject matter experts (SME) who utilized the scale of subjective probability evaluation of events. These procedures have been described in the previous chapters. As can be seen, the considered event tree has six branches which demonstrate the probability of above-listed events.

Let us now present the morphological analysis of the third version of the inventory receiving task after improvement. Table 7.8 shows the algorithmic and time-structure description of this version of the task.

The total performance time of the third version of the task after improvement is $T_3 = 37.56$ sec

There is a new option in Version 3 of the task after improvement that begins when the item is accepted. In the algorithm, it starts with O_{34}^{α}.

As can be seen from Figure 7.15, an item is either on order from the workshop, then the message "WORK-IN-PROCESS" (field 8) is displayed. Otherwise, a bin location in the warehouse appears on the screen (field 7). The member of the algorithm O_{34}^{α} shows this situation.

If the item has been ordered by the workshop, the system gives a message at the bottom of the screen that states "WORK-IN-PROCESS" (WIP) and an operator can print out a label with the workshop number and put the item in the tote marked with the corresponding workshop number. If the item has not been ordered by the workshops, it is going to be stored in the warehouse. The receiver checks if the message WIP is displayed or information about the bin location is presented. Based on the presented information, there are three outputs for the logical condition l_{10}: (1) if WIP, go to O_{35}^{ε} ($P_1 = 0.2$); (2) if bin number is assigned (bin type number plus shelf number), go to O_{44}^{ε} ($P_2 = 0.72$); (3) if only bin type is assigned for a new item by the system, go to $O_{40}^{\alpha\mu}$ ($P_3 = 0.08$). Analysis of logical condition l_{10} demonstrates that one out of two mutually exclusive alternatives should be extracted that are important for our consideration at this step of analysis. The first one is "if WIP, go to O_{35}^{ε} ($P = 0.2$)", and the other one is when an operator has to determine a bin location ($P_2 + P_3 = 0.72 + 0.08 = 0.8$). These alternatives are presented by the event tree (see Figure 7.18). Probabilities P_2 and P_3 are shown by the other branches of the same figure. As can be seen, the output of WIP has probability $P = 0.2$. This means that only in 20% of the cases is the WIP message presented.

O_{35}^{ε} has the probability $P = 0.2$ and O_{36}^{ε} through O_{39}^{ε} which directly follow it, have the same probability because there are no logical conditions, or arrows that are addressed to them from other logical conditions that could change their probability.

The third output of the logical condition l_{10} has the probability $P = 0.08$ and therefore $O_{40}^{\alpha\mu}$ also has the same probability. Similarly, l_{11}^{μ} has the probability $P = 0.08$.

Let us consider how we can calculate the probability of O_{44}^{ε}. This member of the algorithm accumulates several probabilities. According to the outputs of $l_{10}\uparrow^{10\,(2)}$ "if bin number is assigned, go to O_{44}^{ε}" (see also arrow with number 10 but in reverse direction $\downarrow^{10\,(2)}$ in front of O_{44}^{ε}). The probability of this event is

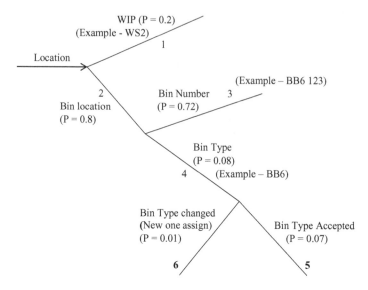

FIGURE 7.18
Event tree for Work-In-Process (WIP) and bin location assignment for Version 3 after improvement.

$P_1 = 0.72$. Similarly, according to the output of l_{11}^μ, the second accumulated probability is $P_2 = 0.07$. There is also a scenario when a receiver needs to change the bin type and according to the event tree this occurs with probability $P_3 = 0.01$. Based on this decision, a receiver performs motor actions described by O_{41}^ε; O_{43}^ε; and O_{44}^ε. This stage of performance also includes the member of the algorithm O_{42}^α, which depicts perceptual actions. Thus, these four members of the algorithm are performed with the probability $P_3 = 0.01$. Therefore, the accumulated probability of O_{44}^ε is $P = P_1 + P_2 + P_3 = 0.72 + 0.07 + 0.01 = 0.8$.

As we have mentioned above, the coefficient 1.2 is used when evaluating the performance time of motor components of activity such as O_{35}^ε, O_{36}^ε, and so on. For example, in order to determine the performance time of O_{35}^ε one would first calculate the time of motor action R15 + AP1; then this time is multiplied by 1.2, and finally the obtained result, according to the output from l_{10}, is multiplied by $P = 0.2$.

Actions performed by an operator when she/he uses the drop-down menu are described by the members of the algorithm O_{41}^ε - O_{43}^ε. The drop-down menu is another innovation that reduces the mental workload. When comparing members of the algorithms that describe the stage of task performance associated with assigning the bin type before and after improvement, it becomes obvious that after innovation, the memory workload is reduced and this part of the task becomes simpler (See Tables 7.7 and 7.8).

In the improved version, an operator can use the drop-down menu to choose the bin type instead of trying to recall it or referring to the hard copy

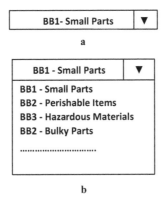

FIGURE 7.19
Drop-down menu: (a) before drop-down menu is activated; (b) after drop-down menu is activated.

on the desk. This relatively simple example demonstrates that changes in the presentation of information on the screen changes the system of cognitive and motor actions utilized by a computer user. It is necessary to stress here that the need for such improvements is not often obvious. Morphological analysis helps to determine such needs. Hence, we can evaluate the usability of a screen design based on analysis of cognitive and behavioral actions performed by users. In spite of the additional functions performed in Version 3 of the task after innovation, and specifically the WORK-IN-PROCESS step, the total performance time for the improved version is less than for the before improvement version of the task.

The member of the algorithm ω_1 (always false logical condition) means that after performance of O_{39}^ε, there is the automatic transfer to the execution of O_{49}^α. An "always false logical condition" is not an action that is performed by an operator but a symbol that is used in the algorithmic description when it is necessary to switch to another member of the algorithm without taking any action. Version 3 is completed when a receiver places a label on the item, goes to the base unit 11, finds the corresponding tote, and puts an item in the tote. When the tote is filled up, a receiver places it on the base unit 7 for work-in-process or on unit 6 for the put-away process, and pushes it into belt 9 for delivering items into the workshop or belt 8 to deliver the tote to the stocking area.

Below we list all the innovations that have been implemented based on the performed analysis that utilized principles of studying human work proposed in SSAT:

1. Scanning PO number utilizing the barcode scanner instead of keying it in.

2. Scanning the item number instead of finding it on the PO list on the computer screen.

3. Software calculates the price difference, giving a warning when the price increase is over 10% instead of mentally calculating the price difference.

4. Introduce the Drop down menu for Bin Types and their descriptions instead of recalling this information.

5. The last and most important enhancement is to check if the item is on order from the workshop. If it is on order, the item is put in the Work-in-Process area and is delivered directly to the workshop instead of putting it on the shelf in the warehouse first. This innovation impacts the whole operation. For the items that are delivered through this new process, put away and pickup processes are eliminated. The cost-effectiveness of all these enhancements leads to a reduction of processing time of 20%.

Figure 7.7 demonstrates that the considered task has three versions. Version 1 is completed when an item is rejected according to the quantity criterion. Version 2 is completed when an item is accepted based on this first criterion but is still rejected at the following stage according to the price criterion. The third version is completed when the bin location is assigned or the item is going to be delivered to the workshop as per WIP request. Version 3 is most probable one. All three versions of the task have a probabilistic logical organization.

Here, for the first time, we give an example of a description of a task with a very complex logical and probabilistic structure. Using this example, we demonstrated the new method of description of the probabilistic structure of the task utilizing the specially developed scale for estimating the probabilities of individual members of the algorithm. We also presented the new method of building the event tree using the conditional probabilities. A new method of morphological analysis requires not only new steps in describing the structure of activity but also changes in their order. Let us consider this using as an example the description of the member of the algorithm O_{44}^{ε} from Table 7.8. At the first step, we describe this member without determining its probabilistic characteristics.

11 10(2) $\downarrow\ \downarrow\ O_{44}^{\varepsilon}$	Press ENTER to print the label for corresponding bin	One motor action: (R15B + AP1)	$0.58 \times 1.2 = 0.696$

We want to remind the reader that the MTM-1 system utilizes the pace of task performance for mass production. Moreover, the analysis of energy expenditure demonstrates that in MTM-1 the pace of task performance exceeds physiological norms (Bedny and Karwowski, 2007). Hence the coefficient 1.2 of pace is more suitable for the considered task. Only after developing the algorithm without specifying the probability of occurrence of each member, we then determined probabilistic characteristics of each member and the time of its performance.

We widely utilize the MTM-1 system in our studies. However, MTM-1 does not always contain the required temporal data, especially when cognitive activity is a complex one. Such data can be obtained by using a chronometrical analysis of the execution time of individual members of the algorithm and other sources that have the necessary information.

11 10(2) $\downarrow \quad \downarrow O_{44}^{\varepsilon}$	Press ENTER to print the label for corresponding bin. $P = 0.72 + 0.07 + 0.01 = 0.8$	One motor action: (R15B + AP1)	$0.58 \times 1.2 \times 0.8 = 0.7 \times 0.8 = 0.56$

Our study demonstrates that morphological analysis includes psychological units of analysis (columns 1 and 3) that are compared with technological units of analysis (column 2). Utilization of technological units of analysis assumed that the task description is performed in technological terms or by using common language. When employing psychological units of analysis, the activity is described using standardized psychological terminology. The combination of psychological and technological units of analysis makes it possible to clearly define what an operator does when she/he performs the described member of the algorithm. The units of analysis in column 1 and 3 have a hierarchical description. The description of the member of the algorithm in column 1 is further decomposed in column 3. Activity as an object of study is described as a system. Hierarchical units of analysis are: *members of algorithm* → *cognitive and behavioral actions* → *psychological operations (motions and cognitive operations that are components of actions)*. Each member of the algorithm is studied as a quasi-system that consists of smaller units. Strategies of performance of such quasi-systems and their time structure is considered. Logical organization and probability of appearance of such subsystems in the holistic structure of activity is described.

Thus, morphological analysis developed in SSAT presents an opportunity to describe extremely complex tasks that have probabilistic and logical organization using analytical procedures.

The presented material demonstrates that the framework of SSAT can be used to generate an analytical model for a complex task with a highly complex probability structure of activity. In this chapter, we discussed an example of analyzing a complex computerized task consisting of three relatively independent tasks. Such production processes are usually considered as a number of different tasks. This new approach presents a more unified analysis. The three versions of the task occur with different probabilities in the work process, and together comprise the overall unitary task activity

References

Bedny, G. Z. and Karwowski, W. (2007). *A Systemic-Structural Theory of Activity. Application to Human Performance and Work Design.* Boca Raton, FL: CRC, Taylor and Francis.

Beregovoy, G. T., Zavalova, N. D., Lomov, B. F., and Ponomarenko, V. A. (1978). *Experimental-Psychology in Aviation and Aeronautics*. Moscow, Russsia: Science Publishers.

Card, S., Moran, T., and Newell, A. (1983). *The Psychology of Human-Computer Interaction*. Hillsdale, NJ: Lawrence Erlbaum Associates Publishers.

Dobrolensky, U. P., Zavalova, N. D., Ponomarenko, V. A., and Tuvaev, V. A. (1975). In U. P. Dobrolensky (Ed.). *Methods of Engineering Psychological Study in Aviation*. Moscow, Russia: Manufacturing Publishers.

Petrov, V. P. and Myasnikov, V. P. (Eds.) (1976). *Aviation Digital Systems of Control and Management*. Leningrad, Russia: Manufacturing Publishers.

Zarakovsky, G. M. and Pavlov, V. V. (1987). *Laws of Functioning Man-Machine Systems*. Moscow, Russia: Soviet Radio.

8

Quantitative Assessment of the Complexity of Computerized Tasks

8.1 Basic Principles of Measuring the Task Complexity

Complexity has been studied in various fields of science, including economics, biology, engineering, and so on. Some economists, such as Rosser (2010a,b), apply a transdisciplinary approach when considering complexity. According to this approach, complexity can be studied from a unified single position in economics, physics, and biology. Assessment of the complexity of human labor is not considered by these scientists. We consider the concept of complexity from a psychological perspective and are going to demonstrate that this approach is important for studying the efficiency of human labor and that the analysis of complexity is useful for economics, work psychology, and ergonomics.

Simon (1999) postulated that complexity is a basic property of a system. As we have demonstrated in previous chapters, human work activity should be considered as a system. Hence, we have to describe activity as a system and only after that can we evaluate its complexity. Now ergonomists, work psychologists, and economists need to estimate, first of all, not the physical but the cognitive workload of a task performer because physical efforts for manual work are significantly reduced and cognitive components are increased. Cognitive effort can be estimated if task complexity is evaluated. The more complex the task, the more cognitive effort for its performance is required. The task complexity should be considered as the most important characteristic of a task. It affects the strategies of task performance, mental fatigue, errors, and the failure of task performance.

Activity has a complex structure that unfolds in time, which creates significant obstacles for task complexity evaluation. The main problem is associated with the selection of adequate units of task-complexity measurements. Some scientists, such as Mirabella and Wheaton, 1974; Venda, 1975; Galwey and Drury, 1986; Payne, 1976; and so on, suggest utilizing such units of measure as the number of controls and indicators, the number of alternatives in multiple-choice tasks, and so on. For evaluation of the complexity of computer-based tasks, such measures as task-solving time, the number of different transitions, and the total number of states of the system that describe the task-solving process were suggested by Rauterberg (1996). The example that demonstrates an attempt to utilize non-commensurable units of measure is the GOMS system. This system suggests using *units of roughly equal size* as units of complexity measures. One cannot perform mathematical operations using such units of measure. Recently, the quantitative evaluation of task complexity from a reliability-engineering perspective has been described by Park (2009, 2011). For this purpose, he suggested decomposition of the task into procedural steps. The complexity of a task depends on the basic characteristics of the steps and their quantity. However, all of these steps are not adequate units of measurement of the complexity of human activity (Bedny, 2015). One step of a task can be more complex than another step. A task can have fewer steps but be more complex than the one that has more steps. In general, an analysis of task complexity evaluation methods clearly demonstrates that all authors utilize incommensurable units of measure.

A prerequisite for task complexity evaluation is the development of an activity time structure based on morphological analysis of activity and selection of adequate units of measure. The time structure of activity is a new concept in psychology and task analysis. It is not a separate temporal data of activity such as time of motions, time of task performance, reaction time, reserve time, and so on. The time structure of activity describes a process that unfolds in time and determines the duration of various elements of activity, their logical, sequential, or simultaneous organization specific to their emergence and disappearance in the structure of activity. The main purpose of the morphological analysis is description of the time structure of activity. It includes two basic methods of study: the algorithmic description of activity and determining the temporal parameters of the algorithmic description of activity that unfold in time. The description of the activity time structure is needed when studying human-computer interaction, equipment design, skill acquisition process, and so on. Any changes in the configuration of equipment or software alter the structure of activity in a probabilistic manner. Specialists can evaluate design solutions based on an analysis of activity time structure. This was illustrated using our studies of the inventory receiving task. Consideration of the preferable strategies of task performance based on studying the mechanisms of self-regulation is an important preliminary stage of task analysis (Bedny, 2015; Bedny et al., 2015).

One important aspect of the time structure of activity development is determining what elements of activity can be performed simultaneously and which ones can be performed only in sequence. Depending on preferable strategies, in a particular situation an operator can change the sequence of performance of activity elements to some degree. In cognitive psychology, the possibility of performing activity elements in sequence or simultaneously is discussed, while a human is considered to be a single or a multiple-channel processor of information (see, for instance, Pushler, 1998; Welford, 1967). From an SSAT perspective, a person is not a channel for information processing. He/she is a subject who, depending of his/her goal, motivation, significance of the task, level of concentration of attention on required elements of activity, and so on can develop various strategies of task execution and perform task elements in sequence or simultaneously. A critically important factor that determines the possibility of performing actions in sequence or simultaneously is their complexity and the required level of attention concentration for their performance. The more complex performed actions and the task in general are, the more mental effort they require and the higher the level of attention concentration is during their performance. In SSAT, the possibility of performing elements of activity simultaneously depends on the level of attention concentration. The more complex activity elements are, the higher the attention concentration is, and therefore it is more difficult to perform them simultaneously. This idea is in agreement with data in the MTM-1 system that studied the possibility of performing various motions simultaneously depending on the level of control they require. Motions in the MTM-1 system have a level of control which can be compared with the attention concentration level. This system distinguishes between three levels of control: low, average, and high, where the easiest motions are associated with the low level of control. For example, in motion *Reach (RA)*, an object in a fixed location requires a minimum level of control. Such motion can be performed without visual control by using kinesthetic sense. The motion *Reach (RB)* is more complex. It involves reaching for an object in a location that can vary and requires some visual control. The motion RC (to reach an object that is mixed among other objects) requires a high level of visual and muscular control and a decision in order to select one object out of a bunch of other objects. (UK MTMA, 2000). From a cognitive and activity theory perspective, the term "control" should be substituted by the term "attention concentration". So, RA requires a minimal level of attention concentration and RC requires the highest level of attention concentration in this ordered scale. In our further analysis of motions, we utilize the term "level of attention concentration".

In SSAT, the concept of attention is used for evaluation of the complexity of not just separate motions but also of the whole motor actions, and of cognitive elements combined with motor components of activity. The time structure of activity includes various combinations of activity elements. Depending on the complexity of separate elements, and the specifics of their combinations, the level of attention concentration and therefore the complexity of activity can change.

According to Bloch (1966), attention from neuropsychological perspectives should be considered as activation of neural centers of the brain. The higher the level of activation, the higher the level of wakefulness. Attention per Bloch is one of the levels of wakefulness. Utilizing Bloch's data, we determined ranges of wakefulness and nonspecific activation that can be related to work activity. Attention concentration and nonspecific activation of the brain associated with it is connected with the difficulty to perform particular types of activity (Bloch, 1966; Lazareva et al., 1979, etc.). An analysis of ranges of nonspecific activation of neural centers and wakefulness allowed us to distinguish five levels of activity complexity depending on criteria such as activation of neural centers and concentration of attention (Bedny and Meister, 1997; Bedny, 2015). Elements of activity that require a minimal level of wakefulness (attention) are the simplest, whereas activity elements requiring a maximum level of wakefulness are the most complex ones. We understand activity elements not only as separate motions or actions but also as possible combinations. The fact that activity is a structure that unfolds in time as a process allows selecting qualitatively different intervals of time that include qualitatively different elements of activity as the units of complexity measurements.

Analysis of the obtained theoretical data facilitated development of the order scale for elements of activity. SSAT identifies three categories of complexity of motor components (1–3) and three categories of cognitive components of activity (3–5) (see Figure 8.1).

This figure shows that the simplest motor actions belong to category 1. The most complex motor activity belongs to category 3. The simplest cognitive action has the third category of complexity and the most complex one has the fifth category of complexity. However, in stressful situations, less complex elements of activity can become more complex. For example, a motor action that has the third level of complexity can be transformed into the fifth category of complexity.

We can make a general conclusion. The quantitative evaluation of task complexity is referred to choosing adequate units of measurement that would permit a comparison of different elements of activity. Activity is a structure that unfolds in time, which suggests using time intervals for qualitatively different elements of activity as units of task complexity measurements.

The complexity of activity can be evaluated using the concept of *complexity of a time interval*. Such a unit of measure allows evaluation of the complexity of various elements of an activity that is performed in certain time intervals based on the level of attention concentration.

Activity elements (cognitive and behavioral actions and their constituting elements) can overlap in time, and this factor also influences the complexity

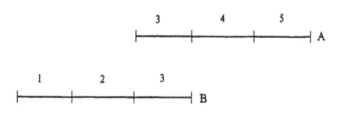

FIGURE 8.1
Five categories order scale for evaluation of complexity of actions or activity in general.

of a considered time interval. Basic rules, which determine the possibility of performing actions simultaneously or in sequence based on the level of concentration of attention, have been developed (Bedny, 2015). Below we present some of these rules.

1. Two motor actions that require a high level of concentration of attention and visual control along all of the action trajectories and are performed outside of the normal visual field should be performed only in sequence.
2. Two motor actions when one of them requires a high level of concentration of attention and the other requires a low or average level of concentration of attention can be performed simultaneously.
3. Cognitive actions require a high level of attention concentration (the third level of complexity) and therefore should be performed in sequence.
4. The decision-making actions and motor actions that require a high level of attention concentration should be performed sequentially.

The short theoretical analysis of assessment of task complexity given above allows us to proceed to the justification of units of measure considered below and to describe formalized procedures of complexity evaluation. All attempts outside of SSAT to measure task complexity found it impossible to select commensurable units of measure. The selection of such units would allow evaluating complexity quantitatively by using adequate mathematical procedures. Activity is a structure that unfolds in time, which suggests using time intervals for qualitatively different elements of activity as units of complexity measurement (Bedny, 1987; Bedny and Meister, 1997; Bedny, 2015). If the time structure of activity during task performance is known, and the time for qualitatively different elements of activity and time of performance of a holistic task is known, one can calculate a mathematical mean to evaluate the fraction of qualitatively different elements of a task.

The following criteria are used to classify qualitatively different units of measure of task complexity: (1) qualitative content of activity elements during a particular interval of time; (2) complexity of these elements of activity according to a five-point scale; (3) possibility of their performance simultaneously or sequentially; (4) probability of appearance of activity elements. For example, based on the first criterion, one should distinguish between time intervals devoted to cognitive activity and those that are devoted to motor activity. A time interval devoted to mental activity is classified based on the dominant cognitive process, such as perceiving a weak signal in the sensory threshold area, perceiving various signals, keeping information in working memory, decision-making, and so on. Each qualitatively different interval of time is related to a particular category of complexity according to the five-point order scale of complexity. At the next step, one should determine the possibility of performing elements of activity simultaneously or sequentially because simultaneous performance of elements of activity changes the complexity of the considered time interval. Then one should evaluate the probability of the appearance of these time intervals during task performance. Finally, a specialist would calculate the mathematical mean and fraction of time for every qualitatively different element of activity. This would allow estimation of the amount of mental effort for the task performance.

Let us consider another aspect of task complexity evaluation that requires development of procedures of evaluation of time intervals when elements of activity are performed simultaneously. What elements of activity could be performed either sequentially or simultaneously depends on the complexity of each element, on strategies of performance, on the significance of the situation, and so on. For example, in a dangerous situation, when each element of activity has a high level of significance, an operator performs them sequentially, even if they are simple. However, in normal situations, if an operator has well-developed skills, these elements of activity could be performed simultaneously. There is also a need to know preferable strategies of task performance. In order to determine what elements of activity can be performed simultaneously, it is necessary to identify the complexity of each element or the level of concentration of attention during their performance. Below we present some formalized rules that have been developed based on the above-described theoretical data and on the data obtained in cognitive psychology, activity theory, and in the MTM-1 system.

Rule 1. Time intervals for motions requiring either a low (A), average (B), or high (C) level of concentration of attention (see the MTM-1 system) can be related according to our rules to the first, second, or third category of complexity, respectively.

Rule 2. If an activity is performed in a stressful situation, then time intervals related to the third and fourth categories of complexity should be classified as the fifth category of complexity intervals. Time intervals related to the first and second categories of complexity should be elevated as the third category. For instance, if an operator performs a simple decision-making action (third level of complexity) but this action is taking place under stressful conditions, the action should be considered as being of the fifth level of complexity.

Rule 3. A time interval when a performer makes a simple "yes-no" or "if-then" decision (simultaneous perceptual action) can be evaluated by using the MTM-1 system microelement *EF*. SSAT distinguishes two situations when microelement *EF* can be utilized if a decision is made based on visual information

Case A. In MTM-1, microelement *EF* means recognition of the stimulus and performance of "yes-no" or "if-then" decisions. According to the MTM-1 system, it takes 0.3 sec to perform *EF*. The first scenario is when in the mass production a worker performs the same production operation multiple times. In such cases, the pace of task performance is approximately the same as in the MTM-1 system.

If an operator recognizes an object and makes a simple "yes-no"/"if-then" decision selecting one out of two possible actions, this is considered as a decision-making action at the sensory-perceptual level, which, according to the data obtained in psychophysics (Green and Swets, 1966), should be divided into two separate cognitive operations. One of them is a *sensory-perceptual operation* and the other one is a *decision-making operation*. In the considered situation, the term operation means element of cognitive or behavioral actions. In the algorithmic description of the task, the first operation is defined as $1/2EF$ and is shown by an afferent operator (O^α), and another one is $1/2EF$, which should be considered as the simplest decision "what should be done based on obtained information" (logical condition *l*). Hence, when analyzing an activity element *EF* for the algorithmic description of activity and defining its time structures, it is important to pay attention to the relationship between the detection or recognition stage and the decision-making stage. Both considered stages require 0.3 sec. According to MTM-1, this simplest cognitive element of work activity requires a high level of concentration of attention and thus should be related to the third category of complexity. This rule is applied when an operator works in time-restricted conditions and performs a well-known task.

Case B. In the second situation, the performer is an operator in man-machine or HCI systems. In such scenarios, the pace of perceiving information and performing the following decisions is slower. An operator performs not tightly interconnected cognitive operations but two interdependent cognitive actions. The first is a simultaneous perceptual action that takes 0.3 sec., and the second is a simultaneously performed decision-making action that also requires 0.3 sec. Thus, performance of these two actions requires 0.6 sec. A specialist should decide when to use Rule 3a or 3b.

Rule 4. If cognitive activity coincides with motor activity, the complexity of such a time interval depends on the specificity of the motor and cognitive elements. Cognitive and motor elements of such an activity interval should be assigned the same level of complexity.

Rule 5. If there are two types of time intervals, one period of time occurs during an actual performance of the task, and the other period of time is associated with the active waiting period, in which an operator observes an ongoing production process and is not directly involved in performance, then the preceding rule is used.

Rule 6. The period of time when two elements of activity that have different categories of complexity are performed simultaneously should be evaluated based on the complexity of the more difficult element.

As an example, we consider Figure 8.2, which depicts simultaneous performance of two elements of activity. Element *A* has longer duration, and the second level of complexity according to the order scale in Figure 8.1. Element *B* has the third level of complexity. The bold-lined rectangle C shows the complexity of the time interval. The time interval t_1 has the third level of complexity, and time interval t_2 has the second category of complexity.

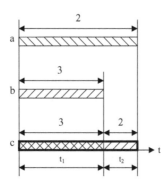

FIGURE 8.2
Graphical interpretation of time interval complexity according to rule 6.

Sometimes it is possible to perform two elements of activity that belong to the third category of complexity (a high level of attention concentration) simultaneously. Such a period of time should be assigned the fourth category of complexity. The combination of these kinds of elements requires the highest level of resource mobilization and often leads to an increase in performance time.

Rule 7. The period of time when two elements of activity that require a high level of concentration of attention (the third category of complexity) are performed simultaneously should be associated with the fourth category of complexity.

Rule 8. When two elements of activity that require a low or average level of concentration of attention are performed simultaneously (motor elements of activity of the first or second category of complexity), the complexity of the overlapping time interval remains unchanged.

Rule 9. Time intervals dedicated to decision-making have three categories of complexity: (a) a time interval with the simplest decision-making actions (when an operator knows in advance how to react to a particular situation) is related to the third category of complexity; (b) a time interval for the more complex decisions, when an operator does not know in advance how to react to a changing situation or decision-making requires extraction of information from memory, such time interval should be assigned the fourth category of complexity; (c) similarly, if choosing from various alternatives is performed when stereotypical decision-making processes are not possible and there are no alternatives given in advance, such time intervals belong to the fourth category of complexity.

There are some additional rules for complexity evaluation (Bedny and Meister, 1997; Bedny and Karwowski, 2007) that are not considered here. Complexity measures and their application will be discussed further when we consider some examples.

In some cases, complex motor activity components can be combined with simple cognitive components. The temporal structure of such activity can be very complex. Such structure can be better understood when the graphical method of its representation is used. In Chapter 6 we described the time structure of a task that involves installation of cylindrical pins into the holes when pins had various shapes. This is a complex manual task that includes simultaneous perceptual actions and a number of decisions. We also described in detail the members of the algorithm O_3^ε - O_6^ε (see Figure 8.3).

Here, we consider evaluation of the complexity of activity during performance of the O_6^ε member of the algorithm that has the most complicate structure when both pins are fluted. After developing the time structure of the activity during its performance, we evaluate the complexity of various time intervals that have different combinations of elements. When assessing the complexity of the elements of the activity, it was taken into account that the pace of the task execution exceeded the optimal one, and that recollection of the rules for the installation of the pins presents a sufficient load on the memory.

A- time structure; B - category of complexity.

The example presented demonstrates that adjacent intervals have different complexity (see bold line B on Figure 8.3). In this figure, element EF is designated by two mental operations: P means perceptual operation and D.M. means decision-making operation. The cognitive action EF (P.D.M.) belongs to the fourth category of complexity because during simultaneous movements of two hands a subject needs to quickly extract a rule about how to turn a pin in the left or right hand. This makes decision-making more complex, and cognitive action should be assigned the fourth category of complexity instead of the third category. In the considered time interval, two motions M22B are performed simultaneously by the left and right hands. Their simultaneous performance does not change the complexity of the time interval. However, combining them with a mental action (P. D. M.) which has

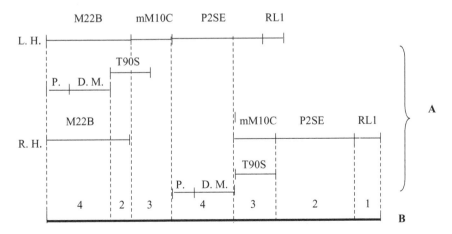

FIGURE 8.3
Time structure of activity and category of complexity activity's elements.

the fourth category of complexity elevates this time interval into the fourth category of complexity (see bold bottom line B, the first time interval from the left). The small time interval that is second from the left has the second category of complexity because this period of time includes elements of the second and the first category of complexity (two elements M22B for the left and right hand of the 2nd category of complexity, and T90S of the first category of complexity). According to the existing rules, a combination of elements of activity of the first or the second level of complexity does not change complexity of a considered interval. Therefore, the second interval (line B) has the second category of complexity. The third interval from the left (see bold line B) is a combination of elements of activity with the third (mM10C) and the first (T90S) categories of complexity. Therefore, this interval has the third category of complexity. The fourth interval includes cognitive action with the fourth (P. D. M.) and second (P2SE) categories of complexity. Hence, this time interval belongs to the fourth category of complexity. The fifth interval from the left has the third level of complexity because an element mM10C according to the level of attention concentration belongs to the third category of complexity and T90S to the first category. PSE belongs to the second level of complexity and T90S to the first level of complexity. The complexity of the two time intervals on the very right of line B is determined similarly.

This example demonstrates how we can evaluate different time intervals when their activity elements can be performed not only in sequence but also simultaneously. Depending on the combined complexity of the described intervals, it is possible to assess the overall complexity of the stage of activity. Although we don't consider the overall activity complexity in this example, we'd like to indicate that according to the ratio between intervals with various complexities, the complexity of the considered stage of activity can also be evaluated (Bedny and Karwowski, 2007). The time structure of activity depends on the constructive features of the equipment of software design and on the methods of performance. Hence it can be utilized not only for time study and assessment of complexity of task performance but also for assessment of equipment and software design solutions.

8.2 Quantitative Assessment of Task Complexity

Analysis of contemporary work clearly demonstrates that tasks are increasingly cognitively demanding, and physical components of work no longer play a critical role in job analysis and design. The design, analysis, and assessment of cognitively demanding tasks play a central role in the study of human performance by ergonomists, work psychologists, and economists. This became especially relevant with the introduction of computers in various types of industries, including aviation and the military. Cognitive task analysis as a new area of studying human performance emerged in response to these most recent demands for job analysis and design. There are a number of critical remarks that can be made for this area of task analysis. First of all, this area of applied science is considered as a number of independent techniques that are not integrated into a unitary theoretical framework. Cognitive components of work are considered to be relatively independent from behavioral components. The strategies of task performance that depend on emotionally evaluative and motivational factors, and principles of goal-directed self-regulation of activity are not consider in this area of study. Precise motor actions that require complex cognitive regulation are eliminated from cognitive task analysis. Unified and standardized principles of formalized and quantitative methods of task analysis and design are not suggested in this area of study. Analysis of material presented in previous chapters demonstrates that SSAT overcomes these shortcomings. In the previous chapters, we have presented qualitative and formalized methods of task analysis and now we are going to describe basic quantitative methods of analyzing cognitively demanding tasks. Two of the most important characteristics of tasks are their complexity and reliability.

We begin our discussion by describing general principles of task complexity evaluation. Complexity is one of the most critical cognitive characteristic of a task that might affect the strategies and efficiency of task performance. The more complex the task, the more mental effort is required for its performance. In cognitive psychology, cognition is separated from motor activity. Complexity is a concept that is used for evaluation of cognitive demands of tasks. According to the cognitive approach, motor components of a task cannot be evaluated by using a concept of complexity. In SSAT, the concept of complexity can be used for evaluation of motor components of activity. The more precise a motor action, the more concentration of attention it requires and the more complex the action is. In general, the more complex motor components of a task are, the more rigorous the requirements are to motor actions regulation and coordination. Thus, motor actions are analyzed not so much based on the positions of the applied physical efforts but based on their cognitive regulation.

We distinguish between a concept of task complexity and a concept of task difficulty. Complexity is an objective characteristic of the task, and difficulty is the performer's subjective evaluation of the task complexity. Depending on the past experience and individual features of a person, the same complex task can be subjectively evaluated as relatively more or less difficult by this person. Complexity and difficulty are often considered to be the same for an idealized and/or average subject. However, such subjects do not exist as we are all different in our experiences and abilities. The concept of complexity can be used for evaluation of design solutions of equipment or interface and evaluation of complexity of methods of task performance. This concept is useful for safety analysis, training process, time study, analysis of wages and efficiency in economics, and so on. Analysis of ergonomic, psychological, and economic literature demonstrates that there are four basic characteristics of human task performance: physically demanding tasks (heavy work), intensity of work, complexity, and difficulty of work. These basic concepts are discussed in great detail by G. Bedny (2015).

Evaluation of task complexity can be performed only after morphological analysis of activity. The activity is a multidimensional system. So the complexity measurement cannot be reduced to one unified measure of complexity. Multiple measures of task complexity should be used. Selection of particular measures of complexity depends on specifics of a considered task. There are differences in measuring task complexity when we evaluate computerizes or computer-based tasks and tasks that do not involve computers (Bedny et al., 2015). The analysis of the possibility of simultaneous and sequential performance of activity elements and evaluation of probability of their occurrence are important features of tasks when measuring their complexity. For a complexity evaluation, it is necessary to identify typical intervals of time that are devoted to the cognitive and behavioral components of activity. Selected elements, with the interval of time for their performance, are classified based on dominant psychic preprocesses; for example, intervals of time for perceiving information, memorization or extraction of information from memory, time intervals for thinking or decision-making, or for performance of motor components. Intervals of time for actual performance should be distinguished from the periods of time when an operator actively waits for a possible signal. Actively waiting periods are classified depending on the level of attention concentration and emotional pressure. After this abbreviated analysis we present below some measures of complexity for rule-based tasks.

The first measure determines the duration of task performance. It should be taken into account that more complex tasks can require more time than simple ones. Calculation is performed according to the following formula:

$$T = \sum_{j=1}^{n} P_i t_i,$$
(8.1)

where:

P_i is the probability of occurrence of the i-th member of the algorithm

t_i is the duration of the i-th member of the algorithm

The performance time of all logical conditions (decision-making actions that determine the logic of the transition from one member of an algorithm to another) can be evaluated according to the following formula:

$$L_g = \sum_{i=1}^{k} P_i^l t_i^l,$$
(8.2)

where:

P_i^l is the probability of occurrence of the i-th logical condition

t_i^l is the duration of the i-th logical condition

Time for logical conditions in the rule-based tasks are usually shorter than in the problem-solving tasks and in most cases they take less than a second. However, in decisions with more than two outputs, they can take longer than that.

Knowing T and L_g, we can determine the fraction of time for decision-making as

$$N_l = L_g / T,$$ (8.3)

where:
L_g is the time for performing logical conditions
T is the time for the entire task performance

This measure characterizes the complexity of the decision-making process during task performance. There are several other measures that characterize the decision-making process from various aspects of activity performance.

The time taken for afferent components of activity and the executive (response) components of activity can be determined as

$$T_\alpha = \sum P_r^\alpha t_r^\alpha; \quad T_{ex} = \sum P_j t_j,$$ (8.4)

where:
P^α and P_j are probabilities of r-th afferent and j-th efferent components of activity
t_r^α, t_j^o are the performance time of r-th afferent and j-th efferent components of activity

Time related to recognizing and identifying weak (i.e., approaching the threshold range) signals can be determined utilizing the following formula:

$$T_\alpha = \sum{}' P_{r'} t_{r'}$$ (8.5)

where:
$P_{r'}$ probability of r'-th afferent operator
$t_{r'}$ performance time of r'-th afferent operator, characteristics of which approach the threshold of the Recognition Differential

The fraction of time for afferent components of activity in the time for executive activity is evaluated as follows:

$$N_\alpha = T_\alpha / T_{ex}$$ (8.6)

The proportion of time for afferent components of activity that is related to recognizing and identifying weak signals (i.e., approaching the threshold) to time for executive activity is evaluated using the following formula:

$$Q_\alpha = {}'T_\alpha / T_{ex}$$ (8.7)

The fraction of time for logical components of activity that depends largely on information selected from long-term memory rather than externally presented information is identified as

$$L_{ltm} = l_{ltm} / L_g ,$$
(8.8)

where l_{ltm} – time for logical components of activity which operational nature is predominantly governed by information retrieved from the long-term memory.

The measure that characterizes the workload on the working memory can be determined according to the following formula:

$$N_{wm} = t_{wm} / T,$$
(8.9)

where:

N_{wm} is the fraction of time for retaining information in working memory during the entire task performance

t_{wm} is the time for storing current information in working memory during task performance

T is the task performance time

Activity can be either stereotypical (repetitive) or changeable (variable). The performance of a stereotypical activity is normally easier; if procedures always take place in a set order, or a given procedure always follows some particular member of an algorithm, these logical components of activity are stereotypical. When procedures and the transition from one action to another have probabilistic features, these procedures are considered variable. Members of the algorithm that are always performed in the same sequence can be considered as stereotypical components of activity. Their sequence is subjectively perceived by an operator as the habitual performance of the same order of actions. If the habitual performance of a stereotypical efferent operator is always followed by the same afferent operator and its associated logical condition, then the afferent operator and its logical condition also belong to the stereotyped activity. It can be hypothesized that the more time in a process is devoted to variable procedures, the more complex this process is. It is possible to calculate measures of stereotypical and variable (changeable) components for executed activity and logical conditions. The time devoted to stereotypical and variable operators and logical conditions can be determined according to the following formulas:

$$t_{st} = \Sigma P_{jst}^o \, t_{jst}^o \, ; \, t_{ch} = \Sigma P_{jch}^o \, t_{jch}^o$$
(8.10)

$$l_{st} = \Sigma P_{ist}^l \, t_{ist}^l \, ; \, l_{ch} = \Sigma P_{ich}^l \, t_{ich}^l ,$$
(8.11)

where:

P^o_{jst}, P^o_{jch} are the probability of the appearance of *j*-th stereotyped and variable operators

t^o_{jst}, t^o_{jch} are the performance time of *j*-th stereotyped and variable operators

P^l_{ist}, P^l_{ich} are the probability of the appearance of *j*-th stereotyped and variable logical conditions

t^l_{ist}, t^l_{ich} are the performance time of *i*-th stereotyped and variable logical conditions

The fraction of stereotypical and variable logical components of activity (decisions) can be determined using the expressions:

$$L_{st} = l_{st} / L_g \tag{8.12}$$

$$L_{ch} = l_{ch} / L_g, \tag{8.13}$$

where l_{st} and l_{ch} - mathematical mean of performance time of stereotypical and variable logical activity.

The fraction of stereotypical and variable executive components of activity can be determined similarly.

Let us now consider methods for evaluation of complexity of a time interval for active waiting periods. The complexity of an active waiting period can be evaluated based on the level of concentration of attention during the waiting period, in accordance with the following rules:

a. If waiting periods require a low, average, or high level of concentration of attention, they are assigned the first, second, and third categories of complexity respectively.

b. When waiting periods convey emotional stress (that is, there is danger of trauma or accident), they belong to the fourth category of complexity.

c. If waiting periods of any level of complexity require continuously keeping information in working memory, their category of complexity should be increased by one.

The existence of an active waiting period in a task requires introduction of additional measures of complexity. One such measure is the "fraction of active waiting periods in the entire task execution time" that is calculated using the following formula:

$$\Delta T_w = t_w / T, \tag{8.14}$$

where:

t_w is the entire time for active waiting periods

T is the total task execution time

If an active waiting period consistently occurs following a particular element of the activity or task, it is considered to be a stereotypical active period that can be determined by the following formula:

$$W^{st} = t_{wst} / t_w,$$ (8.15)

where:
   t_{wst} is the time for stereotypical waiting components
   t_w is the total duration of waiting periods

A task can include several active waiting periods. If the internal psychological content of waiting periods is identical for all of them, then they are repetitive. The fraction of time for repetitive waiting periods in the entire time for active waiting period can be calculated in a similar fashion. All developed in SSAT measures of task complexity evaluation and their psychological interpretation are presented in Table 8.1.

The presented approach to the quantitative assessment of complexity allows, if necessary, to develop new measures of complexity that reflect the specifics of the task under consideration. Thus, it is possible to select measures of complexity that satisfy a wide range of tasks and more specific measures that can be used for a narrower range of tasks. Table 8.1 shows the measures for determining the complexity of computerized tasks. They are used for analyzing the efficiency of information presentation on the computer screen where different areas of the screen have a specific purpose.

As has been stated above, the complexity of motor or cognitive components of activity can be evaluated utilizing the five-point scale of complexity depending on the level of concentration of attention. Development of the principles for estimating complexity of logical conditions (decision-making) using a five-point scale of complexity requires special consideration. Here we need to distinguish rule-based and problem-solving tasks. Decision-making analysis is usually associated with the assessment of problem-solving tasks. Wickens et al. (1998) gave the following description of decision-making tasks: (a) a person needs to select one choice from a number of choices, (b) there is some amount of information available with respect to the choices, (c) the time frame is relatively long (longer than a second), and (d) the choice is associated with uncertainty; that is, it is not clear which is the best choice. The definition presented demonstrates that decision-making involves risk. We can add that in some situations the subject may not know the possible alternatives and may formulate them independently. Decisions that are included in problem-solving tasks may be static or dynamic, well or poorly defined, risky or not risky, and involving or not involving diagnosis. In the last case, the operator does not know all the alternatives, the situation is very unpredictable and dynamic, and he or she may not have enough information to

TABLE 8.1

Measures of the Complexity of Task Performance and Their Psychological Interpretation

No	Name of Measure	Formula for Calculation	Variables	Psychological Meaning
1	Algorithm (task) execution time	$T = \sum P_i t_i$	P_i - probability of occurrence, t_i - duration of i-th member of the algorithm	Duration of task performance
2	Sum of the performance time of all afferent operators	$T_\alpha = \sum P^\alpha t^\alpha$	P^α - probability of occurrence, t^α - duration of r-th afferent operator	Duration of perceptual components of activity
3	Sum of the performance time of all thinking operators	$T_{th} = \sum P^{th} t^{th}$	P^{th} - probability of occurrence, t^{th} - duration of r-th thinking operator	Duration of thinking components of activity
4	Sum of the performance time of all operators which requires keeping information in working memory	$T_{wm'} = \sum P_{wm} t_{wm}$	P_{wm} - probability of occurrence, t_{wm} - duration of wm -th operator associated with keeping information in working memory	Time for retaining current information in working memory
5	Sum of the performance time of all logical conditions	$L_g = \sum P_i t_i$	P_i - probability of occurrence, t_i - duration of i-th logical condition	Duration of decision-making components of activity
6	Sum of the performance time of all efferent operators	$T_{ex} = \sum P_j t_j$	P_j - probability of occurrence, t_j - duration of j-th efferent operators	Duration of executive components of activity
7	Sum of the performance time of all cognitive components of the task (including perceptual activity)	$T_{cog} = T_\alpha + T_{th} + L_g + T_{wm}$	T_α; T_{th}; L_g; (see above); T_μ –duration of keeping information in working memory; (some cognitive components can be equal zero in a particular task)	Total duration of cognitive components of activity
8	Sum of time spent on discrimination and recognition of perceptual distinctive characteristics that are approaching threshold of sensory receptors	$'T_\alpha = \sum P_r t_r$	P_r- occurrence probability, t_r - duration of r'-th afferent operators, characteristics of which approach threshold value (required additional EF)	Duration of perceptual process connected with weak stimuli (approaches threshold characteristics)

(Continued)

TABLE 8.1 (CONTINUED)

Measures of the Complexity of Task Performance and Their Psychological Interpretation

No	Name of Measure	Formula for Calculation	Variables	Psychological Meaning
9	Fraction of time for afferent operators in the time for the entire task performance (N_α)	$N_\alpha = T_\alpha/T$	T_α - time performance of afferent operator; T - time of the entire task performance	Fraction of perceptual components of activity in performance of the entire task
10	Fraction of time for thinking operators in the time for entire task performance (N_{th})	$N_{th} = T_{th}/T$	T_{th} - time performance of thinking operators; (see above)	Fraction of thinking components of activity in performance of the entire task
11	Fraction of time for logical conditions in the time for the entire task performance	$N_l = L_g/T$	L_g - time for performance of logical conditions; T - time for the entire task performance	Fraction of decision-making components of activity in performance of the entire task
12	Proportion of time for cognitive components of task (including perceptual activity) to the performance time for all efferent operators	$N_{cog} = T_{cog}/T_{ex}$	T_{cog} - time for performance of cognitive components; T_{ex} - performance time for all efferent operators	Relationship between cognitive and external behavioral (executive) components of task
13	Fraction of time for logical components of work activity, that largely depends on information selected from long-term memory rather than from external sources of information, in entire time for logical conditions performance	$L_{ltm} = I_{ltm}/L_g$	I_{ltm} - time for logical components of activity whose operational nature is predominantly governed by information retrieved from the long-term memory	Level of memory workload and complexity of decision-making process
14	Fraction of time for retaining current information in working memory in the time for the entire task performance	$N_{wm} = T_{wm}/T$	T_{wm} - time for storing information related to task performance in working memory	Level of working memory workload during the task performance
15	Fraction of time for performance all efferent operators in the time for the entire task performance	$'N_{beh} = T_{ex}/T$	T_{ex} - time for external behavioral (executive) components	Fraction of external behavioral (executive) components of activity in the entire task

(Continued)

TABLE 8.1 (CONTINUED)

Measures of the Complexity of Task Performance and Their Psychological Interpretation

No	Name of Measure	Formula for Calculation	Variables	Psychological Meaning
16	Fraction of time for cognitive activity in the time for the entire task performance	$'N_{cog} = T_{cog}/T$	T_{cog} - time for cognitive components	Fraction of cognitive components of activity in performance of the entire task
17	Fraction of time spent on discrimination and recognition of perceptual distinctive characteristics that are approaching threshold of sensory receptors in the time for the entire task performance	$Q = T_{\alpha}/T$	T_{α} - time for discrimination and recognition of various perceptual distinctive characteristics that are approaching threshold of sensory receptors	Characteristics of complexity, sensory, and perceptual processes
18	Measure of stereotypy of logical processing of information	$L_{st} = l_{st}/L_g$	l_{st} - time for stereotypy of logical processing of information, L_g - time for performance of logical conditions	Characteristic of inflexibility or rigidity of decision-making process
19	Measure of changeability of logical processing of information	$L_{ch} = l_{ch}/L_g$	l_{ch} - time for changeable logical processing of information	Characteristic of irregularity or flexibility of decision-making process
20	Measure of stereotypy of executive components of work activity	$N_{st} = t_{st}/T_{ex}$	t_{st} - time for stereotypy of executive components of activity, T_{ex} - time for executive components of activity	Characteristic of inflexibility or rigidity of executive components of activity
21	Measure of changeability of executive components of work activity	$N_{ch} = t_{ch}/T_{ex}$	t_{ch} - time for changeable executive components of activity	Characteristic of irregularity or flexibility of executive components of activity

(Continued)

TABLE 8.1 (CONTINUED)

Measures of the Complexity of Task Performance and Their Psychological Interpretation

No	Name of Measure	Formula for Calculation	Variables	Psychological Meaning
22	Scale of complexity a. algorithm b. member of algorithm	X_i - level of complexity $(1, 2,\dots5)$	Level of concentration of attention during task performance $(1$-minimum concentration, 5-maximum$)$	Level of mental effort during task performance and performance of various elements. Unevenness of mental effort and critical points of task performance
23	Fraction of time for repetitive logical components of work activity in the performance time of all logical conditions	$Zl = t_{rep}/L_g$	t_{rep} - time for performance of identical logical conditions	Characteristic of habitualness of information processing
24	Fraction of time for repetitive afferent components of work activity in the performance time of all afferent operators	$Z^\alpha = t_{rep}/T_\alpha$	t_{rep} - time for performance of identical afferent components	Characteristic of habitualness of perceiving process
25	Fraction of time for repetitive efferent components of work activity in the performance time of all efferent operators	$Z^{ef} = t_{exrep}/T_{ex}$	t_{exrep} - time for performance of identical efferent components	Characteristic of habitualness of executive components of activity
26	Fraction of active waiting period in the entire work process	$\Delta T_w = t_w/T$	t_w - entire time for active waiting period in work process	Relationship between active waiting period and performance
27	Category of complexity of active waiting periods	$X_w - 1\dots4$	Concentration of attention during waiting period $(1$-minimum, 4-maximum$)$	Level of mental effort during active waiting period
28	Fraction of time for repetitive waiting periods of work activity in the entire time for active waiting period in work process	$Z^w = t_{wrep}/t_w$	t_{wrep} - time for repetitive waiting periods	Characteristic of habitualness of waiting periods
29	Measure of changeability of waiting periods	$W^{ch} = t_{wch}/t_w$	t_{wch} - time for changeable waiting periods	Characteristic of irregularity of waiting periods
30	Measure of stereotypy of waiting periods	$W^{st} = t_{wst}/t_w$	t_{wst} - time for stereotypy waiting periods	Stereotypy of waiting periods

evaluate the consequences of alternative decisions. In the production process, problem-solving tasks that include decision-making are not common.

In most situations, operators of a complex system perform rule-based tasks, which means that decision-making is carried out based on known rules that are derived from existing instructions and regulations. The relationships between rule-based and problem-solving tasks are not sufficiently understood and clearly defined. Only in unexpected, complex, and abnormal situations can tasks that are performed based on existing rules and prescriptions be transformed into problem-solving tasks and knowledge-based decisions associated with them. It is not unusual that some authors who study the work of operators in complex systems discovered that most decisions are made based on prescribed rules. For example, in the study of decision-making in the Navy Anti-Air Warfare Systems, 78% of decisions were performed without any deliberation, based on condition-action rules (Kaempf et al., 1992). However, Kaempf et al. and other researchers who study various type of decisions do not discuss what the production process is and what the relationship is between rule-based and knowledge-based decisions in such a process.

In the past, in the Age of Mechanization, skill-based tasks were the dominant ones. Cognitive components of tasks were reduced, and in most cases motor components of activity were performed in the same sequence. When a worker had to make some decisions, these decisions were simple and of the skill-based type. For example, "if the green bulb lights up, press the green button; if the red bulb is on, press the red button". In the contemporary production process, rule-based tasks, when an operator makes decisions based on prescribed rules, dominate. Problem-solving tasks emerge in unexpected and abnormal situations and only in such situations do knowledge-based decisions play a leading role. Suppose a military pilot is engaged in a fight. He must assess the situation quickly and make adequate decisions in a time-limited and stressful situation. This means that problem-solving situations should be minimized. For such situations, it is very important to develop various training methods and develop special instructions and recommendations in order to transform problem-solving tasks into adequate rule-based tasks.

There are also cases when a subject does not know the rules, forgets them, is distracted or not quite accurately interprets a task, and so on. The listed reasons can lead to transformation of the rule-based tasks into problem-solving ones, which is an unforeseen and undesirable situation. Thinking and memory play the most significant role in problem-solving tasks. Thought processes should be as simplified and as shortened as possible, and the memory workload should be reduced.

Rule-based tasks that include multiple decisions with multiple outputs that occur with various probabilities can be very complex. Inadequate design of such tasks that transfers them into problem-solving, decision-making tasks can lead to undesirable results with serious consequences in the production process or for operator performance, including computer-based tasks.

In order to overcome the above-discussed issues, it is necessary to develop formalized rules for assessment of complexity of decisions (logical conditions) in the rule-based tasks. This type of task plays an important role in the activity of operators who serve complex technical systems or work in a modern industry where a large number of tasks are computer-based or computerized tasks.

The analysis of decisions made by the operator demonstrates that logical conditions or decisions in rule-based tasks can have two or more outputs. The simplest one has only two outputs. More complex ones have three or more outputs with different probabilities. Usually they are performed on the basis of perceptual information and do not require preliminary analysis. We can extract two basic groups of decisions: the simplest ones involve decisions on a sensory-perceptual level, whereas the second group involves the verbally thinking level. For classification of decisions according to their complexity, such criteria as involvement of thinking and memory, existence of emotional stress, and the resulting increased attention concentration during implementation of decisions should also be added.

SSAT offers an order scale with five levels of complexity depending on the attention concentration (see Section 8.1). For *motor components* of activity in normal conditions, there are three levels: I - with minimal concentration of attention; II - with average concentration; and III -with maximum concentration. In stressful situations, complexity can increase up to the fifth level. The same rules can be applied to the complexity of cognitive components based on the level of attention concentration. We can also similarly determine the complexity of only perceptual, only thinking, or only decision-making components of activity. According to the level of concentration of attention, *cognitive components* of activity have the following three levels of complexity: III - minimal concentration of attention; IV- average concentration; and V - maximum concentration. The third and the fourth level of complexity of cognitive activity can shift in stressful situations to the fourth and the fifth level of complexity.

This theoretical analysis can be used for evaluation of various decisions or logical conditions according to the following rules: (1) the simplest logical conditions are related to the third category of complexity, and the most complex decisions are of the fifth category, (2) the simplest third-category decisions are the ones that are based on perceptual information and are of a "yes-no" type. Such decisions do not require keeping information in working memory and are not accompanied by emotional stress. The duration of such decisions usually does not exceed one second. (three logical conditions with performance time exceeding one second and decision-making actions that include mnemonic and thinking operations that make them more complex belong to the fourth category of complexity; (4) if time limit and stress are added to the list of factors, then logical conditions belong to the fifth category of complexity. In rule-based tasks, such a category of complexity of logical conditions is rare.

There are some additional rules for the evaluation of the complexity of logical conditions.

As an additional factor, it is necessary to take into account the total number of logical conditions in the task performance or the variety of tasks performed by an operator. Such factors can lead to an increase in the number of memorized rules and to memory overload. They are especially relevant when the logical conditions have a duration of less than one second. Given these factors, logical conditions that are shorter than one second can be transferred from the third to the fourth category of complexity.

8.3 Evaluation of the Computerized-Task Complexity before and after Innovation

8.3.1 Evaluation of the First Versions of the Computerized-Task Complexity before and after Innovation

This chapter is dedicated to task-complexity evaluation of three versions of the computerized inventory receiving task before and after improvements. We remind the reader that computerized tasks include manual work and utilization of a computer with software specially designed for this task. At the first stage, it is necessary to evaluate the first version of the task before and after innovation. For this purpose, we select the most relevant measures of complexity for the considered version of the task. The comparative analysis of the complexity of the corresponding versions before and after improvement will give us an idea of the efficiency of the implemented innovations based on quantitative criteria. The performance time of the members of the algorithm (absolute measures of task complexity) is presented in Table 8.2. The times are taken from the corresponding algorithmic descriptions of the three versions of the task performance that are presented in Chapter 7. The execution time for each member of the algorithm has been determined by taking into account its probabilistic structure. This simplifies the calculation of quantitative measures. Therefore, we can further determine the duration of the considered elements of activity simply by summarizing their performance time. When the performance time of each member of the algorithm and their probabilities are determined, the performance time of the whole task can be calculated. In order to streamline the calculation of all measures of complexity for the first version of task before improvement, the following table has been developed (see Table 8.2).

From Table 8.2 we can see that the total performance time of the considered version of the task before improvement is $T = 39.18$ sec. At the next step, it is necessary to determine the duration of the separate components of activity that are classified as perceptual components, thinking components, the total performance time for keeping information in working memory, the performance time for decision-making components (logical conditions), the

TABLE 8.2

Absolute Measures of Task Complexity of the First Version of Task before Improvement

Member of Algorithm	Version 1 Before Improvement					
	Performance Time (secs)	T_{ex}	T_α	L_g	T_{wm}	T_{th}
O_{17}^ε	5.7	5.7				
$O_{18}^{\alpha\mu}$	6		6		6	
l_4	0.3			0.3		
O_{19}^ε	0.62	0.62				
↓O_{20}^ε	2.64	2.64				
$O_{21}^{\alpha\mu}$	1.64		1.64		1.64	
l_5	0.4			0.4		
O_{22}^ε	1.7	1.7				
O_{23}^α	1.14		1.14			
**$O_{24}^{\mu th}$	4.2				4.2	4.2
l_6	1.5			1.5		
O_{25}^ε	0.84	0.84				
O_{26}^ε	12.5	12.5				
Total time	39.18	24	8.78	2.2	11.84	4.2

performance time for executive components of activity (motor components), and the performance time of all cognitive components of activity. Table 8.1 shows the following absolute measures of complexity: T_{ex} - the performance time of executive (motor) components of activity; T_α - the performance time of perceptual components; L_g - the performance time of logical conditions (decision-making); T_{wm} - the time for keeping information in working memory; T_{th} - the time for performing thinking operations.

This data allows calculation of the fraction of time for all the above-listed components (relative measures of task complexity). These measures and several others presented in Table 8.1 make it possible to evaluate the complexity of the considered version of the task.

The duration of perceptual components of activity is determined utilizing the formula:

$$T\alpha = O_{18}^{\alpha\mu} + O_{21}^{\alpha\mu} + O_{23}^\alpha = 6 + 1.64 + 1.14 = 8.78 \text{ sec,} \qquad (8.16)$$

where $O_{18}^{\alpha\mu}$, $O_{21}^{\alpha\mu}$, O_{23}^α are members of algorithms that are involved in receiving information.

There is only one thinking member of the algorithm, $O_{24}^{\mu th}$, and therefore $T_{th} = 4.2$ sec.

The duration of time when an operator has to keep information in working memory can be determined by summarizing the performance time of the members of the algorithm presented in the following formula:

$$T_{wm} = O_{18}^{\alpha\mu} + O_{21}^{\alpha\mu} + O_{24}^{\alpha\mu} = 6 + 1.64 + 4.2 = 11.84 \text{ sec} \tag{8.17}$$

The duration of the decision-making components (logical conditions) in task performance is:

$$L_g = l_4 + l_5 + l_6 = 0.3 + 0.4 + 1.5 = 2.2 \text{ sec} \tag{8.18}$$

The duration of the executive components of activity (the performance time of efferent operators) is:

$$T_{ex} = O_{17}^{\varepsilon} + O_{19}^{\varepsilon} + O_{20}^{\varepsilon} + O_{22}^{\varepsilon} + O_{25}^{\varepsilon} + O_{26}^{\varepsilon}$$
$$= 5.7 + 0.62 + 2.64 + 1.7 + 0.84 + 12.5 = 24 \text{ sec} \tag{8.19}$$

The duration of all cognitive components of the task are determined by utilizing the formula below:

$$T_{cog} = T_{\alpha} + T_{th} + L_g = 8.78 + 4.2 + 2.2 = 15.18 \text{ sec} \tag{8.20}$$

If we know total task performance time and duration of its separate components, we can now determine the fractions of time for the above-listed components in the total task performance time. These fractions are presented in Table 8.3.

As discussed above, in SSAT there is an order scale with five levels of complexity depending on attention concentration (see Section 8.1). In normal conditions, the simplest motions require minimal attention concentration and can be related to the first category of complexity. The more complex motions that require maximum concentration can be related to the third category of complexity. For example, in normal conditions we can distinguish the following standardized motions according to the level of concentration of attention. R-A; M-A; GIA are examples of the first category of complexity; R-B; M-B; GIB belong to the second category; and R-C; M-C; G4C belong to the third category of complexity. In stressful conditions, according to SSAT, the complexity category of the above-listed motions can be increased by one.

In the inventory receiving task, there are no motions that belong to the third category of complexity. In spite of this fact, we present measures 15 and 16 in order to demonstrate how to evaluate the complexity of motor components of activity. These measures of complexity are based on an analysis of the level of concentration of attention during the performance of motor activity. They determine the level of mental effort when motor components of activity are performed. The general principles of obtaining such measures are that if any one category of complexity is predominant and exceeds 70% of the performance time, the general complexity of this component of activity belongs to this category of complexity. For example, if 70% of the

TABLE 8.3

Measures of Complexity for Version 1 of the Task before and after Innovation

Order Number of Measures	Measures (Time Measured in Seconds)	Before Innovation	After Innovation	Psychological Meaning
1	Algorithm (task) execution time (T) $T = \Sigma P_i t_i$	39.18	33.42	Duration of task performance
2	Total of the performance time of all afferent operators (T_α) $T_\alpha = \Sigma P \alpha t_\alpha$	8.78	2.78	Duration of perceptual components of task
3	Total of the performance time of all thinking operators (T_{th}) $T_{th} = \Sigma P^{th} t^{th}$	4.2	4.2	Duration of thinking components of task
4	Total of the performance time of all operators that requires keeping information in working memory (including mnemonic operations that combined with other components of activity) (T_{wm}): $T_{wm} = \Sigma P_{wm} t_{wm}$	11.84	5.84	Time for retaining current information in working memory
5	Total of the performance time of all logical conditions (L_g) $L_g = \sum P^l t^l$	2.2	1.9	Duration of decision-making components of task
6	Total of the performance time of all efferent operators (T_{ex}) $T_{ex} = \Sigma P_j t_j$	24	24.54	Duration of executive components of task

(Continued)

TABLE 8.3 (CONTINUED)

Measures of Complexity for Version 1 of the Task before and after Innovation

Order Number of Measures	Measures (Time Measured in Seconds)	Before Innovation	After Innovation	Psychological Meaning
7	Total of the performance time of all cognitive components including perceptual activity (T_{um} is not considered as independent component of cognitive activity in this formula; $T_{um} = 0$), (T_{cog}) $T_{cog} = T_\alpha + T_{th} + L_g$	15.18	8.88	Duration of all cognitive components of task
8	Fraction of time for afferent operators in time for entire task performance (N_α) $N_\alpha = T_\alpha / T$	0.22	0.08	Perceptual workload in the task performance
9	Fraction of time for thinking operators in the time for the entire task performance (T_{th}) $N_{th} = T_{th} / T$	0.11	0.13	Problem solving workload in the task performance
10	Fraction of time for logical conditions in the time for the entire task performance (N_l) $N_l = L_g / T$	0.06	0.06	Decision-making workload in the task performance
11	Fraction of time for performance of cognitive components of task to the entire task performance, including perceptual components (N_{cog}) (combined in time with other components of activity mnemonic operations are not considered) $'N_{cog} = T_{cog} / T$	0.39	0.27	General cognitive workload in the task performance

(Continued)

TABLE 8.3 (CONTINUED)

Measures of Complexity for Version 1 of the Task before and after Innovation

Order Number of Measures	Measures (Time Measured in Seconds)	Before Innovation	After Innovation	Psychological Meaning
12	Fraction of time for logical conditions, which largely depends on information selected from long-term memory rather than from external sources of information, in the entire time for logical conditions performance ($'N_{ltm} = l^u / L_g$)	0	0	Level of memory workload and complexity of the decision-making process
13	Fraction of time for retaining current information in working memory in the time for the entire task performance (N_{wm}) $N_{wm} = T_{wm}/T$	0.3	0.17	Memory workload in the task performance
14	Fraction of time for performance of all efferent operators in the time for the entire task performance $'N_{beh} = T_{ex}/T$	0.61	0.73	Behavioral or motor workload in the task performance
15	Scale of complexity of motor components of activity based on level of concentration of attention (three ordered scale of complexity; 1 – minimum and 3 maximum concentration). In stressful or complex situation complexity of motor components can be increased	1	1	Level of mental effort when motor components are performed
16	Fraction of time (t_{ex}) for more complex components of motor activity to entire executive activity (T_{ex}) $F_{ex} = (t_{ex}II/T_{ex})$	$t_{ex}II = 5.4$ sec-II; 5.4/24 = 0.22 (I – 0.78) (II – 0.22)	$t_{ex}II – 6.6$ sec-II; 6.6/24.54 = 0.27 (I -0.73; II – 0.27)	Level of mental effort when motor components are performed

(Continued)

TABLE 8.3 (CONTINUED)

Measures of Complexity for Version 1 of the Task before and after Innovation

Order Number of Measures	Measures (Time Measured in Seconds)	Before Innovation	After Innovation	Psychological Meaning
17	Scale of complexity of cognitive components of activity based on level of concentration of attention (three order scale of complexity; 3- minimum and 5-maximum concentration). In stressful situation cognitive components of the third or fourth category can increase their complexity a. in the whole task b. logical conditions (decision-making) c. thinking components	For the whole task only IV	For the whole task only IV	Level of mental effort when cognitive components are performed
18	Time (t_{cog}) and fraction of the more complex component of cognitive activity (t_{cog}/T_{cog}) in entire executive activity Fraction of time (t_{cog}) for more complex components of cognitive activity to entire cognitive activity (T_{cog}) $F_{cog} = (t_{cog}/T_{cog})$	$t_{cog} = 13.34$ sec -IV; $13.34/15.18 = 0.88$ (IV – 0.88; III – 0.12)	$t_{cog} = 7.34$ sec -IV; $7.34/8.88 = 0.83$ (IV – 0.83; III – 0.17)	Level of mental effort when cognitive components are performed
19	Scale of complexity of logical conditions (decision-making)	IV	IV	

Here comparison before and after improvement Version 1.

performance time of a particular member of the algorithm is associated with the second category of complexity and the rest belongs to the first category, the total category of this member of the algorithm is of the second category of complexity. If the considered activity element consists of 20% of the first category, 20% of the second, and 60% of the third, the total category of complexity will be of the second order. As a result of stressful situations, some elements of the third category should be shifted to the fourth category of complexity. In order to evaluate the fraction of the fourth category of complexity, the performance time of this interval should be multiplied by the weight coefficient 1.5. If multiplying the performance time allocated to the fourth category of complexity results in 70% of the time for the element performance or more, the total complexity of the considered element of activity should be related to the fourth category of complexity. If the result is less than 70%, it should be added to the time for elements that belong to the third category of complexity. If time for the elements of the third category exceed 70% of total performance time for a member of the algorithm or the entire motor activity, then the considered activity belongs to the third category of complexity. Let us consider a hypothetical example. The motor activity of the task takes 100 sec and 20% of it is of the second category of complexity, 30% belongs to the third category, and 50% belongs to the fourth category. Multiplying 50 sec by the coefficient 1.5 gives 75 sec or 75%. This means that the considered motor activity belongs to the fourth category of complexity.

Now we are going to calculate this measure for our task. Analysis of the motions in Version 1 of the task shows that motor activity has only the first and the second level of complexity because there are no precise motor components that require the third level of attention concentration and there are no stressful situations when motor activity is performed. In order to determine the 15th measure, all motions that can be related to the *second* category of complexity are selected from Table 7.3.

Member of algorithm O_{17}^{ε} contains $(R60B + G1B)$ and $M40B$; O_{19}^{ε} has $R10B$; O_{20}^{ε} has two $R30B$, one $M5B$; O_{22}^{ε} consists of $R20B$ and $R12B$; O_{25}^{ε} has $R26B$; O_{26}^{ε} includes $R6B$ and $G1B$.

Knowing the list of standard motions and taking into account that the pace of execution is used applying the coefficient 1.2, the total time for the motor component of the first and the second category of complexity can be calculated. The second level of complexity is slightly more difficult in this task. Motor components of the activity can be performed simultaneously. However, according to the SSAT rules, combination of the first and second categories of complexity of the motor components does not change the complexity of their time intervals. So the factors of simultaneous performance of motor components can be ignored.

Let us consider measures of complexity 15–18. Their numbers are designated by a bold font. These measures are utilized relatively seldomly. They normally are used in situations when motor and cognitive components of activity can become very complex under stressful conditions. In our example, such conditions do not exist. We utilize these measures just to illustrate the method.

So let us consider measures 15–18 based on the data presented in Table 8.3. Based on the above-described method, we determine the performance time of the motor activity of the second category of complexity (more complex category). This time is $t_{ex}II = 5.4$ sec. So the fraction of time for the motor activity of the second category of complexity is 0.22 and for the first category it is 0.78 (measure 16). Therefore, the fraction of time for the motor activity of the first category of complexity exceeds 70%. Hence, the entire motor activity in this case belongs to the first category of complexity. This measure reflects the mental effort for the performance of the motor components of the activity. The 16th measure defines the fraction of time ($t_{ex}II$) for the complex components of the motor activity in the entire executive activity (T_{ex}). It is determined according to the formula $F_{ex} = (t_{ex}II/T_{ex})$. There are no motor components that would require a high level of attention concentration (the third category) in this task. The second category of complexity is the highest one for the motor actions for this task. If we take into consideration the above statement that measures 15 and 16 can be useful only when motor activity belongs predominantly to the third and higher categories when activity is accompanied by stressful factors, it becomes obvious that these measures are presented for illustration only. Some simplified rules for evaluation of the third and higher categories of complexity for motor activity are the following: determine the duration of motor activity of the third category of complexity; then determine the performance time of the motor activity of the fourth and fifth category; multiply the performance time of the fourth and fifth category by the coefficient 1.5; and add this time to the time for the motor activity that belongs to the third category. If this time exceeds 70% of the overall motor activity, then it belongs to the third or even a higher category. Motor activity randomly belongs to the category that is higher than the second category of complexity.

The same rules can be applied to the analysis of the complexity of the cognitive components of activity. As has been discussed above, in normal conditions the simplest cognitive components belong to the third category of complexity. In the task we are analyzing, the most complex cognitive components can be related to the fourth category of complexity. In our study, the performance time of members of the algorithm that belong to the cognitive components has been obtained by using experimental procedures or data from engineering psychology handbooks. Therefore, the coefficient 1.2 should not be applied in these calculations. Let us determine the complexity of the entire cognitive activity during the task performance and the fraction of the most complex cognitive components of the activity according to the level of attention concentration. In Version 1 of the task, members of the algorithm depicting perceptual or thinking actions that are performed in combination with mnemonic operations should be assigned the fourth category of complexity.

The following members of the algorithm for cognitive activity are related to the fourth category of complexity: $O_{18}^{\alpha\mu}$; $O_{21}^{\alpha\mu}$; $O_{24}^{\mu th}$; l_6. Hence, the performance time of the cognitive components that are related to the fourth category is

$$t_{\text{cog}}IV = O_{18}^{\alpha\mu} + O_{21}^{\alpha\mu} + O_{24}^{\mu th} + l_6$$

$$= 6 + 1.64 + 4.2 + 1.5 = 13.34 \text{ sec}$$

(8.21)

F_{cog} can be determined utilizing the below formula.

$$F_{\text{cog}}IV = (t_{\text{cog}}IV / T_{\text{cog}}) = 13.34 / 15.18 = 0.88 \tag{8.22}$$

Therefore, the fraction of time for the cognitive activity of the fourth category of complexity is 0.88, and is 0.12 for the third category (minimum level of attention concentration). These results bring us to the conclusion that cognitive activity for Version 1 of the task before innovation belongs to the fourth category of complexity. Measures 15 and 16 for the motor components of the activity and measures 17 and 18 for the cognitive activity complement each other and present an accurate picture of the complexity characteristics of the considered activity.

The simplified method of the evaluation of the complexity of the cognitive components can be performed similarly to the evaluation of the motor components. However, we want to remind the reader that the simplest level of complexity of the cognitive components of activity is the third one.

We have completed the analysis of the quantitative measures of complexity of Version 1 before improvement. Before we start our analysis of Version 1 after improvement, it is necessary to remind the reader that after introduction of the barcode scanner, the quantity of members of the algorithm has been reduced. Subtask 0, which precedes Version 1 after improvement, has ten members of the algorithm instead of 16. Hence, the order number of the members of the algorithm O_{17}^{ε} has changed to O_{11}^{ε} after innovation.

The performance time of the members of the algorithm or absolute measures of task complexity of the first version of the task after innovation are presented in Table 8.4.

All measures after improvement have been obtained the same way as for Version 1 before improvement. Therefore, we do not include mathematical calculations for these measures. It is simply necessary to compare the obtained data for Version 1 before and after improvement.

Measures 1 through 7 present the performance time of qualitatively different components of the task. The first measure (task execution time) demonstrates that before improvement, the task performance time was 39.18 sec. After improvement, this time is 33.42 sec. Hence, the performance time for the entire task has been reduced by 5.66 sec (14%). The perceptual workload has been significantly reduced (see measures 2 and 8). Measure 2

TABLE 8.4

Absolute Measures of Complexity for Version 1 of the Task after Improvement

Member of Algorithm	Version 1 After Improvement					
	Performance Time	T_{ex}	T_α	L_g	T_{wm}	T_{th}
↓↓ O_{11}^{ε}	5.7	5.7				
O_{12}^{ε}	2.9	2.9				
O_{13}^{ε}	0.9	0.9				
$O_{14}^{\alpha\mu}$	1.64		1.64		1.64	
l_3	0.4			0.4		
O_{15}^{ε}	1.7	1.7				
O_{16}^{α}	1.14		1.14			
*$O_{17}^{\mu th}$	4.2				4.2	4.2
l_4	1.5			1.5		
O_{18}^{ε}	0.84	0.84				
O_{19}^{ε}	12.5	12.5				
Total	33.42	24.54	2.78	1.9	5.84	4.2

(T_α) for the performance time of the perceptual components of the activity before improvement was 8.78 sec, and performance time of the whole task "T" is 39.18 sec. Therefore, $N_\alpha = T_\alpha/T = 0.22$ before innovation, and $N_\alpha = T_\alpha/T = 2.78/33.42 = 0.08$ after improvement.

The problem-solving workload (thinking process) before and after innovation remained the same in absolute value, but the relative value increased insignificantly (measures 3 and 9). This is due to the fact that the total duration of the task has decreased, and the duration of the thinking component remained the same ($T_{th} = 4.2$ sec). So N_{th} has slightly increased. Analysis of the examples with perceptual and thinking components of activity shows that the corresponding absolute and relative measures should be compared during their interpretation.

The memory workload after implementation of improvements has significantly decreased. Before innovation, time for retaining information in working memory was 11.84 sec and after innovation 5.84 sec (measure 4). The relative value of the memory workload was 0.3 before and 0.17 after innovation (measure 13). The duration of decision-making processes has been slightly reduced after innovation, but the relative value of the decision-making workload has not changed (measures 5 and 10). The duration of all cognitive components in the task performance after innovation has been significantly reduced (see measure 7). Measure 11 also demonstrates that the general cognitive workload has been reduced. The duration of motor components in the task performance (duration of efferent operators) is practically the same (measure 6). The difference, especially for motor activity, is negligible. However, the fraction of behavioral or motor components in the task performance increased slightly (measure 14). Such changes are due to the fact that the time for the motor activity in both versions of the task are practically unchanged because using the barcode scanner after innovation involves some additional motor actions. However, before innovation, the worker can often perform some additional motor actions for scanning an item (see the member of the algorithm $O\,{}^{\varepsilon}_{20}$ repeat if required"). As can be seen, in our calculations the performance time of these additional motor actions before innovation is not included because the quantity of these repetitions can vary.

The time for cognitive components of activity and their complexity are reduced after introduction of the barcode scanner. Usage of the barcode scanner eliminated the following members of the algorithm from Version 1 before improvement in the task performance after innovation (see $O^{\alpha\mu}_{18}$ –6 sec; l_4 – 0.3 sec).

We can observe the following changes in motor activity. Before improvement, after decision-making (l_4), an operator performs O^{ε}_{19} (press arrow key, repeat if required – motor action "R10B + AP1"). The performance time of this action is 0.62 sec. O^{ε}_{20} has a complicated structure, and its performance time is 2.64 sec (see Version 1 before improvement, Table 7.3). In Version 1 after improvement, O^{ε}_{19} is eliminated and O^{ε}_{13} has the same purpose as the member of the algorithm O^{ε}_{20} before innovation. At the same time, O^{ε}_{13} after innovation requires 0.9 sec (Table 7.4).

It is interesting to compare members of the algorithm O^{ε}_{19} and O^{ε}_{20} before innovation with O^{ε}_{12} and O^{ε}_{13} after innovation. The duration of the motor activity after innovation for the last two members of the algorithm has slightly increased. The performance time of O^{ε}_{12} and O^{ε}_{13} after innovation is 3.8 sec. The performance time of O^{ε}_{19} and O^{ε}_{20} before innovation was 3.26 sec (difference 0.54 sec). An operator has to press the arrow key (perform *AP1*) in order to move the cursor to the required line. If the item at hand is not on the list on the first page of the screen, an operator has to browse the item list and press the down key more times. Such browsing requires some perceptual actions and motor actions that include *AP1*. This can increase the duration of the motor activity, and the difference in 0.54 sec might be eliminated. For simplicity, we do not consider this possible situation. Thus, when we compare measure 6 (T_{ex}) before and after improvement, the increase in duration of the executive component from 24 to 24.54 sec (difference 0.54 sec) should not be taken into account due to the high probability of repetition of motion *AP1*. This motion is included in O^{ε}_{19} that depicts scanning of the list of items for the task before improvement.

As a result of the innovations, the cognitive activity and its complexity are reduced, and the behavioral components of activity are shorter on average. Measure 16 shows that before improvement, the fraction of time for the more complex second category of complexity motor components was 0.22 and 0.78 for the first category. After improvement, it was 0.27 for the second category and 0.73 for the first category. Hence, the fraction of time for the more complex category (the second category) of motor activity has decreased.

The meaning of relative measures (fractions) can be understood only by comparison with absolute measures (the performance time of motor components of activity).

Let us evaluate the data in measures 17 and 18 for the cognitive components of activity. A comparison of measure 18 demonstrates that the fraction of time for the cognitive components that belongs to the fourth category of complexity in the entire cognitive activity has been reduced from 0.88 to 0.83.

It is interesting to compare the complexity of the logical conditions before and after improvement. Before improvement there were three logical conditions: l_4, l_5, and l_6. The first two conditions l_4, l_5, belong to the third category of complexity and l_6 to the fourth category of complexity. The performance time of l_4 and l_5 is 0.7 sec, and the performance time of l_6 is 1.5 sec. The total performance time for logical conditions is 2.2 sec. The fraction of time for the logical conditions of the fourth category is $1.5/2.2 = 0.68$. It is close to 70%. So even without taking into account the weight coefficient 1.5 (for the fourth category of activity elements) decision-making can be related to the fourth category of complexity. After improvement, there are only two logical conditions, l_3 and l_4. The first one requires 0.4 sec and belongs to the third category of complexity. The second one requires 1.5 sec and belong to the fourth category of complexity. Therefore, the fraction of time for logical conditions of the fourth category of complexity is 0.79, and 0.21 is of the third category of complexity. Hence, the complexity of the decision-making before and after improvement has the same category of complexity. However, the performance time for the decision-making L_g after innovation has been reduced from 2.2 to 1.9 sec.

Based on the analysis of various measures of complexity, we can make the following conclusions. After innovation, the complexity of the task is reduced in general due to the optimization of the cognitive components of activity. The time of the task performance is reduced. The cognitive workload, and especially the absolute value of the measures, has been significantly reduced.

8.3.2 Evaluation of the Second Versions of the Computerized-Task Complexity before and after Innovation

Absolute measures of the task complexity of Version 2 of the task before implementation of innovations are presented in Table 8.5.

Let us consider the absolute complexity measures for Version 2 of the task before and after innovation (see Table 8.6) for the cognitive components of activity. This would help us to understand how all other measures have been

TABLE 8.5

Absolute Measures of Complexity for Version 2 of the Task before Innovation

Member of Algorithm	Performance Time	T_{ex}	T_{α}	L_g	T_{wm}	T_{th}
O_{17}^{ε}	5.7	5.7				
$O_{18}^{\alpha\mu}$	6		6		6	
l_4	0.3			0.3		
O_{19}^{ε}	0.62	0.62				
$\downarrow O_{20}^{\varepsilon}$	2.64	2.64				
$O_{21}^{\alpha\mu}$	1.64		1.64		1.64	
l_5	0.4			0.4		
O_{22}^{ε}	0.17	0.17				
O_{23}^{α}	0.11		0.11			
$**O_{24}^{\mu th}$	0.42				0.42	0.42
l_6	0.15			0.15		
O_{25}^{ε}	These members of the algorithm are performed in Version 1 only					
O_{26}^{ε}						
O_{27}^{α}						
l_7						
$\downarrow O_{28}^{\varepsilon}$	0.126	0.126				
$\downarrow O_{29}^{\varepsilon}$	1.01	1.01				
$***O_{30}^{\alpha}$	2.36		2.36			
l_8	0.4			0.4		
$O_{31}^{th\mu}$	3.6				3.6	3.6
$O_{32}^{\varepsilon\mu}$	2.98	2.98			2.98	
$O_{33}^{\alpha\mu}$	1.14		1.14		1.14	
l_9^{μ}	4			4	4	
O_{34}^{ε}	These members of the algorithm are performed in Version 3 only					
$\downarrow O_{35}^{\varepsilon}$	0.89	0.89				
O_{36}^{ε}	12.5	12.5				
Total	47.156	26.636	11.25	5.25	19.78	4.02

Version 2 Before Improvement

TABLE 8.6

Absolute Measures of Complexity for Version 2 of the Task after Innovation

Member of Algorithm	Performance Time	T_{ex}	T_α	L_g	T_{wm}	T_{th}
O_{11}^ε	5.7	5.7				
O_{12}^ε	2.9	2.9				
O_{13}^ε	0.9	0.9				
$O_{14}^{\alpha\mu}$	1.64		1.64		1.64	
l_3	0.4			0.4		
O_{15}^ε	0.17	0.17				
O_{16}^α	0.11		0.11			
$*O_{17}^{\mu th}$	0.42				0.42	0.42
l_4	0.15			0.15		
O_{18}^ε	These members of the algorithm are performed in Version 1 only					
O_{19}^ε						
O_{20}^α						
l_5						
↓ O_{21}^ε	0.126	0.126				
↓ O_{22}^ε	1.01	1.01				
$**O_{23}^\alpha$	2.36		2.36			
l_6	0.4			0.4		
O_{24}^ε	2.98	2.98				
O_{25}^α	1.14		1.14			
l_7	3.5			3.5		
↓ O_{26}^ε	0.84	0.84				
O_{27}^ε	12.5	12.5				
Total	37.246	27.126	5.25	4.45	2.06	0.42

obtained. The performance time of all afferent operators before (8.23) and after innovation (8.24) in the second version of the task are:

$$T_\alpha = O_{18}^{\alpha\mu} + O_{21}^{\alpha\mu} + O_{23}^{\alpha\mu} + O_{30}^\alpha + O_{33}^{\alpha\mu}$$
$$= 6 + 1.64 + 0.11 + 2.36 + 1.14 = 11.25 \text{ sec}$$

(8.23)

$$T_\alpha = O_{14}^{\alpha\mu} + O_{16}^\alpha + O_{23}^\alpha + O_{25}^\alpha = 1.64 + 0.11 + 2.36 + 1.14 = 5.25 \text{ sec,}$$

(8.24)

where $O_{14}^{\alpha\mu}$; O_{16}^{α}; ... O_{25}^{α} are members of the algorithms that are involved in receiving information.

Before improvement, the thinking component is associated with two members of the algorithm $O_{24}^{th\mu}$ and $O_{31}^{th\mu}$. After improvement, there is only one thinking component $O_{17}^{th\mu}$. So T_{th} before improvement is determined using formula 8.25 and after improvement using formula 8.26.

$$T_{th} = O_{24}^{th\mu} + O_{31}^{th\mu} = 0.42 + 3.36 = 4.02 \text{ sec} \tag{8.25}$$

$$T_{th} = O_{17}^{th\mu} = 0.42 \text{ sec} \tag{8.26}$$

The performance time of all operators (members of the algorithm and the logical conditions) that require keeping information in working memory before improvement is depicted in formula 8.27 and after improvement in formula 8.28.

$$T_{wm} = O_{18}^{\alpha\mu} + O_{21}^{\alpha\mu} + O_{24}^{th\mu} + O_{31}^{th\mu} + O_{32}^{\varepsilon\mu} + O_{33}^{\alpha\mu} + l_9^{\mu}$$
$$= 6 + 1.64 + 0.42 + 3.6 + 2.976 + 1.14 + 4 = 19.78 \text{ sec} \tag{8.27}$$

$$T_{wm} = O_{14}^{\alpha\mu} + O_{17}^{th\mu} = 1.64 + 0.42 = 2.06 \text{ sec} \tag{8.28}$$

The performance time of all logical conditions (decision-making actions) before (equation 8.29) and after improvement (equation 8.30) are shown below.

$$L_g = l_4 + l_5 + l_6 + l_8 + l_9^{\mu} = 0.3 + 0.4 + 0.15 + 0.4 + 4 = 5.25 \text{ sec} \tag{8.29}$$

$$L_g = l_3 + l_4 + l_6 + l_7 = 0.4 + 0.15 + 0.4 + 3.5 = 4.45 \text{ sec} \tag{8.30}$$

The total performance time of all cognitive components, including perceptual activity before improvement, is determined by formula 8.31 and after improvement by formula 8.32. T_{wm} is not included in this formula as an independent component of cognitive activity.

$$T_{cog} = T_{\alpha} + T_{th} + L_g = 11.25 + 4.02 + 5.25 = 20.52 \text{ sec} \tag{8.31}$$

$$T_{cog} = T_{\alpha} + T_{th} + L_g = 5.25 + 0.42 + 4.45 = 10.12 \text{ sec} \tag{8.32}$$

The purpose of the 12th measure presented in Table 8.1 is determining the complexity of decision-making components of activity when information is extracted from memory ($'N_{ltm} = l_{ltm}/L_g$). There are the following five logical conditions in Version 2 before improvement: $l_4 + l_5 + l_6 + l_8 + l_9^{\mu}$ and only one of them is based on extracting information from memory. This is l_9^{μ}, which

TABLE 8.7

Measures of the Task Complexity for Version 2 before and after Innovation

Order Number of Measures	Measures (Time Measured in Seconds)	Before Innovation	After Innovation	Psychological Meaning
1	Algorithm (task) execution time (T) $T = \Sigma P_i t_i$	47.16	37.25	Duration of task performance
2	Sum of the performance time of all afferent operators (T_α) $T_\alpha = \Sigma P^\alpha t^\alpha$	11.25	5.25	Duration of perceptual components of task
3	Sum of the performance time of all thinking operators (T_{th}) $T_{th} = \Sigma P^{th} t^{th}$	4.02	0.42	Duration of thinking components of task
4	Sum of the performance time of all operators that requires keeping information in working memory (including mnemonic operations that combined with other components of activity) (T_{um}); $T_{um} = \Sigma P_{um} t_{um}$	19.78	2.06	Time for retaining current information in working memory
5	Sum of the performance time of all logical conditions (L_g) $L_g = \Sigma P_g t_g$	5.25	4.45	Duration of decision-making components of task
6	Sum of the performance time of all efferent operators (T_{ex}) $T_{ex} = \Sigma P_j t_j$	26.64	27.13	Duration of executive components of task
7	Sum of the performance time of all cognitive components including perceptual activity (combined in time mnemonic operations with other activity components are not considered in this formula, $T_{um} = 0$), (T_{cog}) $T_{cog} = T_\alpha + T_{th} + L_g$	20.52	10.12	Duration of all cognitive components of task

(*Continued*)

TABLE 8.7 (CONTINUED)

Measures of the Task Complexity for Version 2 before and after Innovation

Order Number of Measures	Measures (Time Measured in Seconds)	Before Innovation	After Innovation	Psychological Meaning
8	Fraction of time for afferent operators in time for entire task performance (N_a) $N_a = T_a/T$	0.24	0.14	Perceptual workload in the task performance
9	Fraction of time for thinking operators in the time for the entire task performance (T_{th}) $N_{th} = T_{th}/T$	0.09	0.01	Problem solving workload in the task performance
10	Fraction of time for logical conditions in the time for the entire task performance (N_l) $N_l = L_g/T$	0.11	0.12	Decision-making workload in the task performance
11	Fraction of time for performance of cognitive components of task to the entire task performance, including perceptual components (N_{cog}) (T_{um} is not considered as independent component of cognitive activity in this formula; $T_{um} = 0$) $N_{cog} = T_{cog}/T$	0.44	0.27	General cognitive workload in the task performance
12	Fraction of time for logical components of work activity, which largely depends on information selected from long-term memory rather than from external sources of information, in the entire time for logical conditions performance ($'N_{ltm} = 'N_l = l^u/L_g$)	0.76	0	Level of memory workload and complexity of the decision-making process
13	Fraction of time for retaining current information in working memory in the time for the entire task performance (N_{um}) $N_{um} = T_{um}/T$	0.42	0.06	Memory workload in the task performance

(Continued)

TABLE 8.7 (CONTINUED)

Measures of the Task Complexity for Version 2 before and after Innovation

Order Number of Measures	Measures (Time Measured in Seconds)	Before Innovation	After Innovation	Psychological Meaning
14	Fraction of time for performance of all efferent operators in the time for the entire task performance $^*N_{beh} = T_{ex}/T$	0.56	0.73	Behavioral or motor workload in the task performance
15	Scale of complexity of *motor* components of activity based on level of concentration of attention (three ordered scale of complexity; 1 – minimum and 3 maximum concentration). In stressful or complex situation complexity of motor components can be increased	II	I	
16	Fraction of time (t_{ex}) for more complex components of motor activity to entire executive activity (T_{ex}) $F_{ex} = (t_{ex}/T_{ex})$	t_{ex}II = 7.21; $9.01/26.64 = 0.34$ t_{ex}III = 1.17; $1.17/26.64 = 0.04$ (II+III – 0.38 (II); I – 0.62)	t_{ex} = 7.8; $7.8/26.67$ 0.29 (II – 0.29; I – 0.71)	Level of mental effort when motor components are performed
17	Scale of complexity of *cognitive* components of activity based on level of concentration of attention (three order scale of complexity; 3- minimum and 5- maximum concentration). a. in the whole task b. logical conditions (decision-making) c. thinking components	a. 4 b. 4 c. 4	a. 3 b. 4 c. 4	
18	Fraction of time (t_{cog}) for more complex components of cognitive activity to entire cognitive activity (T_{cog}) $F_{cog} = (t_{cog}/T_{cog})$	t_{cog} IV = 16.95; $F_{cog} = 16.95/20.7 = 0.82$	t_{cog}IV = 5.56 $F_{cog} = 5.56/10.12 = 0.55$	Fraction of cognitive workload with higher level of mental effort. Unevenness of cognitive workload
19	Scale of complexity of logical conditions (decision-making)	IV	IV	Mental effort during decision-making

describes decision-making actions from three alternatives and is based on a syllogistic conclusion that involves utilizing extracting information from memory. The duration of this decision is 4 sec and the probability of its performance is $P=1$. From above formula 7 one can see that total performance time for L_g is 5.25 sec. Hence, $'N_{ltm}=4/5.25=0.76$. This makes decision-making components of activity during the performance of Version 2 of the task before improvement more complex.

We do not discuss here how time for the executive activity (the motor components of activity) is calculated and just state that $T_{ex}=26.64$ sec for Version 2 before improvement and 27.13 sec after.

Knowing the absolute measures of complexity, it is easy to calculate the relative measures of complexity (see Table 8.7).

Measures of complexity and the method of their calculation has been discussed above. Hence, we now consider measures presented in Table 8.7. It should also be noted that in our calculation we already took into consideration probabilistic characteristics of activity and therefore instead of calculating mathematical means by using various probabilistic data, we simply carry out the final results. As before, the coefficient 1.2 has been utilized to determine performance time of motor activity.

The first measure of complexity demonstrates that task performance time after improvement is reduced by 9.91 sec (21%). The duration of the perceptual processes is reduced approximately two times. The duration of the thinking process also is significantly reduced after innovation. Critically important for this task is measure 4, which demonstrates demands for keeping information in working memory. This measure is reduced from 19.78 to 2.06 sec. The performance time of all logical conditions is also lower after improvement. Thus, all measures that present the performance of various cognitive components of the task demonstrate that the cognitive workload has been significantly reduced. At the same time, the duration of motor components of activity in this task before and after improvement remains practically unchanged. More precisely, the duration of motor components of activity increased by 0.49 sec. However, similarly to Version 1 where we had to compare O_{19}^ε and O_{20}^ε before innovation with O_{12}^ε and O_{13}^ε after innovation in the second version of the task, we compare the same members of the algorithm. Such comparison demonstrates that before improvement, a receiver had to move the cursor down by using an arrow key and look for the item numbers she/he is going to process. After improvement, this activity is replaced by utilizing the barcode scanner. Hence, the difference of 0.49 sec in duration of motor activity before and after innovation for Version 2 can be ignored.

We have considered all measures that have an absolute value. Let us now discuss measures that utilize relative value and compare them with the above-described measures. Before innovation, the fraction of time for afferent operators (the perceptual components of activity) N_α has a value of 0.24 and after innovation 0.14. A comparison of the performance time of all afferent operators T_α and the fraction of time for afferent components (N_α)

before and after improvement demonstrates that the perceptual workload has been significantly reduced. The fraction of time for thinking operations in the entire task performance time (N_{th}) has been reduced from 0.09 to 0.01, and the duration of thinking operations has decreased from 4.02 sec to 0.42 sec (90% reduction). The total time for thinking operations does not present a significant workload, but at the same time, thinking operations belong to the fourth category of complexity because they are performed based on information extracted from memory (see measure 17). This factor can be a reason for further computerization of this operation in the future.

The comparison of time for the logical conditions before and after innovation demonstrates that the total time for decision-making L_g was reduced 15% and the fraction of time for logical conditions (N_l) has increased insignificantly from 0.11 to 0.12 due to the reduction of the total performance time of the task after innovation. The fraction of time for the logical components of work activity, which largely depend on information selected from long-term memory rather than from external sources of information ($'N_{ltm}$), in the total time for logical conditions decreased from 0.76 to 0 (see measure 12). Therefore, decision-making components of activity, according to the comparison of L_g, N_l and $'N_{ltm}$, are simpler after innovation.

The working memory workload is a critical factor for the considered task performance because it can result in fatigue and errors. A comparison of the fourth measure (the duration of keeping information in working memory T_{wm}) and measure number 13 (the fraction of time for retaining current information in working memory in the entire task performance N_{wm}) shows that the workload on working memory has been significantly reduced. In fact, the duration of the active involvement of working memory before improvement was 19.78 sec and only 2.06 sec after improvement (see T_{wm}). The fraction of time when working memory is involved in task execution before improvement was 0.42, and after improvement it has been reduced to just 0.06 (see N_{wm}).

Measure 11 (N_{cog}) demonstrates that the fraction of time for performing the cognitive components of the task in the entire task performance, including the perceptual components, has been reduced, and the fraction of time for motor components (measure 14) has increased because the total task performance time has been significantly decreased.

The fraction of time t_{ex} for the most complex components of motor activity in the entire executive activity (measure 16) is determined as $F_{ex}=t_{ex}/T_{ex}$. All executive members of the algorithm, excluding $O_{32}^{\varepsilon\mu}$, belong to the first or second category of complexity. For this member of the algorithm, the motor actions coincide with the memory workload. This factor raises the level of the attention concentration and the category of complexity of $O_{32}^{\varepsilon\mu}$.

Let us consider the scenario when we ignore this fact and consider $O_{32}^{\varepsilon\mu}$ simply as O_{32}^{ε}.

Then *R20B*, *R6B*, and *R12B* belong to the second category of complexity.

In such a situation, the fraction of time t_{ex}II for the most complex components of motor activity in the entire executive activity (measure 16) is determined as $F_{ex} = t_{ex}II/T_{ex}$. The most complex executive (motor) components of activity belong to the second category of complexity according to the level of attention concentration. The performance time of these motor components is $t_{ex}II = 7.21$ sec. The rest of the motor components belong to the first category of complexity (the lowest level of attention concentration for motor components). The fraction of time of the second category of complexity in the entire time of the motor activity is $F_{ex} = t_{ex}II/T_{ex} = 7.21/26.64 = 0.27$ (measure 16). Hence, the fraction of the first category in the entire motor activity is 0.76. So, motor activity is not complex according to this criterion.

However, if we take into consideration that the receiver has to key in the new price and continue keeping it in the working memory when performing $O_{32}^{\varepsilon\mu}$ in order to be able to perform the next members of the algorithm, this member of the algorithm should be upgraded to the next category of complexity. It includes $(R20B + AP1) + 2 \times (R6B + AP1) + (R12B + AP1)$.

As per SSAT rules and MTM-1 data, motions $R20B$, $R6B$, and $R12B$ belong to the second category of complexity, and three motions $AP1$ are of the first category. The memory workload brings the first group up from the second to the third category and $AP1$ from the first to the second category of complexity. The calculation shows that $t_{ex}III = 1.17$ sec and the performance of $AP1$ that occurs four times takes 1.8 sec when applying the coefficient of pace 1.2.

Now the time for the second category is $t_{ex}II = 7.21$ sec + 1.8 sec = 9.01 sec, accounting for the memory workload and $F_{ex} = t_{ex}II/T_{ex} = 9.01/26.64 = 0.34$.

The performance time of all motions that belong to the third category of complexity in the considered member of the algorithm is 1.17 sec and $F_{ex} = t_{ex}III/T_{ex} = 1.17/26.64 = 0.04$.

The fraction of the motor components of the third category in the entire motor activity is very small. So we can simply integrate the third category with the second category of complexity. Then, the performance time of the second category of the motor activity is 9.01 sec + 1.17 sec = 10.18 sec, and the fraction of the second category is $F_{ex} = t_{ex}II/T_{ex} = 10.18/26.64 = 0.38$. Hence, the fraction of the first category of complexity is 0.62.

This shows that the first category of complexity takes less than 70%, and we should consider the motor components of Version 2 of the task before improvement to be of the second category of complexity. If we don't take into account the memory workload, it belongs to the first category of complexity (see prior calculations).

In general, estimation of the complexity of the majority of the motor activity based on the assessment of the level of concentration of attention before and after the innovation belongs to the first category (measure 15). This is mainly due to the fact that considerable time is spent on leg movement while an operator moves around in the workplace, and hand movements are relatively simple.

Leg movements are complex only in specific situations, such as walking on a slippery surface, walking where there are obstacles, walking where there is a possibility of getting injured, and so on. Otherwise, this kind of motor activity does not require a considerable concentration of attention, and it belongs to the first category of complexity (a lower level of complexity according to the level of attention concentration). Only certain motor actions of the hands require a high level of accuracy and are of the second category of complexity. Hence, in general, motor components of activity are relatively simple and belong to the first or the second category of complexity according to level of attention concentration. We have presented some aspects of motor movement assessment in order to illustrate the method, because the qualitative analysis of the standardized motions demonstrates that there are practically no precise motions in the entire task. We recommend using this measure only if the task has very complex motor components or if these components are performed under stress, which increases the difficulty of their performance.

The simplest cognitive activity requires the third level of attention concentration. More complex cognitive components can have the fourth or even fifth category of complexity. As can be seen in Version 2 of the task, even after improvement some cognitive components belong to the fourth category of complexity. For example, $O_{14}^{\alpha\mu}$ and $O_{17}^{th\mu}$ after improvement are of the fourth category of complexity because the first one combines perceptual actions with mnemonic operations, and the second one combines thinking actions with mnemonic operations.

Hence, the fraction of time t_{cog} for the most complex components of cognitive activity in the entire cognitive activity before improvement is $F_{cog} = t_{cog}II/T_{cog} = 16.95/20.7 = 0.82$, and after improvement it is $F_{cog} = t_{cog}II/T_{cog} = 5.56/10.12 = 0.55$. This measure demonstrates that cognitive efforts are reduced after implementation of innovations.

Measure 18 is involved in evaluation of the complexity of the entire cognitive activity. For this version of the task after improvement, 55% of the time in cognitive activity is dedicated to the perceptual process and 45% is dedicated to thinking and decision-making. The last cognitive components belong to the fourth category of complexity. Hence, coefficient 1.5 should be applied when we evaluate the fraction of time for the fourth category of complexity. So, $45\% \times 1.5 \approx 70\%$ of time during the performance of cognitive activity is of the fourth category of complexity. The perceptual components of the activity are not always optimal because visual information from the computer screen is frequently not located in the optimal visual field. Thus, cognitive activity before innovation should be related to the fourth category of complexity. After innovation, all cognitive activity belongs to the third category of complexity because cognitive component of the fourth category are eliminated.

It is interesting to consider the logical conditions (decision-making). Before improvement, an operator has to make a decision l_9^μ, not by using externally presented information but by using information extracted from memory. Such decisions are more complex than the ones that utilize externally presented information.

After innovation, the need to keep information in memory during decision-making is eliminated, and the same logical condition is described as l_7. This decision-making requires less time for performance than l_9^μ.

Let us now evaluate the complexity of decision-making components for the whole task. Before improvement, there are two logical conditions l_6 and l_9^μ, that are of the fourth category of complexity. The performance time of l_6 is 1.5 sec, and for l_9^μ it is 4 sec. Taking into consideration that the probability of performance of l_6 is $P=0.1$, time of performing these logical conditions is $1.5 \times (0.1+4) = 4.15$ sec. The performance time of all logical conditions is 5.25 sec. So the fraction of time for more complex logical conditions to the time for all logical conditions is $4.15/5.25 = 0.8$. This means that an operator makes decisions that belong to the fourth category of complexity 80% of the time (more than 70%). This means that the whole decision-making process should be related to the fourth category of complexity.

After improvement, the logical conditions l_4 and l_7 are of the fourth category of complexity.

Taking into consideration that l_4 has probability $P=0.1$ and its performance time is 1.5 sec, and l_7 has duration 3.5 sec, the fraction of the logical conditions that are related to the fourth category is 0.82. Hence, after innovation, the logical conditions also belong to the fourth category of complexity.

Summarizing the analysis of the second version of the task before and after innovation, we can conclude that, thanks to innovation, its complexity is decreased and that is demonstrated in the analysis of the cognitive components of the task.

8.3.3 Evaluation of Complexity of Third Versions of the Computerized Task before and after Innovation

Before analyzing Version 3 we would like to remind the reader that we are considering a computerized task with a complex logical and probabilistic structure. The purpose of this task is to register all purchases and move the intermediate and final products to the warehouse or one of the workshops. The task includes physical elements and work with a computer. An operator receives items arriving from vendors, fulfills special and emergency orders, and organizes items for further put-away in the warehouse. Parts arrive in boxes that are delivered to the reception area where a receiver performs her/his task. The task begins when an item is taken out of the box, registered in the computer system, and put in a special tote based on the information provided by the software in use. Analysis of this complex task demonstrates that there is the setup subtask and three versions of the main task. We mostly analyzed the main task.

Based on the analysis conducted, we outlined three versions of the main task: Version 1, Version 2, and Version 3. The first version of the task ends when an item is rejected due to the unacceptable quantity. The second version of the task ends when an item is rejected because its price is unacceptable.

We analyzed these versions of the tasks and created their models before and after improvements. By utilizing the original method of analysis of the probabilistic and logical structure of the task, we discovered that the probability of Version 1 of the task is $P_1 = 0.07$, and the probability of Version 2 is $P_2 = 0.03$. Thus, their combined probability is $P = 0.1$. Therefore, the third version of the task has the probability $P_3 = 0.9$. This probability also has been determined by using our theoretically developed method. The third version of the task is the most frequently encountered version when an item is accepted and put into the corresponding tote. These items are later moved into the warehouse. As can be seen, the third version of the task includes all the elements of the second task, but here an item price is accepted, and an item is processed all the way through and received. All improvements that were implemented in the first and second versions of the tasks were also applied in the third version.

Version 3 of the task includes assigning the bin location for an item that has been received before or a bin type for the new items. Bin location is determined based on bin number that includes information about the area in the warehouse and the shelf number. The first two capital letters and number present information about the area and the following numbers define the shelf. For example, BB123456 gives the exact information about the location of the item. If a shelf is already reserved for an item, the system will select it automatically.

The label with the warehouse location is printed out, placed on the item and the item is put in the corresponding tote. If an item is received for the first time, the bin type is assign by the system based on the item description. If this assignment is rejected by an operator, she/he can change it. An operator then needs to recall the bin types and their description or refer to the printed-out table located on the desk. This causes an additional mental workload for an operator.

This short description of the main parts of Version 3 is a reminder of what an operator has to do at this stage of the task performance.

Let us briefly consider measures of complexity for Version 3 of the task before innovation that have absolute value. These measures are presented in Table 8.8.

Absolute measures of task complexity for Version 3 of the inventory receiving task after improvement are presented in Table 8.9.

In order to evaluate the efficiency of the third version of the inventory receiving task improvements, it is necessary to compare measures of complexity of the before and after improvement versions.

Let us consider measures 15 and 16, which are presented in Table 8.10 for before and after improvement versions of the task. In Version 3, the same as in the prior versions, the motor activity includes only standardized motions of the first and the second category of complexity.

SSAT offers new principles of utilizing the MTM-1 system for the description of motor activity that is combined with cognitive components in the

TABLE 8.8

Absolute Measures of Task Complexity for Version 3 of the Inventory Receiving Task before Improvement

Member of Algorithm	Performance Time	T_{ex}	T_α	L_g	T_{wm}	T_{th}
O_{17}^ε	5.7	5.7				
$O_{18}^{\alpha\mu}$	6		6		6	
l_4	0.3			0.3		
O_{19}^ε	0.62	0.62				
$\downarrow O_{20}^\varepsilon$	2.64	2.64				
$O_{21}^{\alpha\mu}$	1.64		1.64		1.64	
l_5	0.4			0.4		
O_{22}^ε	0.17	0.17				
O_{23}^α	0.11		0.11			
$**O_{24}^{\mu th}$	0.42				0.42	0.42
l_6	0.15			0.15		
O_{25}^ε	These members of the algorithm are performed in Version 1					
O_{26}^ε	only					
O_{27}^α						
l_7						
$\downarrow O_{28}^\varepsilon$	0.126	0.126				
$\downarrow O_{29}^\varepsilon$	1.01	1.01				
$***O_{23}^\alpha$	2.36		2.36			
l_8	0.4			0.4		
$O_{31}^{th\mu}$	0.72				0.72	0.72
$O_{32}^{\varepsilon\mu}$	0.59	0.59			0.59	
$O_{33}^{\alpha\mu}$	0.23		0.23		0.23	
l_9^μ	0.8			0.8	0.8	
O_{34}^ε	0.2	0.2				
ω_1	These members of the algorithm are performed in Version 2					
$\downarrow O_{35}^\varepsilon$	only					
O_{36}^ε						
O_{37}^α						
l_{10}						
$\downarrow\downarrow O_{38}^\varepsilon$	0.89	0.89				
$***O_{39}^\alpha$	0.72		0.72			
l_{11}^μ	0.4			0.4	0.4	
O_{40}^ε	0.06	0.06				
$\downarrow O_{41}^\varepsilon$	0.86	0.86				
O_{42}^α	0.3		0.3			
l_{12}	0.3			0.3		
$O_{43}^{\alpha\mu}$	0.18		0.18		0.18	

(Continued)

TABLE 8.8 (CONTINUED)

Absolute Measures of Task Complexity for Version 3 of the Inventory Receiving Task before Improvement

	Version 3 Before Improvement					
Member of Algorithm	Performance Time	T_{ex}	T_α	L_g	T_{wm}	T_{th}
l_{13}^μ	0.06			0.06	0.06	
$O_{44}^{\varepsilon\mu}$	0.02	0.02			0.02	
ω_2						
O_{45}^{ε}	0.03	0.03				
O_{46}^{α}	0.03		0.03			
O_{47}^{ε}	0.067	0.067				
$\downarrow\downarrow\downarrow O_{48}^{\varepsilon}$	0.7	0.7				
O_{49}^{ε}	6	6				
O_{50}^{ε}	6.7	6.7				
O_{51}^{α}	1.71		1.71			
l_{14}	0.4			0.4		
O_{52}^{ε}	1.32	1.32				
O_{53}^{α}	1.71		1.71			
$\downarrow O_{54}^{\varepsilon}$	1.8	1.8				
O_{55}^{α}						
l_{15}						
Total	48.843	29.503	14.99	3.21	11.06	1.14

contemporary tasks. Traditionally, the MTM-1 system is applied for production operation where all elements of activity are performed in the same sequence. We also want to remind the reader that in some situations the performance times of motor actions or their motions can be obtained specific cases, but the symbolic description of motions still should be the same as in the MTM-1 system. So, the uniform description of motor components of activity is preserved.

Let us consider the motor motions of the second category of complexity and their probability of occurrence in Version 3 before innovation. There are also two members of algorithm $O_{32}^{\varepsilon\mu}$ and $O_{44}^{\varepsilon\mu}$ where motor actions are combined with mnemonic operations.

Involvement of mnemonic operations (keeping information in working memory) increases the necessary level of attention concentration. Hence, these standardized motions should be upgraded to a higher category of complexity. $O_{32}^{\varepsilon\mu}$ includes *R20B*; 2×*2R6B* and *R12B*, which according to MTM-1 belongs to the second category of attention control and in accordance the SSAT rules should be transferred into the third category of complexity; 4×*AP1* belongs to the first category of attention control and should be raised to the second category of complexity.

Similarly, $O_{44}^{\varepsilon\mu}$ consists of one motor action that includes two standardized motions: *R37B* of the second category and *AP1* of the first category of

TABLE 8.9

Absolute Measures of Task Complexity for Version 3 of the Inventory Receiving Task after Improvement

Member of Algorithm	Performance Time	T_{ex}	T_α	L_g	T_{wm}	T_{th}
		Version 3 After Improvement				
O_{11}^ε	5.7	5.7				
O_{12}^ε	2.9	2.9				
O_{13}^ε	0.9	0.9				
$O_{14}^{\alpha\mu}$	1.64		1.64		1.64	
l_3	0.4			0.4		
O_{15}^ε	0.17	0.17				
O_{16}^α	0.11		0.11			
$^*O_{17}^{\mu th}$	0.42				0.42	0.42
l_4	0.15			0.15		
O_{18}^ε	These members of the algorithm are performed in Version 1					
O_{19}^ε	only					
O_{20}^α						
l_5						
$\downarrow O_{21}^\varepsilon$	0.126	0.126				
$\downarrow O_{22}^\varepsilon$	1.01	1.01				
$^{**}O_{23}^\alpha$	2.36		2.36			
l_6	0.4			0.4		
O_{24}^ε	0.59	0.59				
O_{25}^α	0.23		0.23			
l_7	0.7			0.7		
$\downarrow O_{26}^\varepsilon$	These members of the algorithm are performed in Version 2					
O_{27}^ε	only					
$^{***}O_{28}^\alpha$						
l_8						
$\downarrow O_{29}^\varepsilon$	0.178	0.178				
O_{30}^ε	0.83	0.83				
$^{***}O_{31}^\alpha$	0.72		0.72			
l_9^μ	0.4			0.4	0.4	
O_{32}^ε	0.06	0.06				
$\downarrow O_{33}^\varepsilon$	0.86	0.86				
O_{34}^α	0.74		0.74			
l_{10}	0.1			0.1		
$\downarrow O_{35}^\varepsilon$	0.14	0.14				
O_{36}^ε	1.19	1.19				
O_{37}^ε	1.1	1.1				

(Continued)

TABLE 8.9 (CONTINUED)

Absolute Measures of Task Complexity for Version 3 of the Inventory Receiving Task after Improvement

Member of Algorithm	Performance Time	T_{ex}	T_α	L_g	T_{wm}	T_{th}
O_{38}^α	0.34		0.34			
O_{39}^ε	0.37	0.37				
ω_1						
$\downarrow O_{40}^{\alpha\mu}$	0.15		0.15		0.15	
l_{11}^μ	0.048			0.048	0.48	
O_{41}^ε	0.00936	0.00936				
O_{42}^α	0.016		0.016			
O_{43}^ε	0.005	0.005				
$\downarrow\downarrow O_{44}^\varepsilon$	0.56	0.56				
O_{45}^ε	4.8	4.8				
O_{46}^ε	4.32	4.32				
O_{47}^α	1.34		1.34			
O_{48}^ε	1.48	1.48				
ω_1						
$\downarrow O_{49}^\alpha$						
l_{15}						
Total	37.56	27.29836	7.646	2.2	2.66	0.42

complexity. Because these motions are combined with mnemonic operations, their category of complexity should be increased as well. As a result, *R37B* is related to the third category of complexity and *AP1* to the second category. Knowing the list of standard motions and their probability, and taking into account that the pace of their execution is used with the coefficient 1.2, we can calculate the total time for the motor component of the first, the second, and the third category of complexity.

The performance time of motions that belong to the second and the third categories of complexity has been determined first. The performance time of motions that belong to the second category of complexity is 8.97 sec. At the next step, we have applied the coefficient 1.2 and 8.97 sec × 1.2 = 10.77 sec. Therefore, $F_{ex} = (t_{ex}II/T_{ex}) = 10.77/29.5 = 0.37$.

The performance time for motion of the third category of complexity has been determined similarly. Before taking into consideration the coefficient 1.2 this time was 0.5 sec and after multiplying it by 1.2 this time is equal 0.6 sec. Hence, this $F_{ex} = t_{ex}III/T_{ex} = 0.6/29.5 = 0.02$. This category has a very small part in the whole motor activity. So, the second and the third categories of complexity have been combined. The fraction of time for the second and the third categories is $0.37 + 0.02 = 0.39$. Knowing this data, it is apparent

TABLE 8.10

Measures of Complexity of Version 3 of the Task before and after Improvement

Order Number of Measures	Measures (Time Measured in Seconds)	Before Innovation	After Innovation	Psychological Meaning
1	Algorithm (task) execution time (T) $T = \Sigma P_i t_i$	48.84	37.56	Duration of task performance
2	Sum of the performance time of all afferent operators (T_α) $T\alpha = \Sigma P^\alpha t^\alpha$	15	7.65	Duration of perceptual components of task
3	Sum of the performance time of all thinking operators (T_{th}) $T_{th} = \Sigma P^{th} t^{th}$	1.14	0.42	Duration of thinking components of task
4	Sum of the performance time of all operators that requires keeping information in working memory (including mnemonic operations that combined with other components of activity) (T_{wm}); $T_{wm} = \Sigma P_{wm} t_{wm}$	11.06	2.66	Time for retaining current information in working memory
5	Sum of the performance time of all logical conditions (L_g) $L_g = \Sigma P_g t_g$	3.21	2.2	Duration of decision-making components of task
6	Sum of the performance time of all efferent operators (T_{ex}) $T_{ex} = \Sigma P_j t_j$	29.5	27.3	Duration of executive components of task
7	Sum of the performance time of all cognitive components including perceptual activity (T_{wm} is not considered as independent component of cognitive activity in this formula; $T_{wm} = 0$), (T_{cog}) $T_{cog} = T_\alpha + T_{th} + L_g$	19.34	10.27	Duration of all cognitive components of task

(Continued)

TABLE 8.10 (CONTINUED)

Measures of Complexity of Version 3 of the Task before and after Improvement

Order Number of Measures	Measures (Time Measured in Seconds)	Before Innovation	After Innovation	Psychological Meaning
8	Fraction of time for afferent operators in time for entire task performance (N_a) $N_a = T_a/T$	0.31	0.2	Perceptual workload in the task performance
9	Fraction of time for thinking operators in the time for the entire task performance (T_{th}) $N_{th} = T_{th}/T$	0.02	0.01	Problem- solving workload in the task performance
10	Fraction of time for logical conditions in the time for the entire task performance (N_l) $N_l = L_g/T$	0.066	0.07	Decision-making workload in the task performance
11	Fraction of time for performance of cognitive components of task to the entire task performance, including perceptual components (N_{cog}) (combined in time with other components of activity mnemonic operations are not considered) $N_{cog} = T_{cog}/T$	0.4	0.27	General cognitive workload in the task performance
12	Fraction of time for logical components of work activity, which largely depends on information selected from long-term memory rather than from external sources of information, in the entire time for logical conditions performance $(L_{lim} = l^y/L_g)$	0.39	0.33	Level of memory workload and complexity of the decision-making process
13	Fraction of time for retaining current information in working memory in the time for the entire task performance (N_{wm}) $N_{wm} = T_{wm}/T$	0.23	0.07	Memory workload in the task performance

(Continued)

TABLE 8.10 (CONTINUED)

Measures of Complexity of Version 3 of the Task before and after Improvement

Order Number of Measures	Measures (Time Measured in Seconds)	Before Innovation	After Innovation	Psychological Meaning
14	Fraction of time for performance of all efferent operators in the time for the entire task performance $^t N_{beh} = T_{ex}/T$	0.604	0.73	Behavioral or motor workload in the task performance
15	Scale of complexity of *motor* components of activity based on level of concentration of attention (three ordered scale of complexity; 1 – minimum and 3 maximum concentration). In stressful or complex situation complexity of motor components can be increased	II	II	
16	Fraction of time (t_{ex}) for more complex components of motor activity to entire executive activity (T_{ex}) $F_{ex} = (t_{ex}/T_{ex})$ $F_{ex} = (t_{ex}\text{II}/T_{ex}) = 10.77/29.5 = 0.37$	$t_{ex}\text{II} = t_{ex}\text{II}+$ $t_{ex}\text{III} = 0.39$ $t_{ex}\text{I} = 0.61$	$tex\text{ II} = 0.46;$ $tex\text{ I} = 0.54$	Level of mental effort when motor components are performed
17	Scale of complexity of *cognitive* components of activity based on level of concentration of attention (three order scale of complexity; 3 - minimum and 5- maximum concentration). a. in the whole task b. logical conditions (decision-making) c. thinking components	a. 4 b. 4 c. 4	a. 3 b. 3 c. 4	Level of mental effort when cognitive components are performed
18	Fraction of time of cognitive components related to the fourth category of complexity in entire cognitive activity $F_{cog} = (t_{cog}/T_{cog})$	$t_{cog}\text{ IV} = 10.6;$ $F_{cog} = 10.6/19.34 * 1.5 = 0.82$	0.12	Fraction of cognitive workload with higher level of mental effort. Unevenness of cognitive workload
19	Scale of complexity of logical conditions (decision making)	III	IV	Mental effort during decision-making

that the fraction of time for the motions of the first category of complexity is 0.61.

The same calculations have been performed for Version 3 after improvements. Motions of the second category of complexity took 10.62 sec. Taking into consideration the coefficient 1.2 makes it 12.74 sec. Then, $F_{ex} = t_{ex}\text{II}/T_{ex} = 12.74/27.3 = 0.46$. In Version 3 after improvement there are no motions of the third category of complexity. Therefore, the fraction of time for the second category of complexity is 0.46 and for the first category it is 0.54.

It is interesting that innovation does not give any advantages in terms of motor activity according to the criteria considered because the introduction of the WIP option requires some additional movements around the workplace and other additional motor activity.

It should be noted that we present these calculations, first of all, to demonstrate the method of obtaining these measures. There are no high-precision motions that require a high level of attention concentration in the considered task. Only two members of the algorithm are of the third category of complexity, and even they have a low probability of appearance in the context of this task. Thus, the qualitative analysis demonstrates that the first and the second category of complexity are the dominating complexity levels of motor activity. At the same time, if the task includes the noticeable quantity of precise motions that require a high level of attention concentration, such as *RC*; *MC*; *G4C*; and so on, measures 15 and 16 are very useful for evaluation of motor components of activity before and after innovation. These measures, the same as the other ones, can be obtained at the design stage when observation or experiment can't be used.

It is necessary to stress that although two members of algorithm $O_{32}^{\varepsilon\mu}$ and $O_{44}^{\varepsilon\mu}$ have low probability, we should pay attention to them because they appear in the activity infrequently and being rather complex, can provoke errors in task performance.

This study demonstrates that SSAT makes it possible to adapt the MTM-1 system not only for time study and such important areas of economics as evaluation of efficiency and complexity of performance, but also for work psychology, engineering psychology, and ergonomics.

It is interesting to compare the L_g (performance time of all logical conditions) for Version 2 and Version 3 before improvement. In spite of Version 3 being much longer, L_g for Version 2 is greater than for Version 3 ($L_g = 5.25$ sec for Version 2 and $L_g = 3.21$ sec for Version 3). Let us compare the performance time of the logical condition l_9^μ in Versions 2 and 3 before improvement. It is the same logical condition when an operator performs the same decision. However, the probability of this element being performed is different in these two versions of the task. In Version 2, the probability of this logical condition is $P = 1$, and in Version 3 it is $P = 0.2$. Therefore, the performance time for this logical condition for these two versions is drastically different.

L_g (the performance time of all logical conditions) for Version 2 after improvement is 4.45 sec and for Version 3 after improvement L_g has a duration of 2.2 sec. And for Version 2 of the task after improvement, l_7 takes 3.5 sec. The performance time for this logical condition is the same if the probability of this logical condition is not taken into account. However, the probability of this logical condition in Version 2 is $P = 1$, and for Version 3 it is $P = 0.2$. This is the reason that the time for this logical condition and for the whole L_g decreases in Version 3.

This difference in probability is due to the fact that in Version 2 of the task, the ordered price of the item and the received price are always not the same. In Version 3, this scenario occurs only for 20% of the items. So, in the third version the price comparison takes less time and is much simpler than in in the second version.

This example demonstrates how sensitive the suggested measures are when they are applied at the analytical stage of evaluation of the task performance methods.

Let us evaluate the complexity of decision-making components for Version 3 of the task. There are four logical conditions that are related to the fourth category of complexity in the before improvement version of the task: l_6 is performed with probability $P = 0.1$; l_9^μ has $P = 0.2$; l_{11}^μ is performed with $P = 1$; and l_{13}^μ has probability $P = 0.1$. Their corresponding performance times are 1.5, 4, 0.4, and 0.6 sec. So their combined performance time is 1.41 sec, and their category of complexity is IV. The fraction of the fourth category in the total performance time of all logical conditions is $1.41/3.21 = 0.44$. The weight coefficient for the fourth category that reflects the significance of this category, is 1.5. If we multiply 0.44 by this coefficient, the result is 0.66, meaning that less than 70% of the time for all decision-making in this version of the task belongs to the fourth category of complexity. Therefore, decision-making in general for Version 3 of the task before improvement should be related to the third category of complexity.

The complexity of decision-making components of the third version of the task after improvement has been determined similarly. There are the following logical conditions that belong to the fourth category of complexity: l_4, l_7, l_9^μ, and l_{11}^μ. Taking into account their probabilities, their combined performance time is 1.3 sec. Knowing that the performance time of all logical conditions in the third version of the task after improvement is L_g is 2.2 sec, the fraction of the fourth category is $1.3/2.2 = 0.59$. Applying the weight coefficient 1.5 for the fourth category results in 0.88. Hence, more than 70% of the time for all decision-making in the third version of the task after improvement belongs to the fourth category of complexity. So the decision-making for Version 3 of the task after improvement should be related not to the third but to the fourth category of complexity. This is due to the fact that the performance time of all logical conditions L_g is reduced in this version of the task. If we compare the absolute measures of the logical conditions, it become obvious that not only the performance time of all logical conditions has been reduced ($L_g = 2.2$ sec instead of 3.21 sec) but that the performance time of the logical conditions of the fourth category also decreased after innovations (instead of

1.41 it became 1.3). This example demonstrates that we have to compare both absolute and relative measures of complexity. Such a comparison gives us a clear understanding of what advantages or drawbacks the proposed innovations have at the analytical stage of the design process.

In this chapter, we have presented the evaluation of the complexity of the computer-based task. Reduction of the task complexity brings multiple tangible and intangible benefits. These are just some of them: the task performance takes less time and effort, the safety of the performance improves, and cognitive effort is reduced. All this makes the task performance much more simple and reliable. Knowing the measures of task complexity allows identifying cognitively demanding aspects of the task and finding the means of optimizing the performance of the task.

References

Bedny, G. Z. (1987). *The Psychological Foundations of Analyzing and Designing Work Processes*. Kiev, Ukraine: Higher Education Publishers.

Bedny, G. Z. (2015). *Application of Systemic-Structural Activity Theory to Design and Training*. Boca Raton, FL and London, UK: CRC Press, Taylor & Francis Group.

Bedny, G. and Meister, D. (1997). *The Russian Theory of Activity: Current Application to Design and Learning*. Mahwah, NJ: Lawrence Erlbaum Associates Publishers.

Bedny, G. Z. and Karwowski, W. (2007). *A Systemic-Structural Theory of Activity. Application to Human Performance and Work Design*. Boca Raton, FL and London, UK: CRC, Taylor & Francis

Bedny, G. Z., Karwowski, W. and Bedny, I. (2015). *Applying Systemic-Structural Activity Theory to Design of Human-Computer Interaction Systems*. Boca Raton, FL and London, UK: CRC Press, Taylor & Francis Group.

Bloch, V. (1966). Level of wakefulness and attention. In Fraisse, P. and Piaget, J. (Eds.), *Experimental Psychology*, Vol. 3 (pp. 97–146). Paris, France: University Press of France.

Gallwey, T. J. and Drury, G. G. (1986). Task complexity in visual inspection. *Human Factor*, 28, 596–606.

Green, D. M. and Swets, J. A. (1966). *Signal Detection Theory and Psychophysics*. New York, NY: Wiley.

Kaempf, G. L., Wolf, S., Thordsen, M. I., and Klein, G. (1992). *Decision Making in the AEGIS Combat Information Center*. Fairborn, OH: Klein Associates. (Prepared under contract N66001- 90 – C – 6023 for the Naval Command, Control and Ocean Surveillance Center, San Diego, CA.)

Lazareva, V. V., Svederskaya, N. E. and Khomskaya, E. D. (1979). Electrical activity of brain during mental workload. In Khomskaya E. D. (Ed.), *Neuropsychological Mechanisms of Attention* (pp. 151–168). Moscow, Russia: Science Publisher.

Mirabella, A. and Wheaton, G. R. (1974). *Effect of Task Index Variation on Transfer of Training Criteria*. Report NAVTRAEQUIPEN 72-C- 0126-1, Naval Training Equipment Center, Orlando, FL. January (AD 773-047/7GA).

Park, J. (2009). *The Complexity of Proceduralized Tasks*. London, UK: Springer-Verlag.

Park, J. (2011). Scrutinizing inter-relationships between performance influencing factors and the performance of human operators pertaining to the emergency tasks of nuclear power plant: An explanatory study. *Annals of Nuclear Energy*, 38, 2521–2532.

Payne, J. W. (1976). Task complexity and contingent processing in decision making: An information search and protocol analysis. *Organizational Behavior and Human Performance*, 16 (2), 366–387, Science Direct.

Rauterberg, M. (1996). How to measure of cognitive complexity in human-computer interaction. In Trappl, R. (Ed.), *Cybernetic and Systems* (Vol. 2, pp. 815–820). Vienna, Austria: Austrian Society for Cybernetic Studies.

Rosser Jr, J. B. (2010a). Introduction to special issue on transdisciplinary perspectives on economic complexity. *Journal of Economic Behavior and Organization, Elsevier*, 75, 1–2.

Rosser Jr, J. B. (2010b). Is a transdisciplinary perspectives on economic complexity possible? *Journal of Economic Behavior and Organization, Elsevier*, 75, 3–13.

Simon, H. A. (1999). *The Sciences of the Artificial* (3rd, rev. ed.). Cambridge, MA: MIT Press.

UK MTMA. MTM-1. Analyst Manual, London: The UK MTM Association, 2000.

Venda, V. F. (1975). *Engineering Psychology and Synthesis of Information Sources*. Moscow, Russia: Manufacturing Publishers.

Welford, A. T. (1967). Single channel operation in the brain. *Acta Psychologica*, 27, 5–21.

Wickens, C. D., Gordon S. E. and Liu, Y. (1998). *An Introduction to Human Factors Engineering*. Boston, MA: Addison-Wesley.

9

Assessment of Complexity and Reliability of the Computer-Based Task with Complex Logical and Probabilistic Structure

9.1 Assessing Reliability of the Computer-Based Task before Innovation

The main purpose of this section is to consider the relationship between complexity and reliability assessment of computer-based tasks and the correlation between these vital characteristics of human performance. In contrast with computerized tasks, these tasks involve only working with computers during task performance. They don't include any other physical work elements.

In this chapter, we also consider analytical methods of assessing complexity and reliability that can be used at the design stage when the new task does not exist yet and experimental methods cannot be applied. It is also extremely beneficial to complement the experimental procedures with the analytic ones.

Complexity evaluates cognitive demands for task performance and helps to optimize human performance. Such optimization includes reconstruction of means for task performance and its method (Jacko et al., 1971; Jacko and Ward, 1996; Jacko, 1997). Optimization of task performance increases efficiency of work, including reducing errors and failures. Reliability is also associated with error and failure analysis. Currently, the difference between such concepts as reliability and precision of task performance is often not clearly distinguished. The main concept for reliability is failure, while for precision it is error. These two basic concepts are tightly interconnected, but they are not the same. Precision refers to the accuracy with which a goal of a task is achieved. Reliability refers to failure in task performance, and how probability of failure in task performance can change over time (Bedny, 2004 and 2006; Bedny et al., 2010). In SSAT, differences between human errors and failure is distinguished based on the criticality of the errors for the system (Bedny and Harris, 2013). This difference is also discussed by specialists in safety analysis (Kotik and Yemelyanov, 1993). Operator actions that negatively affect functioning of the system without shutting it down, or decreasing some criteria of performance that still can be further corrected, are considered as erroneous actions that lead to errors. Operator actions that lead to the errors that render incapability to the system functioning, or shutting it down, or cause inability to achieve a required goal are considered as operator or user failure. Specialists can talk about operator and user errors or failures only when they discover erroneous cognitive or behavioral actions during task performance.

An operator or a user can utilize explorative actions. These actions can be cognitive and behavioral. Such actions provide the information that helps in understanding the system and correct performance strategies. Users can examine consequences of their own actions and intentionally utilize "reversible errors." Such errors perform informative functions (Bedny and Bedny, 2011a,b). Thus, error analysis can be an important component of the task analysis (Bedny et al., 2012).

Cognitive and behavioral actions and their components, such as psychological operations, are used as units of analysis. Qualitative and morphological analysis preceded quantitative analysis of activity during task execution.

In this chapter, we select a real computer-based task *"receiving the orders"* as an object of study. The advanced method of analysis of this task is presented in this chapter. This is a task that does not have independent manual components of work. The whole task is performed by interacting with a computer. The task has a complex logical and probabilistic structure and is critically important in the production process. Below we present the analysis of the task before innovation that includes: (1) qualitative stage; (2) morphological analysis; and (3) assessing reliability of task performance.

In our study, we've select objectively logical analysis for the qualitative stage, which is the most common and relatively simple method of study. It may be reduced to providing a short verbal description of the job and of the task performance. This method also includes description of technological process, equipment, tools, raw materials, and basic technological procedures. Work conditions such as temperature, noise, illumination, and so on should be also considered. Particular importance is given to the analysis of the identified shortcomings in task performance, qualitative analysis of work safety, and the relationship between computerized and non-computerized components of work. Therefore, a brief description of the considered computer-based task and its qualitative analysis is presented below.

The considered "receiving the orders" task is performed just one time at the beginning of a shift. The purpose of this task is to receive the file containing orders into the local computer system from another distant computer. This is the key task at the beginning of the shift that if performed successfully creates work for 50–60 employees for the whole eight-hour shift. Without successful completion of this task, 50–60 employees with hourly pay cannot start their work. Unsuccessful completion of this task leads to a loss of about one hour for these employees. Assuming that the hourly rate is $27.50, it's easy to calculate the financial impact of this task failure for the company. Loss of an hour is due to the fact that re-creation and resending of "the orders" file from another computer system that is located in another state and in a different time zone requires coordination of effort with a number of computer professionals. Such negative event should be considered as a failure. The failures were more frequent on Mondays after two days off. Receiving the orders task is the first task performed by an operator in the early morning shift. At the same time, it is a critically important task. The functional state of an operator is not yet at an optimal level. An optimal state is achieved only after some warm-up period. This factor can have negative effect on the pace of task performance and on the probability of erroneous actions. All of these factors have been considered in our study and particularly when assessing the pace of task performance.

The task includes several stages of performance. The operator checks if the name of the file that contains the orders is on the file list. If the file has been received, the operator checks the date stamp on the file to make sure it is a new order file, not an old one. If the date stamp on the file is from today, then the task "receive the orders" is completed and the file can be processed using the software specifically designed for this purpose, and the orders for the current shift are distributed among all the workers. If the file name is not on the list or its date stamp does not have a current date, then the operator should restart the communication software that facilitates the file transfer process. The qualitative analysis of the task showed that the operator could not understand why the same actions that in most cases have given the desired result in other cases have led to a failure. The operator blamed the computer specialists and insisted that the computer system functioned incorrectly. The cause of the human-computer system malfunctioning can be explained only by uncovering a breakdown in the computer-information system or erroneous cognitive or behavioral actions of the operator.

This brief qualitative analysis helps us to formulate the main purpose of our study. Now we are going to utilize morphological analysis that includes the combination of algorithmic analysis of human activity with time-structure analysis, the new method of analyzing probabilistic structure of activity, analysis of task complexity and its reliability. Finally, we compare complexity and reliability evaluation and examine their relationship. Such task analysis has never been performed before, and the obtained data can be especially useful for analysis of computer-based tasks and computerized tasks. The computerized tasks are the tasks that include a combination of manual work and interaction with a computer.

At the first step of the morphological analysis, an algorithmic description of the task is developed without presenting probabilistic characteristics of the algorithm. The event tree is developed, and the probability of each branch is determined. This step is especially important in determining the reliability of task performance, because such an event tree shows not only probabilities of each step of the algorithm but also probabilities of success and failure of the task performance. Then, probabilities determined in the event tree are utilized in the algorithmic description of the task performance to calculate the performance time of each member of the algorithm. It is also useful to develop the graphic form of the algorithm. The combination of the algorithmic description of the task with its temporal characteristics is called morphological analysis of activity during evaluation of reliability of task performance. All these methods will be described below in this chapter.

In Table 9.1 we present a morphological analysis of the existing method of task performance (before innovation). It includes an algorithmic description of activity and its time structure.

Table 9.1 includes hierarchically organized units of analysis. They describe activity with different levels of decomposition. There are two types of units of analysis. The first type utilizes the psychological units of analysis, and the second one uses the technological units of analysis. This allows a clear reflection of the structure of operator activity. The symbolic standardized description in the left column utilizes combined psychological units of analysis (see Table 9.1, left column). The second column from the left contains the description of the same elements of activity utilizing common language or technological units of analysis. The third column contains a detailed description using psychological units of analysis. The MTM-1 system has been used in the third column to describe motions that the motor actions include. Cognitive actions that are psychological units of analysis are also presented in a standardized manner. The right column reflects the time that is required for performance of the corresponding elements of activity. So Table 9.1 depicts not only algorithmic description but also the time structure of activity. It demonstrates that it is possible to describe extremely variable human activity that includes cognitive and behavioral components. We have to recall that such a model can be created by using purely analytical procedures, without observation of real performance. Experiments are supplemental in design and can be used in a simplified manner if needed. For example, we can select one member of the algorithm, consider preferable strategies of its performance, and conduct a chronometrical analysis to determine the performance time of separate actions included in this member of the algorithm.

Let us briefly consider basic symbols that are utilized in our table. Symbols O_1^ε and O_2^ε mean that the first and the second members of the algorithm are efferent operators (include only motor actions); O_3^α - the third afferent operator (includes only perceptual actions). These members of the algorithm have probability 1. Members of the algorithm that follow the logical conditions (decision-makings) have probabilities that depend on the outputs of the logical conditions. For example, decision-making that is described as logical condition $l_1\uparrow$ has two outputs $P_1 = 0.8$ and $P_2 = 0.2$. The associated arrow with number 1 in the top demonstrates that this arrow belongs to l_1. $O_4^{\alpha\mu}$ is performed according to output P_1. Hence, $O_4^{\alpha\mu}$ has probability $P = 0.8$. $l_2\uparrow$ is the second decision-making action or logical condition. The associated arrow with number 2 at the top demonstrates that this arrow belongs to l_2. This logical condition has two outputs with probabilities $P_4 = 0.76$ and $P_3 = 0.04$. Logical condition l_2 follows after the member of algorithm $O_4^{\alpha\mu}$. The probability of this member

TABLE 9.1

Algorithmic Description of Computer-Based Task and Its Time Structure (Existing Method)

Members of the Algorithm: Psychological Units of Analysis	Description of the Members of the Algorithm (Technological Units of Analysis)	Psychological Units of Analysis (Detailed Method of Description)	Time sec
O_1^ε	Type user name, press "*Enter*". Type password, press "*Enter*"	(1) Type ID (seven signs); (2) Move finger and press "*Enter*" - (AP1); (3) Type password (ten signs); (4) Move finger and press "*Enter*" - (R18B + AP1)	(1) 3.6; (2) 0.38; (3) 3.8; (4) 0.38 $\Sigma 8.16$; $8.16 \times 1.1 = 8.98$
O_2^ε	Type command "*ls –l*" and press "*Enter*"	(1) Move finger and press "*l*" (AP1); (2) Type "*s –l*" (three signs); (3) Move finger and press "*Enter*" (AP1)	(1) 0.38; (2) $1.14 + 0.28 = 1.42$; (3) 0.38 $\Sigma 2.18$ $2.18 \times 1.1 = 2.4$
O_3^α	Check to see if the file "*orders*" is on the list (distance of top-down eye movement is 4 cm)	Simultaneous perceptual action combined with scanning operation of target item; eye movement; time for scanning is calculated using coefficient 1.1 (EF + ET + EF)	$0.3 + 0.53 + 0.3 = 1.13$ $1.13 \times 1.1 = 1.24$
$l_1^1 \uparrow$	If the file is on the list go to $O_4^{\alpha\mu}$. ($P_1 = 0.8$). If the name of the file is absent from the list go to O_5^ε ($P_2 = 0.2$). $P = 1$	Simple decision-making action from two alternatives at sensory-perceptual level ("Yes"–"No" type decision)	$0.3 \times 1.1 = 0.33$
$O_4^{\alpha\mu}$	Check to see if the date stamp of the file "*orders*" has current date (move eyes 3 cm, read the date and compared with the current date that is retrieved from memory). $P = 0.8$	Simultaneous perceptual action combined with mnemonic operation (ET + EF)	$0.45 \times 1.1 = 0.49$ $0.49 \times 0.8 = 0.39$

(Continued)

TABLE 9.1 (CONTINUED)

Algorithmic Description of Computer-Based Task and Its Time Structure (Existing Method)

Members of the Algorithm: Psychological Units of Analysis	Description of the Members of the Algorithm (Technological Units of Analysis)	Psychological Units of Analysis (Detailed Method of Description)	Time sec
$l_2^2 \uparrow$	If the file has today's date then received file is the expected file, go to O_8^ξ $(P_1=0.76)$. If the file has the old date, then go to O_5^ξ. $(P_2=0.04)$. $P=0.8$	Simple decision-making action from two alternatives at sensory-perceptual level ("Yes"-"No" type decision)	$0.3\times1.1=0.33$ $0.33\times0.8=0.26$
$^1\downarrow O_5^\xi$	Type "restore_communication" and press "Enter". Repeat if necessary. $P=0.2+0.04=0.24$	Type "restore_communication" (twenty signs); move finger and press "Enter" (AP1)	(1) 7.6; (2) 0.38; $\approx\Sigma8$; $8\times1.1=8.8$; $8.8\times0.24=2.1$
$O_6^{\alpha w}$	Waite until initial screen comes up. $P=0.24$	Waiting period requires minimum level of attention concentration	$\approx8\times0.24=1.92$
$O_7^{\alpha w}$ 1. $(O_7^{\alpha w})$ 2. $(O_7^{\alpha l w})$ 3. (O_7^{w}) 4. $({}_{ch}O_7^{\alpha w}+l_7^{l w})$ These members of sub-algorithm are repeated 3 times during waiting period	Make a note of the time of completion of $O_6^{\alpha w}$. Waite for $\approx$ 7 minutes. The most preferable strategy of performance includes (1) move the site to the right bottom corner of the screen and read the time; (2) add 7 minutes and remember this time until the time is written down; (3) take a pen, move hand with pen to the paper, write down the target time and put the pen back; (4) checking actions that compare current time with the target time. (The checking actions can be repeated 2–3 times); The rest of the time is pure waiting period. $P=0.24$	Waiting period includes (1) Simultaneous perceptual action" $(ET+EF)$; (2) arithmetic calculation combined with mnemonic operation $(O_7^{al w})=2sec$; (3) (R30B+ G1A)+(M30B+2sec)+(M13A+RL1)=3.33sec; (4) The first perceptual action $(ET+EF)=0.54sec$; the second perceptual action, including mnemonic operation and comparing numbers $(ET+EF+EF)=1.04sec$; Decision-making based on visual information $(EF)=0.5sec$	$O_7^{\alpha w}=7\,min\times0.24=100.8\,sec$ 1. $(0.54\times3)\times1.1=1.78$; $1.78\times0.24=0.43$ 2. $(2\times3)\times1.1=6.6$; $6.6\times0.24=1.58$ 3. $(3.33\times3)\times1.1=10.99$; $10.99\times0.24=2.64$ 4. $(0.54+1.04++0.5)\times3\times1.1=6.86$; $6.86\times0.24=1.65$ $\Sigma\,2.64_{ex}+3.66_{cog}=6.3\,sec$

(Continued)

TABLE 9.1 (CONTINUED)

Algorithmic Description of Computer-Based Task and Its Time Structure (Existing Method)

Members of the Algorithm: Psychological Units of Analysis	Description of the Members of the Algorithm (Technological Units of Analysis)	Psychological Units of Analysis (Detailed Method of Description)	Time sec
$_1O_2^\varepsilon$	Type command "ls -l" and press "Enter" (the same as O_2^ε) P=0.24	Move finger and press "l" (AP1); type "s -l" (three signs); move finger and press Enter (AP1)	(1) 0.38; (2) 1.14+0.28=1.42; (3) 0.38 Σ2.18; 2.18×1.1=2.4; 2.4×0.24=0.576
$_1O_3^\alpha$	Check if the file "order" is on the list (distance of top-down eye movement is 4cm) (the same as O_3^α). P=0.24	Simultaneous perceptual action combined with scanning operation for a target item; eye movement; time for scanning is calculated using coefficient 1.1 (EF+ET+EF)	(0.3+0.53+0.3) ×1.1×0.24=0.3
$l^1_1\,\uparrow_1$	If the file is on the list go to $_1O_4^\alpha$. ($P_1=0.12$). If the name of the file is absent from the list go to $_1O_5^\varepsilon$ (the same as l_1). ($P_2=0.12$). P=0.12+0.12=0.24	Simple decision-making action from two alternatives at sensory-perceptual level ("Yes"-"No" type decision)	0.3×1.1×0.24=0.08
$_1O_4^{\omega_\mu}$	Check if the date stamp of the file "orders" has the current date (move eyes 3cm, read the current date and compared with the current date that is retrieved from memory), the same as $O_4^{\omega_\mu}$. P=0.12	Simultaneous perceptual action combined with mnemonic operation (ET+EF)	0.45×1.1×0.12=0.06
$l^2_2\,\uparrow_2$	If the file has today's date, then received file is the expected file, go to O_8^ε ($P_1=0.06$); If the file has the old date, then go to $_1O_5^\varepsilon$ (the same as l_2) ($P_2=0.06$). P=0.06+0.06=0.12	Simple decision-making action from two alternatives at sensory-perceptual level ("Yes"-"No" type decision)	0.3×1.1×0.12=0.04

(Continued)

TABLE 9.1 (CONTINUED)

Algorithmic Description of Computer-Based Task and Its Time Structure (Existing Method)

Members of the Algorithm: Psychological Units of Analysis	Description of the Members of the Algorithm (Technological Units of Analysis)	Psychological Units of Analysis (Detailed Method of Description)	Time sec
$1\,1_1 O_5^{\varepsilon}$	Type *"restore_communication"* and press *"Enter"*. Repeat if necessary (the same as O_5^{ε}). $P = 0.12 + 0.06 = 0.18$	Type *"restore_communication"* (twenty signs); move finger and press *"Enter"* (AP1)	(1) 7.6; (2) 0.38; $\Sigma 7.98; 7.98 \times 1.1 = 8.8; 8.8 \times 0.18 = 1.6$
$1 O_6^{a w}$	Wait until initial screen comes back (the same as $O_6^{a w}$). $P = 0.18$	Waiting period requires the minimal level of attention concentration	$\approx 8 \times 0.18 = 1.44$
$1 O_7^{a w}$ 1. $(O_7^{a w})$ 2. $(O_7^{a\beta\mu w})$ 3. $(O_7^{a w})$ 4. $(O_7^{a w} + l_7^{a w})$ These members of sub-algorithm are repeated 3 times during waiting period	Make a note of the time of completion of $O_7^{a w}$. Wait for ≈7 minutes. The most preferable strategy of performance includes (1) move the site to the right bottom corner of the screen and read the time; (2) add 7 minutes and remember this time until the time is written down; (3) take a pen, move hand with pen to the paper, write down the target time and put the pen back; (4) checking actions that compare current time with the target time. (The checking actions can be repeated 2–3 times); The rest of the time is pure waiting period. $P = 0.18$	Waiting period includes (1) Simultaneous perceptual action" (ET + EF); (2) arithmetic calculation combined with mnemonic operation $(O_7^{a\beta\mu w}) = 2$ sec; (3) (R30B + G1A) + (M30B + 2 sec) + (M13A + RL1) = 3.33 sec; (4) The first perceptual action (ET + EF) = 0.54 sec; (4) The second perceptual action, including mnemonic operation and comparing numbers (ET + EF + EF) = 1.04 sec; Decision-making based on visual information (EF) = 0.5 sec	αMBED Equation. DSMT4 = 7 min × 0.18 = 75.6 sec (1) (0.54×3) ×1.1 = 1.78; 1.78×0.18 = 0.32 (2) (2×3) ×1.1 = 6.6; 6.6×0.18 = 1.19 (3) (3.33×3) ×1.1 = 10.99; 10.99 × 0.18 × 1.98 (4) (0.54 + 1.04 + 0.5) ×3 ×1.1 = 6.86; 6.86×0.18 = 1.23 $\Sigma = 1.98_{ex} + 2.74_{cog} = 4.72$ sec

(Continued)

TABLE 9.1 (CONTINUED)

Algorithmic Description of Computer-Based Task and Its Time Structure (Existing Method)

Members of the Algorithm: Psychological Units of Analysis	Description of the Members of the Algorithm (Technological Units of Analysis)	Psychological Units of Analysis (Detailed Method of Description)	Time sec
$_2O_2^\varepsilon$	Type command "*ls -l*" and press "*Enter*" (the same as O_2^ε). $P=0.18$	Move finger and press "*l*" (AP1); type "*s -l*" (three signs); Move finger to corresponding key and press "*Enter*" (AP1)	(1) 0.38; (2) $1.14+0.28=1.42$; (3) 0.38 $\Sigma2.18$; $2.18\times1.1=2.4$; $2.4\times0.18=0.43$
$_2O_3^\alpha$	Check to see if the file "*orders*" is on the list (distance from the screen 45 cm; distance of top-down eye movement is 4 cm) (the same as O_3^α). $P=0.18$	Simultaneous perceptual action combined with scanning operation of target item; eye movement; time for scanning is calculated using coefficient 1.1 (EF+ET+EF)	$(0.3+0.53+0.3)\times1.1\times0.18=0.22$
$_2l_1^1\uparrow_2$	If the file is on the list go to $_2O_4^{\alpha\mu}$; ($P_1=0.09$). If the name of the file is absent from the list go to O_9^ε (the same as l_1). ($P_2=0.09$). $P=0.18$	Simple decision-making action from two alternatives at sensory-perceptual level ("*Yes*"–"*No*" type decision)	$0.3\times1.1\times0.18=0.059$
$_2O_4^{\alpha\mu}$	Check to see if the date stamp of the file "*orders*" has the current date (move eyes 3 cm, read the date and compared with the current date that retrieved from memory), the same as $O_4^{\alpha\mu}$. ($P=0.09$)	Simultaneous perceptual action combined with mnemonic operation (ET+EF)	$0.45\times1.1\times0.09=0.044$
$_2l_2^2\uparrow_2$	If the file has today's date then received file is the expected file, go to O_8^ε ($P_1=0.045$); If the file has the old date, then go to O_9^ε (the same as l2), ($P_2=0.045$). $P=0.09$	Simple decision-making action from two alternatives at sensory-perceptual level ("*Yes*"–"*No*" type decision)	$0.3\times1.1\times0.09=0.03$

(Continued)

TABLE 9.1 (CONTINUED)

Algorithmic Description of Computer-Based Task and Its Time Structure (Existing Method)

Members of the Algorithm: Psychological Units of Analysis	Description of the Members of the Algorithm (Technological Units of Analysis)	Psychological Units of Analysis (Detailed Method of Description)	Time sec
$\downarrow_2^2\downarrow_2^1\downarrow O_8^\varepsilon$	Type command "*interface_orders*". The end of the considered task (the goal is achieved) and the beginning of the following task. ($P = 0.76 + 0.06 + 0.045) = 0.865$	Type command with 16 characters; move finger and press "*Enter*" (AP1)	(1) 6.1; (2) 0.38; $\Sigma 6.48$ $6.48 \times 1.1 = 7.3$ $7.3 \times 0.865 = 6.31$
$_2\downarrow_2^1\downarrow O_9^\varepsilon$	Call computer specialist (**Failure**). This member of algorithm is excluded from analysis. $P = 0.09 + 0.045 = 0.135$	Consists of the sequence of verbal and motor actions. The sequence and number of actions are not strictly defined	— —

Total performance time of the whole task is 39.829 sec.

of algorithm is $P=0.8$, and therefore l_2 also has probability $P=0.8$. Member of algorithm $O_4^{\alpha\mu}$ includes perceptual actions in combination with mnemonic operation (μ). Member of algorithm O_5^{ε} has a combined probability $P_2=0.2$ and $P_3=0.04$. Therefore, entire probability of O_5^{ε} is $P=0.2+0.04=0.24$.

Let us consider member of the algorithm $O_7^{\alpha w}$. It reflects the waiting period lasting approximately seven minutes. After that period of time, the worker can start task performance again. Despite the fact that the worker should start performing the task only after the end of the seven-minute waiting period, certain cognitive and behavioral actions take place in this time period. These actions are described in Table 9.1. They include reading the time when the waiting period starts, add to it seven minutes, memorizing the target time until it is written down, performing checking actions, and so on. Thus, $*O_7^{\alpha w}$ includes the passive waiting period and the active waiting period that involves cognitive and motor actions. Some members of the algorithm are performed several times. They have subscripts in front of them. For example, $_1O_2^{\varepsilon}$ has the subscript 1 in front of the symbol O. It demonstrates that $_1O_2^{\varepsilon}$ is the first repetition of O_2^{ε}. If the same member of the algorithm is repeated the second time, the subscript in front of O is 2 ($_2O_2^{\varepsilon}$). It is important to note that each member of the algorithm includes the same type of actions, and sometimes they can be combined with additional mental operations (components of actions). For example, $O_4^{\alpha\mu}$ combines perceptual actions with mnemonic operations. The number of actions that can be integrated by one member of the algorithm is restricted by the capacity of the working memory.

Here we also utilize *technological and psychological units of analysis*. Technological units of analysis describe elements of activity by using common language and technical terminology. For experts who cannot directly observe task performance, it is very difficult to understand what the operator really does during task performance based on considered technological units of analysis. Psychological units of analysis utilize standardized psychological terminology developed in SSAT to describe human cognitive and physical actions. Cognitive actions usually are not divided into constituted elements due to their short duration. So cognitive and behavioral actions are described by utilizing not only technological but also standardized psychological terminology. Thanks to such descriptions, experts who never observed the real task performance but are familiar with SSAT can clearly understand how operators perform their tasks. An algorithm of performance and time structure of activity depend on the method of task performance and configuration of utilized equipment or interface. A specialist can evaluate the method of task performance, interface, or equipment configuration based on the morphological analysis of activity presented here.

After performing a preliminary algorithmic analysis, it is possible to conduct a reliability assessment of the existing method of task performance. Human reliability analysis is an important area of study in ergonomics (Bedny and Meister, 1997; Kirwan, 1994; Kotik and Yemelyanov, 1993). One of the most important aspects of a reliability assessment is analysis of the probabilistic structure of activity. In reviewing Table 9.1, we considered probabilistic characteristics of the members of the algorithm.

It has been demonstrated above that the probability of the previous member of the algorithm can influence the probability of the following members of the algorithm. However, logical conditions (decision-making actions) are the main factors that determine the probabilistic structure of activity. In the following section, we will show how we determine quantitatively the probability of various members of algorithm and how the probabilistic structure of activity can be used for assessment of the reliability of the task performance. Let us consider how we determine the probability of some members of algorithm. In order to determine the conditional probability of each member of the algorithm, we have developed a new event tree method.

Let us consider how the probability of several members of the algorithm are calculated. Figure 9.1 shows the probabilities from P_1 to P_{12}.

All these probabilities have been obtained using subject matter experts' (SME) estimates of the probability of outputs of various logical conditions. The verbal estimates of output have been then transferred into numbers. The method developed involves utilizing a special scale for the evaluation of the subjective probability of the making of various decisions or logical conditions. The suggested method also utilizes an event tree technique adapted for this purpose. The scale developed in SSAT utilizes the notion of scale diapasons. For example, the extremely high diapason has the range $P = 0.9–0.99$, and the extremely low diapason has probability range $P = 0.1–0.01$. When subject matter experts are working on determining the probability of events, they first identify the diapason in which events fall. Only after that can they define the probability of events within this diapason. Probabilities from P_1 to P_{12} have been determined based on this principle.

The first three members of the algorithm are performed with probability 1. Logical condition l_1 has two outcomes, with probabilities 0.8 and 0.2. Considered probabilities fall within two different diapasons ($P = 0.7–0.9$ and $P = 0.1–0.3$) because they are the first probabilities to be determined. These probabilities have been subjectively determined by the experts. Then, using the probability theory rules, we have calculated probabilities of the considered events.

For the first logical condition outputs, the subjectively determined probabilities are $P_1 = 0.8$ and $P_2 = 0.2$. Therefore, the probability of l_2 is also 0.8. The unconditional probability of the outcomes of l_2 are 0.95 and 0.05 ($0.95 + 0.05 = 1$). Let us now determine the probabilities of outcomes for logical condition l_2. If the probability of l_2 is $P_1 = 0.8$ and the probability of the successful outcome $P = 0.95$, then the conditional probability of this outcome is:

$$P_4 = 0.8 \times 0.95 = 0.76$$

The probability of P_4 reflects the successful task completion by reaching the goal-related member of the algorithm O_8^ε by utilizing the main activity

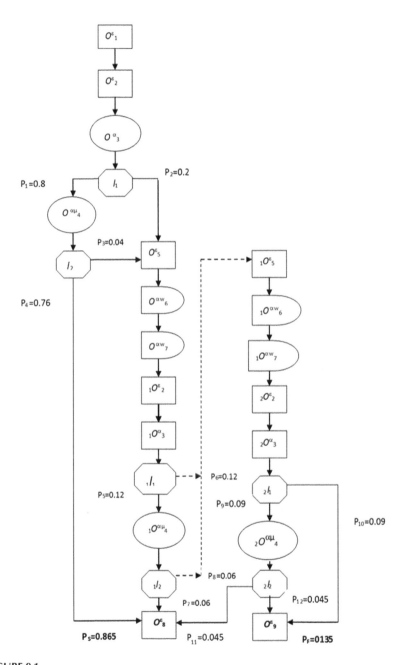

FIGURE 9.1
The graphic form of the algorithm for the existing method of the computer-based task with relative probabilities.

strategy (bottom branch of the graph on Figure 9.1). There are several other strategies that lead to the successful task completion. Therefore, successful completion of the task has a higher probability than 0.76.

The second strategy that leads to success starts with performance of O_5^ε (Figure 9.1). The probability of this member of the algorithm can be determined based on the analysis of logical conditions l_1 and l_2. The probability P_3 of the order file having the wrong date has been determined using expert analysis and is equal 0.05. Knowing P_1 and P_3, we can calculate the second probability of the outcome of the logical condition l_2 on the right-hand side of the graph (see Figure 9.1). The probability of the second outcome of l_2 is:

$$P_3 = 0.8 \times 0.05 = 0.04 \tag{9.1}$$

We can see that the sum of these two outcomes equals the probability of l_2.

$$P_1 = P_4 + P_3 = 0.76 + 0.04 = 0.8 \tag{9.2}$$

So, the probability of $O_4^{\alpha\mu}$ and l_2 is the same as $P_1 = 0.8$. Determining the probability of O_5^ε is more complicated. That probability consists of the combined probabilities of outcomes of l_1 and l_2. The right outcome of l_1 ($P_2 = 0.2$) is added to the probability P_3 of the second outcome of l_2. So the probability of O_5^ε is determined as follows:

$$P = 0.2 + 0.04 = 0.24 \tag{9.3}$$

The probabilities of other members of the algorithm can be determined similarly. These probabilities are presented in Table 9.1 and shown in the graphical form of the algorithm. (see Figure 9.1). They are further utilized for determining the performance time of various members of the algorithm and task performance, and also for the evaluation of the reliability and complexity of the task performance.

Below we demonstrate a new event tree modeling method for evaluation of the probability of successful performance and the probability of failure (see Figure 9.2). The event tree method is used in combination with morphological analysis of task execution. It can present all decision-making actions (logical conditions) performed by an operator in a concise form. Logical conditions are critical points where the probability of performance of various elements of the task are determined. The suggested method of event tree model development helps experts to determine probabilities of events with high precision. The event tree model can also simplify the calculations. As can be seen from Figure 9.2, the outcome of each branch is clearly associated with decision-making actions performed by an operator. Let us consider the event tree for the existing method of task performance. It presents probabilities that were obtained by calculations. It also demonstrates probabilities of failure and successful performance (see Figure 9.2).

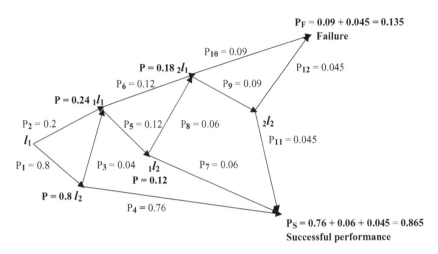

$P_F = 0.09 + 0.045 = 0.135$
Failure

$P_{10} = 0.09$

$P = 0.18 \, _2l_1$

$P_{12} = 0.045$

$P_6 = 0.12$

$P_9 = 0.09$

$P = 0.24 \, _1l_1$

$P_2 = 0.2$

$P_5 = 0.12$ $P_8 = 0.06$ $_2l_2$

l_1

$P_{11} = 0.045$

$P_1 = 0.8$ $P_3 = 0.04$ $_1l_2$ $P_7 = 0.06$

$P = 0.12$

$P = 0.8 \, l_2$

$P_4 = 0.76$

$P_S = 0.76 + 0.06 + 0.045 = 0.865$
Successful performance

FIGURE 9.2
Event tree model for existing method of task performance.

The event tree shows the relationship between logical conditions (decisions made) that determine the probabilistic structure of the considered task. Logical conditions and their probabilities are designated by bold lines. Outcomes of logical conditions (decisions made) are shown by lines with their probabilities. Figure 9.2 shows how outcomes of decision-making influence probabilities of events and the reliability of the task performance. The meaning of each outcome in the event tree is described in Table 9.1.

The first logical condition (l_1) shows the decision when the file name "orders" is browsed for on the list of file names. This logical condition has two outcomes. The first outcome has probability $P_2 = 0.2$ when file is not found on the list. The second outcome has probability $P_1 = 0.8$ and demonstrates the situation when the file name is found on the list. If the file name is not on the list, the operator needs to restore communication. If the file is on the list, the operator has to check the date stamp of the file. This involves the second decision designated by logical condition l_2. This logical condition also has two outcomes. If the file has a date different from the current date ($P_3 = 0.04$), the operator needs to restart communication. This outcome of l_2 ($P_3 = 0.04$) converges with output of l_1 ($P_2 = 0.2$). Thus, these probabilities are combined for the following logical condition $_1l_1$ that has the probability $P = 0.2 + 0.04 = 0.24$. From this example we can see that typically the event tree shows a situation when branches of events only diverge, but in practical situations, branches of events in some cases might converge (see as an example $_1l_1$) and the probabilities of the events are summarized. The logical condition $_1l_1$ has two outcomes with equal probabilities of $P = 0.12$ (P_5 and P_6). One of these outcomes leads to the next logical condition $_2l_1$ with the probability of 0.12. The second outcome leads to the next logical condition $_1l_2$ that has the probability of 0.12. It also has two outcomes with equal probability of 0.06. One of these outputs also leads to the logical condition

$_2l_1$ As a result, the logical condition $_2l_1$ has a probability of $P=0.12+0.06=0.18$. The other probabilities can be described similarly. The probability of failure is a combination of outputs of two logical conditions $_2l_1$ and $_2l_2$ and therefore $P_F=0.09+0.045=0.135$. The probability of the successful result is a combination of outcomes of l_2, $_1l_2$, and $_2l_2$ and is calculated as $P_S=0.76+0.06+0.045=0.865$. Finally, it should be pointed out that decisions (logical conditions) can have more than two outcomes with various probabilities. At the first stage, experts can use the scale of the subjective probability evaluation of events developed in SSAT and described in Chapter 5. The event tree helps to improve the accuracy of determining the probabilistic structure of activity during task performance. The algorithmic and time-structure description of the existing method of task performance (see Table 9.1) demonstrates that the probabilistic structure of activity depends on outcomes of logical conditions (decision-making actions).

Let's as an example consider how various decisions affect the probabilistic structure of the human algorithm. We want to bring readers' attention to the fact that in traditional event tree models, each branch is considered as an independent one and has probability from zero to one. In our model, the probability of preceding branches determines the probability of the following branches. Figure 9.2 demonstrates an event tree model that it is useful for evaluating the probability of failure and of successful performance.

It is also beneficial to use the graphical form of the description of the activity algorithm (see Figure 9.1) along with the event tree model method (see Figure 9.2). Such combination of methods can provide a clear picture of task performance and of its reliability.

The event tree model helps to visualize probabilities of transition from one member of the algorithm to the next and to verify the data already obtained. In our enhanced method of determining probabilities, the SMEs utilize the event tree data, making the analytical part of task analysis more efficient.

Figure 9.1 shows that there are two outcomes: O_8^ε has probability $P_S=0.865$ that shows the successful performance, and O_9^ε that has probability $P_F=0.135$, which represents failure. This approach eliminates the necessity of obtaining the experimental data that is often hard to collect and expensive.

The graphic form of the algorithm clearly demonstrates the strategy when the operator repeats the part of the task starting with operator $_1O_5^\varepsilon$ (type "restore communication" and hit "Enter") after an unsuccessful attempt. This leads to the second performance of O_5^ε and the members of the algorithm following it. If the operator repeats this cycle again, the new line with the same members of the algorithm can appear at the right side of the figure. As we discussed before, such repetitive performance can lead to failure.

The method described in this chapter helps us to determine the reliability of the existing method of task performance with high precision. For the first time, it's demonstrated how the new method of event tree development can be used for reliability assessment of user performance. Furthermore, this method allows determination of the complexity of tasks with a probabilistic structure with high accuracy.

The combination of various methods of task analysis is very useful for the refinement of preliminary algorithmic descriptions of activity.

9.2 Assessment of Reliability of Computer-Based Task after Innovation

At the next stage of task analysis, the new method of task performance is evaluated. Here, the morphological analysis of activity has been utilized as well. The new method of task performance has been suggested based on analysis of the existing method. This method includes some additional stages of task performance. As a result, the algorithm of task performance and time structure of activity has changed. The new method of task performance allowed obtaining information about the file transfer to temporary storage and determining when this transfer is completed. If the file is not found after the first steps, the operator had to type the new command "were_is_file" and check if the name of the order file is on the list and if the time stamp on the file has the current date. The efficiency of the new method of task performance is evaluated similarly to the evaluation of the existing method of task performance. Table 9.2 presents an algorithmic description and time structure analysis of the task performance after innovation.

As can be seen in Table 9.2, we also utilize probabilities of each member of the algorithm. The outcomes of the logical conditions l_1 and l_2 are the same for the new and the existing method of task performance. All other probabilities have been calculated utilizing the expert analysis and the probability theory rules. The graphical form of algorithm for the new method of task performance has also been developed by using the same method. However, we do not show this model here and just present the event tree for the new method of task performance (see Figure 9.3).

The event tree for the new method of performance is built utilizing the same approach as the one used when analyzing the existing method. One can see that the difference between the two event trees starts with l_3 when the new steps are introduced in the performance that allow the operator to monitor the file transfer process. Figure 9.3 shows that the outcomes of l_3 have the probabilities 0.17 and 0.07, meaning that the new method increases the probability of a successful outcome. This is achieved by performing $O_7^{\alpha\mu}$, which allows to check the date stamp of the file in the temporary storage. If the file is absent, the same restore communication command is utilized (O_{10}^{ε}).

The same is true for l_4 where the probability of a successful outcome is much greater as well. As a result of the process enhancement, the final probability of success for the new method is 0.958, which is higher than for the existing method of task performance. The probability of success is calculated as $0.76 + 0.118 + 0.08 = 0.958$, where 0.76 reflects the probability of the file being

TABLE 9.2

Algorithmic Description and Time Structure Analysis of the New Method of Computer-Based Task Performance

Members of Algorithm	Description of Members of Algorithm (Technological Units of Analysis)	Detailed Manner of Description of the Members of the Algorithm (Psychological Units of Analysis)	Time sec
O_1^ε	Type user name, press "Enter" and type password. Press "Enter".	(1)Type ID (seven signs); (2) Move finger and press "Enter" – (AP1); (3) Type password (ten signs); (4) Move finger and press "Enter" – (R18B+AP1)	(1) 3.6; (2) 0.38; (3) 3.8; (4) 0.38; Σ 8.16; $8.16 \times 1.1 = 8.98$
O_2^ε	Type "*ls –l*" command and press "*Enter*".	(1) Move finger and press "*l*" (AP1); (2) Type "*s –l*" (three signs); (3) Move finger and press "Enter" (AP1).	(1) 0.38; (2) $(0.38 \times 3 + 0.28) = 1.42$; (3) 0.38; $\Sigma 2.18$; $2.18 \times 1.1 = 2.4$
O_3^{α}	Check to see if the file "orders" is on the list (distance of top - down eye movement 4cm)	Simultaneous perceptual action combined with scanning operation of target item; eye movement; time for scanning is calculated using coefficient 1.1 (EF +ET +EF)	$0.3 + 0.53 + 0.3 = 1.13$ $1.13 \times 1.1 = 1.24$
$l_1^{\,1}\!\uparrow$	If the file is on the list go to O_4^{α}. ($P_1 = 0.8$). If the name of the file is absent from the list go to O_5^ε ($P2 = 0.2$). Probability of l_1 is: $P = 0.8 + 0.2 = 1$	Simple decision-making action from two alternatives at sensory – perceptual level ("Yes" – "No" type decision)	$0.3 \times 1.1 = 0.33$
$O_4^{\varepsilon\mu}$	Check to see if the date stamp of the file orders has correct date (move eyes at 3 cm, read the date and compared with the current date that retrieved from memory). Probability of $O_4^{\varepsilon\mu}$ is: $P = 0.8$.	Simultaneous perceptual action combined with mnemonic operation (ET +EF)	$0.45 \times 1.1 = 0.49$; $0.49 \times 0.8 = 0.39$

(Continued)

TABLE 9.2 (CONTINUED)

Algorithmic Description and Time Structure Analysis of the New Method of Computer-Based Task Performance

Members of Algorithm	Description of Members of Algorithm (Technological Units of Analysis)	Detailed Manner of Description of the Members of the Algorithm (Psychological Units of Analysis)	Time sec
$l_2^2 \uparrow$	If the file has today's date then received file is the expected file, go to O_{13}^ε ($P_1=0.76$). If the file has the old date, then go to O_5^ε . ($P_2=0.04$). Probability of l_2 is: $P=0.76+0.04=0.8$	Simple decision-making action from two alternatives at sensory – perceptual level ("Yes" – "No" type decision)	$0.3\times1.1=0.33$; $0.33\times0.8=0.26$
$\downarrow O_5^\varepsilon$ (1)	Type "where- are-orders" command and press "Enter". Repeat if necessary. Probability of O_5^ε is: $P=0.2+0.04=0.24$	(1)Type "where-are-orders" (sixteen signs) (2) Move finger and press "Enter" (AP1)	(1) 4.48; (2) 0.38; Σ 4.86; 4.86×1.1=5.35; 5.35×0.24=1.28
O_6^α	Check to see if there is an "orders" file on the list that reflects information stored in computer's temporary storage. Probability of O_6^α is: $P=0.24$	Simultaneous perceptual action combined with scanning operation of target item; eye movement; time for scanning is calculated using coefficient 1.1 (EF+ET+EF)	(0.3+0.53+0.3)=1.13; 1.13×1.1=12.43; 12.43×0.24=0.3
$l_3^3 \uparrow$	If there is an "orders" line on the list, go to $O_7^{\alpha\mu}$ ($P=0.17$); If it is absent, go to O_{10}^ε ($P=0.07$). Probability of l_3 is: $P=0.2+0.04=0.24$ (see event tree, Figure 9.3)	Simple decision-making action from two alternatives at sensory – perceptual level ("Yes" – "No" type decision)	$0.3\times1.1=0.33$; $0.33\times.24=0.08$
$O_7^{\alpha\mu}$	Check the date stamp of the "orders" file stored in computer temporary storage (move eyes at 3cm, read the date and compare with the current date that retrieved from memory). Probability of $O_7^{\alpha\mu}$ is: $P=0.17$	Simultaneous perceptual action combined with mnemonic operation (ET+EF)	0.45×1.1=0.495; 0.495×0.17=0.084
$l_4^4 \uparrow$	If the file in computer temporary storage has today's date, then go to $O_8^{\alpha\mu}$ ($P=0.12$); If the file has the old date, then go to O_{10}^ε ($P=0.05$); Probability of l_4 is $P=0.12+0.05=0.17$	Simple decision-making action from two alternatives at sensory – perceptual level ("Yes" – "No" type decision	$0.3\times1.1=0.33$; $0.33\times0.17=0.056$

(Continued)

TABLE 9.2　(CONTINUED)

Algorithmic Description and Time Structure Analysis of the New Method of Computer-Based Task Performance

Members of Algorithm	Description of Members of Algorithm (Technological Units of Analysis)	Detailed Manner of Description of the Members of the Algorithm (Psychological Units of Analysis)	Time sec
$*O_8^{\alpha w}$	Wait until the order file disappears from computer temporary storage and is transferred into production environment.		$O_8^{\alpha w} = 7$ min
1. $(O_8^{\varepsilon w})$	(1) performs "$ls -l$" and hit "Enter" command that services as a checking function;	(1) Move finger and press "l" (AP1); Type "$s -l$" (three signs); Move finger and press "Enter" (AP1).	(1) 0.38; 1.42; 0.38;
2. $(O_8^{\alpha w})$	The most preferable strategy of performance includes	Waiting period includes	$\Sigma 2.18 \times 1.1 = 2.398_{ex}$
3. $(O_8^{q\alpha w})$	(2) Detect "orders" file on the bottom of the screen	(2) simultaneous perceptual action;	(2) $(0.3 \times 3) \times 1.1 = 0.99$;
4. $(O_8^{q\alpha w})$	(second output of the $ls -l$ command); (3) move sight to	(3) simultaneous perceptual action combined with mnemonic operation;	(3) $(0.47 \times 3) \times 1.1 = 1.55$;
5. $(O_8^{q\alpha w})$	the left, distance 2 sm, look at the size of the file and	(4) simultaneous perceptual action combined with mnemonic operation;	(4) $(0.35 \times 3) \times 1.1 = 1.15$;
6. $(O_8^{th\mu w})$	memorize it; (4) move sight up to the right and detect	(5) simultaneous perceptual action	(5) $(0.45 \times 3) \times 1.1 = 1.5$;
7. (l_8^{w})	position of orders file at the top of the screen (first	combined with mnemonic operation;	(6) $(0.55 \times 3) \times 1.1 = 1.8$;
8. $(_1 O_8^{\varepsilon w})$	output of the $ls -l$ command); (5) move sight to the left	(6) simple thinking action involved in	(7) $(0.3 \times 3) \times 1.1 = 0.99$;
	and read size of the file at the beginning of waiting	detecting differences in numbers;	(8) $(0.71 \times 2) \times 1.1 = 1.56$;
	period; (6) compare current size of the file with	(7) simple decision-making from two	$(2.4_{ex} + 1.56_{ex}) = 3.96_{ex}$;
	existing size before; (7) make a decision to perform	alternatives;	$\sum 8_{cog}$
	$_1 O_8^{\alpha w}$ or O_9^{ε}; (8) Move hand from the start position and	(8) Move finger and press $\uparrow$ key	$(3.96_{ex} + 8_{cog}) \times 11.96$;
	hit $\uparrow$ key to repeat "$ls -l$" command and repeat if	- (R18B + AP1).	$11.96 \times 0.12 = 1.435$
	required until file "orders" disappears (in average		
	three times).		
	$P = 0.12$		
O_9^{ε}	Type "return_to_production", press "Enter" (automatic transition to production environment).	(1) Type twenty signs; (2) Press "Enter" (AP1)	(1) 5.6; (2) 0.38; Σ6
	Probability of O_9^{ε} is: $P = 0.12$		$6 \times 1.1 = 6.6$;
			$6.6 \times 0.12 = 0.79$

(Continued)

TABLE 9.2 (CONTINUED)

Algorithmic Description and Time Structure Analysis of the New Method of Computer-Based Task Performance

Members of Algorithm	Description of Members of Algorithm (Technological Units of Analysis)	Detailed Manner of Description of the Members of the Algorithm (Psychological Units of Analysis)	Time sec
$_1O_2^\varepsilon$	Type command "ls -l" and press "*Enter*" (The same as O_2^ε). Probability of $_1O_2^\varepsilon$ is: $P=0.12$	(1) Press "l" (AP1); (2) Type "$s-l$" (three signs); (3) Press "*Enter*" (AP1)	(1) 0.38; (2) 1.42; (3) 0.38; $\Sigma 2.18$ $2.18\times1.1=2.398$; $2.398\times0.12=0.287$
$_1O_3^\alpha$	Check to see if the file "*orders*" is on the list (distance from the screen 45 cm; distance of top - down eye movement 4 cm) (the same as O_3^α). Probability of $_1O_3^\alpha$ is: $P=0.12$	Simultaneous perceptual action combined with scanning operation of target item; eye movement; time for scanning is calculated using coefficient 1.1 (EF+ET+EF)	$0.3+0.53+0.3=1.13$ $1.13\times1.1=1.24$; $1.24\times0.12=0.15$
$_1^{1(1-2)}l_1\uparrow_1$	If the file is on the list, go to O_{13}^ε ($P=0.118$). If the name of the file is absent from the list, go to O_{14}^ε. ($P=0.002$). Probability of $_1l_1$ is: $P=0.118+0.002=0.12$	Simple decision-making action from two alternatives at sensory – perceptual level ("Yes" – "No" type decision)	$0.3\times1.1=0.33$; $0.33\times0.12=0.04$
$_1\downarrow \downarrow O_{10}^{2(2)3 4}$	Type "restore _ communication". Press "*Enter*". $P=0.05+0.07=0.12$	(1)Type "restore communication" (twenty signs) (2) Move finger and press Enter (AP1)	(1) 7.6; (2) 0.38; Σ 7.98; 7.98 × 1.1=8.8; $8.8\times0.12=1.056$
$O_{11}^{\alpha w}$	Wait until the program that restores communication is completed (initial production screen reappears). $P=0.12$	Waiting period of time requires from worker minimum level of attention concentration	$\approx 8\times0.12=0.96$

(Continued)

TABLE 9.2 (CONTINUED)

Algorithmic Description and Time Structure Analysis of the New Method of Computer-Based Task Performance

Members of Algorithm	Description of Members of Algorithm (Technological Units of Analysis)	Detailed Manner of Description of the Members of the Algorithm (Psychological Units of Analysis)	Time sec
$_1O_{12}^{\alpha w}$	Make a note of the time of completion of the restore. Waite for ≈ 7 minutes (the same as $O_{12}^{\alpha w}$). Make a note of the time of completion of the restore. Waite for ≈ 7 minutes. Waiting period of time includes different content of activity.	Waiting period includes (1) Simultaneous perceptual action" $(ET+EF)$; (2) arithmetic calculative combined with mnemonic operation $(O^{calyw}7) = 2$ sec; (3) (R30B +G1A) + (M30B+2 sec) + (M13A+RL1) $=3.33$ sec;	$O_{12}^{\alpha w} = 7\min \times 0.05 = 21$ sec (1) $(0.54 \times 3) \times 1.1 = 1.78$ $1.78 \times 0.12 = 0.214$ (2) $(2 \times 3) \times 1.1 = 6.6$; $6.6 \times 0.12 = 0.792$
1. $(_1O_{12}^{\alpha w})$	The most preferable strategy of performance includes	(4) the first perceptual action	(3) $(3.33 \times 3) \times 1.1 = 10.99$;
2. $(_1O_{12}^{aλμw})$	(1) move the site to the right bottom corner of the	$(ET+EF) = 0.54$ sec; the second	$10.99 \times 0.12 = 1.319$
3. $(_1O_{12}^{λw})$	screen and read the time; (2) add 7 minutes and remember this time until the time is written down;	perceptual action, including mnemonic operation and comparison	(4) $(0.54 + 1.04 + 0.5) \times 3 \times 1.1 = 6.86$; $6.86 \times 0.12 = 0.823$
4. $(_1O_{12}^{\alpha w} +_1 I_{12}^{\mu w})$	(3) take a pen, move hand with pen to the paper, write down current time and put the pen back;	numbers $(ET+EF+EF) = 1.04$ sec; Decision based on visual information	$1.319_{ex} + 1.829_{cog} = 3.148$ sec
These members of sub-algorithm are repeated 3 times during waiting period	(4) checking actions that compare a current time with the target time. (these checking actions can be repeated 2 times); The rest period of time is purely waiting period. $P = 0.12$	$(EF) = 0.5$ sec.	
$_2O_2^\varepsilon$	Type command "$ls -l$" and press "$Enter$" (The same as O_2^ε). $P = 0.12$	(1) Press "l" (AP1); (2) Type "$s - l$" (three signs); (3) Press "$Enter$" (AP1)	(1) 0.38; (2) 1.42; (3) 0.38; $\Sigma\ 2.18$ $2.18 \times 1.1 = 2.398$; $2.398 \times 0.12 = 0.288$
$_2O_3^\alpha$	Check to see if the file "orders" is on the list (distance from the screen 45 cm; distance of top - down eye movement 4 cm) (the same as O_3^α). $P = 0.12$	Simultaneous perceptual action combined with scanning operation of target item; eye movement; time for scanning is calculated using coefficient $1.1 (EF+ET+EF)$	$0.3 + 0.53 + 0.3 = 1.13$; $1.13 \times 1.1 = 1.24$; $1.24 \times 0.12 = 0.149$

(Continued)

TABLE 9.2 (CONTINUED)

Algorithmic Description and Time Structure Analysis of the New Method of Computer-Based Task Performance

Members of Algorithm	Description of Members of Algorithm (Technological Units of Analysis)	Detailed Manner of Description of the Members of the Algorithm (Psychological Units of Analysis)	Time sec
$_2l_1^{\,1}\uparrow_2$	If the ordered file is on the list, go to $_1O_4^\alpha$. If the name of the file is absent from the list, go to O_{14}^ε (the same as l_1). $P=0.07+0.05=0.12$	Simple decision-making action from two alternatives at sensory – perceptual level ("Yes" – "No" type decision)	$0.3\times1.1=0.33$; $0.33\times0.12=0.04$
$_1O_4^{\alpha\mu}$	Check the date stamp of the "orders" file stored in computer temporary storage (move eyes at 3 cm, read the date and compare with the current date that retrieved from memory), the same as $O_4^{\alpha\mu}$. P=0.1	Simultaneous perceptual action combined with mnemonic operation	$0.45\times1.1=0.495$; $0.495\times0.1=0.05$
$_1l_2^{\,2\,(1\text{-}3)}\uparrow$	If the file has today's date, then the received file is the expected file, go to O_{13}^ε $(P=0.08)$; If the file has the old date, then repeat $O_{10-1}^\varepsilon O_4^{\alpha\mu}$. $(P=0.01)^b$; If result does not change, go to O_{14}^ε $(P=0.01)$. Probability of $_1l_2$ is: $P=0.1$	Decision-making action from three alternatives at sensory – perceptual level	$0.4\times1.1=0.44$; $0.44\times0.1=0.044$
$_1^{1\,(1)}\downarrow\,_1^{2}\downarrow\,_1 \to O_{13}^\varepsilon$	Type command "interface-orders". The end of the considered task (**the goal achieved**) and the beginning of the following task. $(P=0.76+0.118+0.08)=0.958$	(1) Type command with 16 characters; (2) Move finger and press "Enter" (AP1)	(1) 6.1; (2) 0.38; $\Sigma6.48$ $6.48\times1.1=7.3$; $7.3\times0.958=6.99$
$_1^{1\,(2)}\downarrow\,_1^{1}\downarrow\,_1^{2\,(3)}\downarrow\,_1\,O_{14}^\varepsilon$	Call computer specialist (**Failure**). This member of algorithm is excluded from analysis. $(P=0.02+0.002+0.01=0.032)$	Consist of the sequence of verbal and motor actions. The sequence and number of actions are not precisely defined	— —

* See explanation in the text

** This output is not considered in this algorithmic description in order to simplify the calculations. It has a probability of 0.01.

Total performance time of the whole task is 30.828 sec

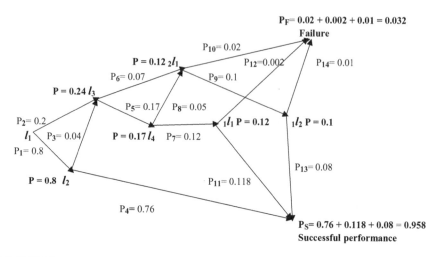

FIGURE 9.3
Event tree model for new method of task performance.

present in the main list right away. 0.118 is a calculated probability of the file being found in the temporary storage, and 0.08 is the probability of the file being received after restoring communication. The corresponding calculations can be found in Table 9.2. The probability of failure is equal to 0.032. This is also a combination of three outcomes $0.02 + 0.002 + 0.01 = 0.032$, where 0.02 is the probability that the file is not on the list after communication has been restored, 0.002 is the probability that the file has been transferred but did not reach the production directory, and 0.01 is the probability that the file has the wrong date. We left out the probability 0.01 for the cases when the operator decides to perform the second attempt to restore communication.

We can see that the probability of the successful performance is significantly higher for the task performance after innovation, which results in much improved economic efficiency of this performance. This task is performed once a day and it's the first task on the list that provides orders for the warehouse employees to pick up and ship out. Failure to receive this file leads to an hour delay. This means that the hourly paid employees are standing around without being able to perform their duties. It's easy to do the math if we know how many employees are affected (50–60), their hourly rate ($27.50) and the duration of the delay (1 hour).

The above-described method of task analysis allows assessment of the economic efficiency of the suggested innovation at the design stage. Some innovation are rather costly and are later proved to be inefficient. So instead of implementing them and then regretting doing so, one can do the preliminary reliability and efficiency analysis to decide if the innovation is going to be cost-effective.

We developed the graphic form of the algorithm for the new method of task performance in the similar fashion. However, this figure is not presented in this book.

9.3 Complexity Evaluation of Computer-Based Task before Innovation

Based on the analysis of Table 9.1, we can choose measures of task complexity evaluation developed in SSAT. Let us first consider some of these measures for the task performance before innovation.

The first measure determines duration of activity during task performance. The time for the algorithm execution (total time of task performance) is determined according to the formula:

$$T = \sum P_i t_i, \tag{9.4}$$

where:

P_i is the probability of the i-th members of the algorithm
t_i is the performance time of the *i*-th member of algorithm

Total performance time for the considered task is $T = 39.829$ sec.

Duration of perceptual components; duration of decision-making components (logical conditions); duration of time that requires retaining current information in working memory; duration of executive components of activity; and total duration of cognitive components of activity are also determined. The time for perceptual components should be determined according to the formula:

$$T_\alpha = \sum P^\alpha t^\alpha, \tag{9.5}$$

where:

P^α is the probability of afferent or perceptual components of activity
t^α is the performance time of afferent components of activity

The right column of Table 9.1 already has the performance time of members of the algorithm taken into account their probability. Therefore, we can further determine the duration of considered elements of activity simply by summarizing their performance times. So duration of perceptual components of activity is determined utilizing the formula:

$$T_\alpha = T\left(O_3^\alpha\right) + T\left(O_4^{\alpha\mu}\right) + T\left(O_6^{\alpha w}\right) + \sum T\left(*O_7^{\alpha w}\right) + T\left(_1 O_3^\alpha\right)$$

$$+ T\left(_1 O_4^{\alpha\mu}\right) + T\left(_1 O_6^{\alpha w}\right) + \sum T\left(*_1 O_7^{\alpha w}\right) + \left(_2 O_3^\alpha\right) + T\left(_2 O_4^{\alpha\mu}\right) \tag{9.6}$$

$$= 1.24 + 0.39 + 1.92 + 1.68 + 0.3 + 0.06$$

$$+ 1.44 + 1.25 + 0.22 + 0.044 = 8.544 \,(\text{sec})$$

where:

$\Sigma T(*O_7^{aw})$

$\Sigma T(*_1O_7^{aw})$ are summarized time of performing perceptual actions during the waiting period. We can similarly determine duration of other components of activity.

Duration of thinking components included in this task of activity is determined as:

$$T_{th} = T\left(O_7^{cal\mu w}\right) + T\left({}_1O_7^{cal\mu w}\right) = 1.58 + 1.19 = 2.77\,(\text{sec}) \tag{9.7}$$

Duration of decision-making components (logical conditions) in task performance is:

$$L_g = (l_1) + (l_2) + \left(l_7^{\mu w}\right) + ({}_1l_1) + ({}_1l_2) + \left({}_1l_7^{\mu w}\right) + ({}_2l_1) + ({}_2l_2)$$

$$= 0.33 + 0.26 + 0.4 + 0.08 + 0.04 + 0.3 + 0.059 + 0.03 = 1.499\,(\text{sec}) \tag{9.8}$$

Duration of retaining current information in working memory is determined according to the following formula:

$$T_{wm} = T\left(O_4^{\alpha\mu}\right) + T\left(O_7^{cal\mu w} + l_7^{\mu w}\right) + T\left({}_1O_4^{\alpha\mu}\right) + T\left({}_1O_7^{cal\mu w} +_1 l_7^{\mu w}\right) + T\left({}_2O_4^{\alpha\mu}\right)$$

$$= 0.39 + (1.58 + 0.4) + 0.06 + (1.19 + 0.3) + 0.044 = 3.964\,(\text{sec}) \tag{9.9}$$

Total duration of cognitive components in task performance is calculated as follows:

$$T_{cog} = T_\alpha + L_g + T_{th} + \left(T_{wm}\right) = 8.544 + 1.499 + 2.77 + (3.964)$$

$$= 12.813\,(\text{sec}) \tag{9.10}$$

Duration of executive components of activity (performance time of efferent operators) is calculated below:

$$T_{ex} = T\left(O_1^\varepsilon\right) + T\left(O_2^\varepsilon\right) + T\left(O_5^\varepsilon\right) + T\left(O_7^{\varepsilon w}\right)$$

$$+ T\left({}_1O_2^\varepsilon\right) + T\left({}_1O_5^\varepsilon\right) + T\left({}_1O_7^{\varepsilon w}\right) + T\left({}_2O_2^\varepsilon\right) + T\left(O_8^\varepsilon\right)$$

$$= 8.98 + 2.4 + 2.1 + 2.64 + 0.576 \tag{9.11}$$

$$+ 1.6 + 1.98 + 0.43 + 6.31$$

$$= 27.016\,(\text{sec})$$

And $T_{cog} + T_{ex} = 12.813 + 27.016 = 39.829$ sec is the execution time of the whole task that consists of these two components. Some elements of activity are counted in several formulas. For example, the performance time of such members of the algorithm as $O_7^{cal\mu w}$ and $_1O_7^{cal\mu w}$ is counted in T_{th} and in T_{wm}. These members of the algorithm include various cognitive operations such as calculations (thinking component) and mnemonic component. Similarly, there are several other members of the algorithm that include different cognitive operations. Obtained data allowed calculation of other measures of complexity.

The first one of them is N_α. This measure presents the fraction of perceptual components of activity in the entire task performance.

$$N_\alpha = T_\alpha / T = 8.544 / 39.829 = 0.215 \tag{9.12}$$

The fraction of thinking components of activity in the entire task performance is calculated as follows:

$$N_{th} = T_{th} / T = 2.77 / 39.829 = 0.07 \tag{9.13}$$

The fraction of decision-making components of activity is presented below:

$$N_J = L_g / T = 1.499 / 39.829 = 0.0376 \tag{9.14}$$

The next measure demonstrates the level of the working memory workload:

$$N_{wm} = T_{wm} / T = 3.964 / 39.829 = 0.0995 \tag{9.15}$$

The fraction of external behavioral (executive) components of activity in the entire task is:

$$N_{ex} = T_{ex} / T = 27.016 / 39.829 = 0.678 \tag{9.16}$$

The next measure determines the fraction of cognitive components of activity in the entire task performance:

$$N_{cog} = T_{cog} / T = 12.813 / 39.829 = 0.322 \tag{9.17}$$

In the considered task, an operator cannot predict the success and/or failure of some of her/his actions. The same cognitive and behavioral actions to restore communication can result in success or failure. Such uncertainty produces emotional tension and an increased level of concentration of attention or level of mental effort. In normal conditions, measures that are associated with attention concentration during performance of computer-based tasks are related to the third category of complexity. However, the period of time when an operator attempts to restore communication for the second time produces emotional tension because the possibility of failure sharply increases. To carry out the necessary calculations, it is necessary to determine the time that is needed for achieving the goal or for encountering failure after performing $_1l_2$. This is the period of time when an operator attempts to restart communication for the second time. We calculate this time that is required for cognitive and motor components of activity separately because it belongs to the fourth category of complexity.

The performance time for perceptual components that are related to the fourth category of complexity are calculated first:

$$T_{\alpha(4)} = T\left(_1O_6^{\alpha w}\right) + \Sigma T\left(^*_1O_7^{\alpha w}\right) + \left(_2O_3^{\alpha}\right) + T\left(_2O_4^{\alpha \mu}\right)$$
$$= 1.92 + 1.25 + 0.22 + 0.044 = 3.434 \ (\text{sec}),$$

(9.18)

At the next step, time for decision-making (logical conditions) that is related to the fourth category of complexity is calculated:

$$L_{g(4)} = \left(_1l_2\right) + \left(_1l_7^{\mu w}\right) + \left(_2l_1\right) + \left(_2l_2\right)$$
$$= 0.04 + 0.3 + 0.059 + 0.03 = 0.429 (\text{sec})$$

(9.19)

Finally, time for thinking components that are related to the fourth category is defined:

$$T_{th(4)} = T\left(_1O_7^{cal\mu w}\right) = 1.19 \text{ sec}$$

(9.20)

The time for mnemonic operations that are performed simultaneously with other cognitive components is not considered in our calculations.

The time for all cognitive components of activity that are related to the fourth category of complexity is calculated according the formula:

$$T_{cog(4)} = T_{\alpha(4)} + L_{g(4)} + T_{th(4)} = 3.434 + 1.19 + 0.429 = 5.053 \ (\text{sec})$$

(9.21)

The time for mnemonic operations that are performed simultaneously with other cognitive components is not included separately.

The time for executive (motor) components of activity that are related to the fourth category of complexity is determined similarly:

$$T_{ex(4)} = T\left(_1O_5^{\varepsilon}\right) + T\left(_1O_7^{\varepsilon w}\right) + T\left(_2O_2^{\varepsilon}\right) + T\left(O_8^{\varepsilon}\right)$$
$$= 1.6 + 2.64 + 0.43 + 6.31 = 10.98 \ (\text{sec})$$

(9.22)

Next, we determine the ratio of time spent performing cognitive component related to the fourth category of complexity to the time for all cognitive components of the task:

$$N_{cog(4)} = T_{cog(4)} / T_{cog} = 5.053 / 12.813 = 0.394 \qquad (9.23)$$

Finally, we determine the ratio of performance time of executive (motor) component related to the fourth category of complexity to the time for all motor components of the task:

$$N_{ex(4)} = T_{ex(4)} / T_{ex} = 10.98 / 27.016 = 0.406 \qquad (9.24)$$

There is an order scale of complexity, with five levels of attention concentration (Bedny, 2015). The simplest one is related to the first category, and the most complex one is of the fifth category of complexity. The last two formulas demonstrate that according to this criterion, the fraction of cognitive components that is related to the fourth category is 0.36, and the fraction of the third category of complexity is 0.64.

The fraction of the fourth category of complexity of motor components is 0.39, and the fraction of the third category is 0.61. Thus, there is a sufficiently high level of mental effort during the task performance, according to this criterion. Specifically, it is true for the executive or motor components of activity. This is due to the fact that for the cognitive components, the minimal level of concentration of attention is of the third category, and the maximum level is of the fifth category. In contrast, motor activity in normal conditions for various manual work requires the first level of attention concentration for the simplest type of motor activity, and the most complex motor activity is related to the third category. The motor activity of a user during interaction with a computer, when he/she utilizes the keyboard, usually is of the third category of complexity. However, because of the stress factor and the emotional tension derived from it, a significant part of the motor activity is transformed into the fourth category. Therefore, the interaction of users with the keyboard in stressful conditions requires a high level of mental effort.

9.4 Complexity Evaluation of Computer-Based Task after Innovation

At the next stage, we conduct a task complexity evaluation of the new method of task performance. This gives us an opportunity to compare task complexity before and after innovation.

Therefore, duration of considered elements of activity can be determined similarly here. The time for the task execution (total time of task performance) is determined according to the following formula:

$$T = \sum P_i t_i = 30.828 \text{ sec,} \qquad (9.25)$$

where:

P_i is the probability of the considered members of the algorithm
t_i is the performance time of i-th member of the algorithm

It means that performance time of the task after innovation decreased by 9 seconds or by 22%.

Duration of perceptual components of activity is determined according to the following formula:

$$T_\alpha = T\left(O_3^\alpha\right) + T\left(O_4^{\alpha\mu}\right) + T\left(O_6^{\alpha w}\right) + T\left(O_7^{\alpha\mu}\right) + T\left(O_8^{\alpha w}\right) + T\Sigma\left(O_8^{\alpha\mu w}\right)$$

$$+ T\left({}_1O_3^\alpha\right) + T\left(O_{11}^{\alpha w}\right) + \Sigma T\left({}_1O_{12}^{\alpha\mu}\right) + T\left({}_1O_{12}^{\alpha\mu w}\right) + T\left({}_2O_3^\alpha\right) + T\left({}_1O_4^{\alpha\mu}\right)$$

$$= 1.24 + 0.39 + 0.3 + 0.084 + 0.99 \times 0.12 + (1.55 + 1.15 + 1.5) \times 0.12 \quad (9.26)$$

$$+ 0.15 + 0.96 + 0.214 + (0.54 + 1.04) \times 3 \times 1.1 \times 0.12 + 0.149 + 0.05$$

$$= 4.786 \text{ sec,}$$

where:

$\Sigma T(O_8^{\alpha\mu w})$ is summarized time of performing perceptual actions during the waiting period ($*O_8^{\alpha w}$)
$\Sigma T({}_1O_{12}^{\alpha\mu})$ is summarized time of performing perceptual actions during the waiting period (${}_1O_{12}^{\alpha w}$)

Before improvement duration of perceptual activity was 8.544 and therefore this time reduced by 3.758 sec or by 44%.

Duration of thinking components of activity is determined as:

$$T_{th} = \left(O_8^{th\mu w}\right) + \left({}_1O_{12}^{cal\mu w}\right) = 1.8 \times 0.12 + 0.792 = 1.008 \text{ sec} \qquad (9.27)$$

The difference in duration of thinking components before and after innovation is 1.762 sec or 63%.

Duration of decision-making components (logical conditions) in the task performance is:

$$L_g = \left(l_1\right)+\left(l_2\right)+\left(l_3\right)+\left(l_4\right)+\left(l_8^{\mu w}\right)+\left(_1l_1\right)+\left(_1l_{12}^{\mu w}\right)+\left(_2l_1\right)+\left(_1l_2\right)$$

$$= 0.33+0.26+0.08+0.056+\left(0.99\times0.12\right)+0.04$$

$$+\left(0.5\times3\times1.1\times0.12\right)+0.04+0.044$$

$$= 1.167\,(\text{sec})$$

(9.28)

The difference in duration of decision-making or logical conditions before and after innovation is 0.332 sec or 22%.

The time for retaining current information in working memory during task performance is determined according to the formula:

$$T_{wm} = T\left(O_4^{\alpha\mu}\right)+T\left(O_7^{\alpha\mu}\right)+\Sigma T\left(O_8^{\alpha\mu w}\right)+T\left(O_8^{th\mu w}\right)+$$

$$T\left(l_8^{\mu w}\right)+T\left(_1O_{12}^{cal\mu w}\right)+T\left(_1O_{12}^{\alpha\mu w}\right)+\left(_1l_{12}^{\mu w}\right)+T\left(_1O_4^{\alpha\mu}\right)$$

$$= 0.39+0.084+\left(\left(1.55+1.15+1.5\right)\times0.12\right)+\left(1.8\times0.12\right)$$

$$+\left(0.99\times0.12\right)+0.792+\left(1.04\times3\times1.1\times0.12\right)$$

$$+\left(0.5\times3\times1.1\times0.12\right)+0.05$$

$$= 2.765$$

(9.29)

Before innovation, $T_{wm}=3.964$ and therefore this time has reduced by 1.199 sec or by 30%.

We need to take into account that T_{wm} is already included into other cognitive components of task and therefore does not affect T_{cog} and performance time of the whole task.

Total duration of cognitive components in task performance can be calculated as follows:

$$T_{cog} = T_\alpha + L_g + T_{th} + \left(T_{wm}\right)$$

$$= 4.786+1.167+1.008+\left(2.765\right)$$

$$= 6.961\,\text{sec}$$

(9.30)

The difference in the total duration of cognitive processes before and after improvement is $12.813-6.961=5.852$ sec or 46%.

Duration of executive or motor components of activity during task performance (time for efferent operators) is:

$$T_{ex} = T\left(O_1^\varepsilon\right) + T\left(O_2^\varepsilon\right) + T\left(O_5^\varepsilon\right) + T\left(O_8^{\varepsilon w}\right) + T\left({}_1O_8^\varepsilon\right) + T\left(O_9^\varepsilon\right)$$

$$+ T\left({}_1O_2^\varepsilon\right) + T\left(O_{10}^\varepsilon\right) + T\left({}_1O_{12}^{\varepsilon w}\right) + T\left({}_2O_2^\varepsilon\right) + T\left(O_{13}^\varepsilon\right)$$

$$= 8.98 + 2.4 + 1.28 + \left(2.398 \times 0.12\right) + \left(1.56 \times 0.12\right) \tag{9.31}$$

$$+ 0.79 + 0.287 + 1.056 + 1.319 + 0.288 + 6.99$$

$$= 23.865 \; \left(\text{sec}\right)$$

The difference in the duration of motor or executive components of task performance before and after innovation is 27.016–23.865 = 3.151 sec or 12%.

Analysis of the above-described measures indicates that the most significant reduction of time was observed during performance of perceptual and thinking components of activity. It is also essential that there is a reduction of time in keeping information in working memory and reduction of time for decision-making. Comparison of time reduction for executive or motor components of activity with reduction for cognitive components demonstrates that the cognitive workload has reduced more significantly than the workload for motor components of activity.

Obtained data allowed calculation of several other important measures of task complexity. The first of them is N_α. This measure presents the fraction of perceptual components of activity in the performance of the entire task.

$$N_\alpha = T_\alpha / T = 4.786 / 30.828 = 0.155 \tag{9.32}$$

The fraction of thinking components of activity in the performance of the entire task is calculated as follows:

$$N_{th} = T_{th} / T = 1.008 / 30.828 = 0.03 \tag{9.33}$$

The fraction of decision-making components of activity in the performance of the entire task is shown below:

$$N_l = Lg / T = 1.167 / 30.828 = 0.038 \tag{9.34}$$

The next measure demonstrates the level of the working memory workload for this version of task performance. This measure is determined as the fraction of time for retaining current information in working memory in the time for the entire task performance.

$$N_{wm} = T_{wm} / T = 2.765 / 30.828 = 0.09 \tag{9.35}$$

The fraction of external behavioral (executive) components of activity in the entire task is:

$$N_{ex} = T_{ex} / T = 23.865 / 30.828 = 0.77 \qquad (9.36)$$

The next measure determines the fraction of cognitive components of activity in the performance of the entire task:

$$N_{cog} = T_{cog} / T = 6.961 / 30.828 = 0.225 \qquad (9.37)$$

Let us consider the last measure of complexity that we use in our analysis. This measure is related to the evaluation of the level of concentration of attention during task performance (1, minimum concentration; 5, maximum) after the innovation. In the first version of the task, this period of time starts when the operator attempts to restore communication for the second time, and it belongs to the fourth category of complexity due to the fact that an operator is unable to predict consequences of her/his own actions. Sometimes an operator performs these actions and they lead to the desired result, and the other times these actions lead to errors. This produces emotional tension and an increasing level of attention concentration from the third to the fourth category of complexity (see this measure before innovation). Let us consider this measure after innovation.

If the file is not present on the screen, the operator now has a new command called "where-are-orders". This is the script that allows the operator to switch to the temporary storage and check if the file is in the process of being transferred. The output of this command is a list of the files in the temporary storage directory. The operator is checking if the file "orders" is on the list and if the date stamp of this file has the current date. If the file is present and the date is right, then the operator types an "ls –l" command to refresh the file list. The operator then compares the size of the file on the last list with the prior file size to make sure the file is growing in size, which means that it is in the process of being transferred. The operator hits the ↑ key a few times to repeat the "ls –l" command until the file "orders" disappears from the file list. This means that the file transfer has been completed and the file has now been moved to the production environment. Such information was not available to the operator before innovation. Now, as a result of innovation, the operator has information about the state of the file and can clearly understand whether the required file has the current date and transfer continues or this process is completed. The uncertainty about the state of the file is removed. Also, now the operator can make an informed decision about whether or not he/she should perform actions that are necessary for restarting communication. Unpredictable consequences of actions cause emotional tension and increasing level of attention concentration, which are eliminated after innovation. So due to this factor, all measures that are associated with attention concentration should be transformed from the fourth category of complexity to the third category of complexity for the enhanced version of the task. This period of time starts when the worker attempts to restore communication for the second time. This period of time starts with negative outcome of $_1l_2$. However, uncertainty of the situation is eliminated, and the last two members of the algorithms $_1l_2$ and O_{13}^ε cannot be related to the fourth category of complexity according to the level of attention concentration. Both of them, according to considered criterion, stay in the third category of complexity. Therefore, after innovation, there is no period of time that can be related to the fourth category of complexity, according to the level of attention concentration.

At the final stage of analysis, we present tables that demonstrates all measures of complexity before and after innovation for the comparative analysis. In Table 9.3, we present the duration of separate types of activity during task performance and the absolute measures of complexity.

In Table 9.4 we present relative measures of complexity for the existing and new method of task performance.

Comparing Tables 9.3 and 9.4 gives a clear idea of the benefits of the considered innovation. Table 9.3 demonstrates that performance time (T) of the entire task for the existing and new method has been significantly reduced after the innovation in spite of the fact that the number of the members of the algorithm has increased in the new method of task execution due to the changes in the content and probabilistic structure of the algorithm. The changes start from O_5^ε, where instead of trying to restore communication, the operator now uses the new script where-are-orders that allows him/her to check if the file is in the process of being transferred. This significantly reduces the risk factor of breaking the process and getting a partial file and increases the probability of a successful result. This also allows reduction of the need to perform O_{10}^ε - $_1l_2$ in most cases. We should also emphasis that

TABLE 9.3

Absolute Measures of Task Complexity Computer-Based Task "Receiving the Order" (seconds)

Conditions	T	T_α	T_{th}	L_g	T_{wm}	T_{cog}	T_{ex}	$T_{cog(4)}$	$T_{ex(4)}$
Existing method	39.829	8.544	2.77	1.499	3.964	12.813	27.016	5.053	10.98
New method	30.828	4.786	1.008	1.167	2.765	6.961	23.865	0	0

TABLE 9.4

Relative Measures of Complexity for Existing and New Method of Computer-Based Task "Receiving the Orders"

Conditions	N_α	N_{th}	N_l	N_{wm}	N_{cog}	N_{ex}	$N_{cog(4)}$	$N_{ex(4)}$
Existing method	0.215	0.07	0.037	0.0995	0.322	0.678	0.394	0.406
New method	0.155	0.03	0.038	0.09	0.225	0.77	0	0

the new method of task performance reduces emotional tension and is less difficult than the existing one. It is also important to consider how the performance time of various components of activity also has reduced. As can be seen, the performance time of cognitive components has been reduced, caused by the reduction of the cognitive workload. The perceptual workload decreased even more noticeably. At the same time, the cognitive workload for the other cognitive processes also went down. The components of work that are associated with the fourth level of concentration of attention during performance of cognitive and behavioral components of activity has been totally eliminated. It is important to compare measures of absolute value with the measures that use the relative value. For example, such measure as Lg demonstrates that the time for decision-making (logical conditions) in the existing method of task performance is 1.499 and in the enhanced one is 1.167 seconds, which is a 0.332 sec decrease. At the same time, the relative measure N_l shows that in the old method, its value is 0.037 and for the new method it's 0.038, due to the fact that after innovation, time needed for decision-making has been reduced, but the time for task performance also went down and as a result this measure with relative values has not changed.

Comparison methods of reliability and complexity assessment and analysis of obtained results leads to some conclusions. In this section, we demonstrated modified advanced methods of reliability and complexity evaluation of a computer-based task.

One of the most important aspects of reliability analysis is the analytical evaluation of the probability of failures of the task performance. In this chapter, we have described the first analytical methods of reliability analysis of complex variable tasks.

Complexity and reliability of task performance are interdependent characteristics. An increase in task complexity often results in a decrease of reliability of its performance. Higher reliability of task performance can usually be achieved by introducing new cognitive and behavioral actions and technical components in task performance. Such new methods of performance can be accompanied by new activity or technical components that can cause errors or failure.

When considering the performance reliability issues, it's important to take into account the operator's efficiency when performing the tasks during the whole shift with time limits, stressful situations, and so on. In other words, an operator has to work reliably under stress and when facing a work overload. An increase in the complexity of the task is one of the most important sources of mental fatigue. This factor is particularly evident at the end of the shift. Therefore, the more complex the task is, the less reliable its performance can be toward the end of the shift. Analysis complexity measures allow detection of potentially critical points in task performance which can be the source of such errors. Introduction of innovations that are aimed at improving reliability should also be accompanied by reducing the complexity of existing or new steps of task performance. Complexity measures help to discover critical points of task performance that cause various errors or failures.

Time study is the critically important method for evaluation of the economic efficiency of innovations. This method is usually applied to analysis of tasks performed by blue-collar workers. Such tasks in most cases have the same repetitive sequence of performance. When an operator performs computer-based tasks or tasks in complex technical systems, the tasks are extremely variable. The traditional method of time study is not adapted for this purpose. The method suggested here, of time structure development, facilitates the time studies and applies the obtained information to determining the time of task performance, time lost, nonproductive time, and so on. Knowing labor cost per unit of time, one can evaluate the economic efficiency of innovations. The study demonstrates that the evaluation of the reliability and complexity of the task should be carried out together.

9.5 Economic Effects of Implemented Innovations

Complex work requires greater cognitive effort and causes more tension than simple work. Economists discuss simple and complex labor in the context of assessing the cost of labor per unit of time. This is also reflected in the different wages for the workers in different professions. Labor productivity is defined as "output per units of time" or "output per labor hour" (Kedrick, 1977). In analysis of productivity, economists pay attention to amount of work done per units of time. Time study becomes an important aspect of analysis of the economic efficiency of human work. Currently, physically demanding manual labor is rather rare and variable and complex tasks are more common. The more complex the tasks performed by a worker or operator in a complex computerized system, the more mental effort is necessary to sustain efficiency of the work activity. Not physical but cognitive demands caused by the complexity of tasks is a major factor in the study of human productivity. Traditionally in economics the difference between complex and simple labor was considered as the difference between skilled and unskilled labor. The assignment of workers to one or the other group according to the work complexity is based on the evaluation of the duration of vocational training, which does not often reflect the real state of affairs. The differences between these professions are inaccurate and often subjective. This is especially true for modern professions when a person works as an operator in the complex man-machine computerized systems. The requirements change for not only the cognitive components of the task but also for its manual components. Contemporary tasks include precise motor actions that require minimal physical effort. Motor actions become not heavy but complex, and their performance requires cognitive effort. Currently, various analytical methods of job evaluation are utilized for assessment of labor complexity. However, the methods for direct measurement of complexity of work that requires significant mental effort have not been developed in economics and psychology at this time. Presented here the SSAT method of measurement of complexity of work allows the evaluation of cognitive efforts during task performance.

Also an important area of economics are the evaluation of technological innovations, methods of task performance, and how innovations influence productivity. Productivity is found to be linked to the amount of expenditures on innovation. It is sometimes suggested that innovation is inherently impossible to quantify and to measure (Smith, 2005). Scientists focus on the conceptualization, collection, and analysis of direct measures of innovation. The major purpose of innovation is producing a new performance outcome. Innovation has multidimensional aspects that are difficult to measure. This is specifically important when we try to evaluate innovations of human performance and consequences of such performance in general. It is especially difficult to examine new methods of task performance and possible innovations at the analytical stage of their evaluation. Quantitative methods of analysis of innovations at the analytical stage is a critical factor in the study of the efficiency of work. Development of measurement procedures, units of measurement, and determining their commensurability are important aspects of innovation evaluation from an economical perspective. The problem of commensurability of units of measurement during assessment of human work activity is central not only in ergonomics and work psychology but also in economics. The method offered here for evaluation of the efficiency of performance of complex variable tasks can be utilized for assessment of various innovations in ergonomics, work psychology, and economics.

Let us now consider the business impact of the innovations implemented in each version of the *computerized inventory receiving task* and their economic effect for the company as a whole.

We will first list these improvements:

1. The PO number is scanned instead of being keyed in.
2. The item number is scanned instead of being keyed in.
3. The price difference is calculated by the system and if the increase is over 10%, the receiver gets a warning.
4. The bin type can be corrected by utilizing the drop-down menu instead of recalling the bin types and their descriptions.
5. If the item being received is on order from one of the workshops, the receiver gets a message and follows the Work-In-Process (WIP) procedure.

The above-listed innovations provide both tangible and intangible benefits.

The Work-In-Process innovation has significant economic effects not only for the inventory receiving task but also for the production process as a whole (Figure 9.4). Before introducing the WIP innovation, it was suggested that all parts received from vendors should be placed directly in storage on the warehouse shelves. Only after that can they be delivered upon request to the various workshops. However, it was discovered that in approximately 20% of the cases it was necessary to immediately deliver some items into the workshop for the production process. Before the innovation, to do this was impossible. The task "delivery to workshop" is usually carried out by the operators who perform tasks such as "putting-away" and "picking-up". This results in in producing unnecessary work by workers that are involved in receiving and delivering items, and delay in the production process. The existing situation ignores the situation when items should be immediately sent to the workshop for production. Figure 9.5 demonstrates how "delivering to workshop" production process is organized after WIP innovation has been introduced.

The above figure demonstrates that received parts have two possible routes: one proceeds to the warehouse, while the other is sent to WIP. Information about these processes is presented on the computer screen (see Figure 7.8).

Approximately 80% of items are delivered to the warehouse (putting-away), and upon request they can be moved to the workshop. After completing the production process, intermediate or ready product is delivered back to receiver. He/she uses route 1 for delivering intermediate product for further production, or route 2 for delivering ready product for sales. If the receiving operator gets the "work-in-process?" message and answers "Y" (see Figure 7.8 with the work-in-process option) on the computer screen, then the items are delivered directly to the workshop (department) for the production process. The organization of the workplace and the additional equipment necessary for the implementation of innovation was also shown before in Figure 7.2. Morphological and quantitative analysis of the task complexity have already been described in Chapters 7 and 8. In this section, we want to draw attention to the assessment of the economic effectiveness of the innovation under consideration.

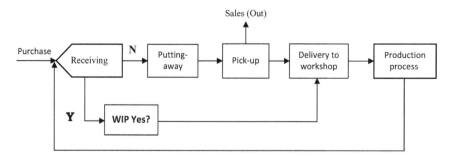

FIGURE 9.4
Production process after improvement. 1 – receiver after questioning "work-in-process?" get answer "N" and items delivered into warehouse; 2 – receiver after questioning "work-in-process?" get answer "Y" and items delivered directly to workshop for production process.

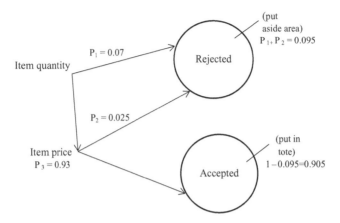

FIGURE 9.5
Relationship between accepted and rejected items' probabilities.

In this case, we can distinguish two aspects of this problem: (1) the effectiveness of the implementation of the "inventory receiving task", which is performed by receiver; and (2) the efficiency of the receiving process as a whole. The first task is performed only by the receiver. The production process as a whole involves the other workers, who are involved in such tasks as "putting-away," "pick-up," and "delivery to workshop." Here it is important to summarize our conclusions. Regarding the first task, we can see that despite the fact that introducing the "work-in-process" innovation that requires additional time for performance, the time of task performance was reduced, and the productivity of worker who performs this task significantly increased. Moreover, this was obtained not on the basis of work intensification but on the basis of task performance optimization, when the complexity of task was reduced and cognitive demand for its performance was reduced. The second economic aspect of the considered innovation is in significantly reducing unnecessary work with delivering items into the warehouse instead of directly delivering them for the production process. This also has a positive impact on the production process. Finally, this innovation can be evaluated at the analytical stage or stage of design, where experiment has only supplementary meaning. Thus, SSAT helps ergonomists and psychologists work with economists together.

Below we present a consolidated time table for computerized inventory receiving tasks before and after improvements that we considered in Chapter 8.

Consolidated time tables are useful because they clearly represent the data for the comparative analysis.

Comparison of the time tables for the computerized inventory receiving task before and after improvement shows a reduction in total performance time after implementation of multiple enhancements to the process. This in itself provides tangible benefits from better productivity. There are also intangible benefits of these innovations manifested in a decrease of cognitive workload.

Table 9.5 A shows all versions of task performance before innovations, and Table 9.5 B shows them after their implementation. Each of these tables is very useful for understanding of the flow of each version's performance. It can be seen from Table 9.5 A that, for instance, O_{25}^g and O_{26}^g are performed in Version 1 only. Table 9.5 A and B demonstrate which members of the algorithm are utilized in different versions of the tasks, and what time is required for each member of the algorithm before and after innovations. These tables are placed next to each other intentionally so that the comparison can also be done for the before and after innovation versions. It can be seen that after implementation of the innovation the algorithm is visibly shorter due to the smaller number of the operators. The performance time of each version in these two tables can be also compared to see the benefits of innovations. This method of comparison can be applied to the various versions of the same task performance. The difference in performance can be caused by the individual style of activity, by different levels of training or experience, or by innovations that lead to more efficient methods of performance.

The complexity measures presented in the previous chapter are the tools for assessing the economic effect of the innovations that can be calculated during the design stage of the new version of the task analysis. Some innovations might result in significant expenses and should be evaluated beforehand. The proposed measures facilitate such evaluation and to choose the most efficient method of task performance.

Below we present Figure 9.5, which demonstrates the relationship between the accepted and rejected items' probabilities. These data are obtained during a morphological analysis of the activity and presented in this figure in an integrative manner.

The figure demonstrates that the items can be rejected based on quantity and price criteria. After considering the item quantity, the item is rejected with $P=0.07$. The probability of the item being accepted based on quantity evaluation is $P=0.93$. After the price of the items is considered, some items also can be rejected. The probability of this event is $P=0.025$. So the probability of the item being rejected is $P=0.07+0.025=0.095$. The probability of the item being accepted, then, is $P=0.905$. When the items are rejected due to quantity or price criteria, it does not mean that the receiver made a mistake. The errors occur when the items are accepted, although they should be rejected according to the business rules or the item is rejected when it should be accepted.

This figure depicts the probabilities and outcomes of three versions of the inventory receiving task. These probabilities can also be inferred from the event trees presented with the morphological analysis of each version. These probabilities have been used to determine the average performance time of the task before and after implementation of multiple innovations. These calculations are presented below.

TABLE 9.5

Consolidate Time Table for the Computerized Inventory Receiving task: A – before and B – after Improvements

A

Consolidate Time Table for Subtask 0 and Versions of Tasks Performance Before Improvement

Members of Algorithm	Subtask 0	Version-1	Version-2	Version-3
Setup Subtask (Subtask 0)				
O_1^α	0.72			
$\downarrow O_2^\varepsilon$	2.01			
O_3^α	0.72			
$l_1\leftarrow$	0.3			→
$\downarrow\downarrow O_4^\varepsilon$	9			→
$O_5^{\alpha\mu}$	2.46			
$O_6^{\alpha\mu}$	0.72			
$O_7^{\varepsilon\mu}$	2.4			
$O_8^{\alpha\mu}$	2.46			
$O_9^{\varepsilon\mu}$	2.7			
O_{10}^ε	1.76			
O_{11}^α	1.02			
$l_2\leftarrow$	0.4			
O_{12}^α	0.18			

B

Consolidate Time Table for Subtask 0 and Versions of Tasks Performance After Improvement

Members of Algorithm	Subtask 0	Version-1	Version-2	Version-3
Setup Subtask (Subtask 0)				
O_1^α	0.72			
$\downarrow O_2^\varepsilon$	2.01			
O_3^α	0.72			
$l_1\leftarrow$	0.3			
$\downarrow\downarrow O_4^\varepsilon$	9			
O_5^α	0.72			
O_6^ε	4.08			
O_7^α	0.72			
$l_2\leftarrow$	0.4			
O_8^ε				
$\downarrow O_9^{\varepsilon\mu}$	0.68			
O_{10}^ε	4.44			
Work with Individual Item				
$\downarrow\downarrow O_{11}^\varepsilon$		5.7	5.7	5.7

(Continued)

TABLE 9.5 (CONTINUED)

Consolidate Time Table for the Computerized Inventory Receiving task: A – before and B – after Improvements

A

Consolidate Time Table for Subtask 0 and Versions of Tasks Performance Before Improvement

Members of Algorithm	Subtask 0	Version-1	Version-2	Version-3
Setup Subtask (Subtask 0)				
O_{13}^ε	0.12			
$I_3 \leftarrow$				
O_{14}^ε				
$\downarrow O_{15}^\varepsilon$	0.68			
O_{16}^ε	4.45			
Work with Individual Item				
$\downarrow\downarrow\downarrow O_{17}^\varepsilon$		5.7	5.7	5.7
$O_{18}^{\alpha u}$		6	6	6
$I_4 \leftarrow$		0.3	0.3	0.3
O_{19}^ε		0.62	0.62	0.62
$\downarrow O_{20}^\varepsilon$		2.64	2.64	2.64
$O_{21}^{\alpha u} \downarrow$		1.64	1.64	1.64
$I_5 \leftarrow$		0.4	0.4	0.4

B

Consolidate Time Table for Subtask 0 and Versions of Tasks Performance After Improvement

Members of Algorithm	Subtask 0	Version-1	Version-2	Version-3
Setup Subtask (Subtask 0)				
O_{12}^ε		2.9	2.9	2.9
O_{13}^ε		0.9	0.9	0.9
$O_{14}^{\alpha u}$		1.64	1.64	1.64
$I_3 \leftarrow$		0.4	0.4	0.4
O_{15}^ε		1.7	0.17	0.17
O_{16}^α		1.14	0.11	0.11
$\rightarrow$		4.2	0.42	0.42
$I_4 \leftarrow$		1.5	0.15	0.15
O_{18}^ε		0.84		
O_{19}^ε		12.5		
O_{20}^α				
$I_5 \leftarrow$				
$\rightarrow O_{21}^\varepsilon$			0.126	0.126
$\rightarrow O_{22}^\varepsilon$			1.01	1.01

(Continued)

TABLE 9.5 (CONTINUED)

Consolidate Time Table for the Computerized Inventory Receiving task: A – before and B – after Improvements

A

Consolidate Time Table for Subtask 0 and Versions of Tasks Performance Before Improvement

Members of Algorithm	Subtask 0	Version-1	Version-2	Version-3
		Setup Subtask 0		
O_{22}^{ε}		1.7	0.17	0.17
O_{23}^{α}		1.14	0.11	0.11
$**O_{24}^{\mu th}$		4.2	0.42	0.42
$l_6 \leftarrow$		1.5	0.15	0.15
O_{25}^{ε}		0.84		
O_{26}^{ε}		12.5		
O_{27}^{α}				
$l_7 \leftarrow$				
$\rightarrow O_{28}^{\varepsilon}$			0.126	0.126
$\rightarrow O_{29}^{\varepsilon}$			1.01	1.01
$***O_{30}^{\alpha}$			2.36	2.36
$l_8 \leftarrow$			0.4	0.4
$O_{31}^{thμ}$			3.6	0.72

B

Consolidate Time Table for Subtask 0 and Versions of Tasks Performance After Improvement

Members of Algorithm	Subtask 0	Version-1	Version-2	Version-3
		Setup Subtask 0		
$**O_{23}^{\alpha}$			2.36	2.36
$l_6 \leftarrow$			0.4	0.4
O_{24}^{ε}			2.977	0.59
O_{25}^{α}			1.14	0.23
$l_7 \leftarrow$			3.5	0.7
$\rightarrow O_{26}^{\varepsilon}$			0.84	
O_{27}^{ε}			12.5	
$***O_{28}^{\alpha}$				
$l_8 \leftarrow$				0.178
$\rightarrow O_{29}^{\varepsilon}$				
O_{30}^{ε}				0.83
$***O_{31}^{\alpha}$				0.72
$l_9^{\mu} \leftarrow$				0.4

(Continued)

TABLE 9.5 (CONTINUED)

Consolidate Time Table for the Computerized Inventory Receiving task: A – before and B – after Improvements

A

Consolidate Time Table for Subtask 0 and Versions of Tasks Performance Before Improvement

Members of Algorithm	Subtask 0 Version-1	Version-2	Version-3
	Setup Subtask (Subtask 0)		
$O_{32}^{\varepsilon\mu}$		2.977	0.59
$O_{33}^{\alpha\mu}$		1.14	0.23
$l_9^{\mu} \rightarrow$		4	0.8
O_{34}^{ε}			0.2
$\omega_1 \rightarrow$	$\rightarrow$		
$\downarrow O_{35}^{\varepsilon}$		0.89	
O_{36}^{ε}		12.5	
O_{37}^{α}			
$l_{10} \rightarrow$			
$\downarrow\downarrow O_{38}^{\varepsilon}$			0.89
$*** O_{39}^{\alpha}$			0.72
$l_{11}^{\mu} \rightarrow$			0.4
O_{40}^{ε}			0.06

B

Consolidate Time Table for Subtask 0 and Versions of Tasks Performance After Improvement

Members of Algorithm	Subtask 0 Version-1	Version-2	Version-3
	Setup Subtask (Subtask 0)		
O_{32}^{ε}			0.06
$\rightarrow O_{33}^{\varepsilon}$			0.86
O_{34}^{α}			0.74
$l_{10} \leftarrow$			0.1
$\downarrow O_{35}^{\varepsilon}$			0.14
O_{36}^{ε}			1.19
O_{37}^{ε}			1.1
O_{38}^{α}			0.34
O_{39}^{ε}			0.37
$\omega_1 \leftarrow$		$\rightarrow$	
$\rightarrow O_{40}^{q\mu}$			0.15
$l_{11}^{\mu} \leftarrow$			0.48

(Continued)

TABLE 9.5 (CONTINUED)

Consolidate Time Table for the Computerized Inventory Receiving task: A – before and B – after Improvements

A

Consolidate Time Table for Subtask 0 and Versions of Tasks Performance Before Improvement

Members of Algorithm	Subtask 0	Version-1	Version-2	Version-3
	Setup Subtask (Subtask 0)			
↓ O_{41}^{ε}				0.86
O_{42}^{α}				0.3
I_{12}				0.3
$O_{43}^{\alpha\mu}$				0.18
I_{13}^{μ}				0.06
$O_{44}^{\varepsilon\mu}$				0.02
ω_2 ↓				
O_{45}^{ε}				0.03
O_{46}^{α}				0.03
O_{47}^{ε}				0.067
↓↓↓ O_{48}^{ε}				0.7
				6

B

Consolidate Time Table for Subtask 0 and Versions of Tasks Performance After Improvement

Members of Algorithm	Subtask 0	Version-1	Version-2	Version-3
	Setup Subtask (Subtask 0)			
O_{41}^{ε}				0.0094
O_{42}^{α}				0.016
O_{43}^{ε}				0.0005
↓↓ O_{44}^{ε}				0.56
O_{45}^{ε}				4.8
O_{46}^{ε}				4.32
O_{47}^{α}				1.34
O_{48}^{ε}				
ω_1 ←				1.48
↓ O_{49}^{α}				
I_{15}				
Total	23.79	33.42	37.25	37.56

(Continued)

TABLE 9.5 (CONTINUED)

Consolidate Time Table for the Computerized Inventory Receiving task: A – before and B – after Improvements

A					B				
Consolidate Time Table for Subtask 0 and Versions of Tasks Performance Before Improvement					Consolidate Time Table for Subtask 0 and Versions of Tasks Performance After Improvement				
Members of Algorithm	Subtask 0	Version-1	Version-2	Version-3	Members of Algorithm	Subtask 0	Version-1	Version-2	Version-3
Setup Subtask (Subtask 0)					Setup Subtask (Subtask 0)				
O_{50}^{ε}				6.7					
O_{51}^{α}				1.71					
$I_{14}\leftarrow$				0.4					
O_{52}^{ε}				1.32					
O_{53}^{α}				1.71					
$\downarrow O_{54}^{\varepsilon}$				1.8					
O_{55}^{α}									
$I_{15}\leftarrow$									
Total	32.1	39.15	47.16	48.84					

The average performance time of the task before of innovations:

$$\textbf{Version 1.}\ T_{\text{ver1}} \times P_1 = 39.18 \times 0.07 = 2.74 \tag{9.38}$$

$$\textbf{Version 2.}\ T_{\text{ver2}} \times P_2 = 47.16 \times 0.025 = 1.18 \tag{9.39}$$

$$\textbf{Version 3.}\ T_{\text{ver3}} \times P_3 = 48.84 \times 0.905 = 44.2 \tag{9.40}$$

$$\textbf{T}_{\text{avaarg}} = T_{\text{ver1}} \times P_1 + T_{\text{ver2}} \times P_2 + T_{\text{ver3}} \times P_3 = 48.12 \tag{9.41}$$

The average performance time of the task after of innovations:

$$\textbf{Version 1.}\ T_{\text{ver1}} \times P_1 = 33.42 \times 0.07 = 2.34 \tag{9.42}$$

$$\textbf{Version 2.}\ T_{\text{ver2}} \times P_2 = 37.25 \times 0.025 = 0.93 \tag{9.43}$$

$$\textbf{Version 3.}\ T_{\text{ver3}} \times P_3 = 37.56 \times 0.905 = 34 \tag{9.44}$$

$$\textbf{T}_{\text{avaarg}} = T_{\text{ver1}} \times P_1 + T_{\text{ver2}} \times P_2 + T_{\text{ver3}} \times P_3 = 37.27 \tag{9.45}$$

Gain in productivity of performance inventory receiving task only can be calculated as $37.27/48.12 = 0.775$ with 22.5% gain in productivity.

It should be noted that WIP stage has been also introduced as a part of improvement of the third version of the task. This requires additional time of task performance. In spite of this fact, general time of task performance in the third version of the task after improvement was 10.85 sec shorter than for the third version of the task before improvement.

It is important to stress that the productivity after implementation of innovations in this case is not achieved by the intensification of performance or an increase of the cognitive or physical workload but by the implementation of the innovations that optimized the work process. Here we want to bring readers' attention to the fact that considered tasks have complex cognitive structure, and the evaluation of improvement is based on comparison of the optimized and initial structure of activity with taking into account utilized material tools, software and other means of work. Such a method is totally different in comparison with traditional simplifications that are derived from time and motion analysis.

At the same time, the above-presented data clearly demonstrate that the field of study that is traditionally known as the time study of the blue-collar jobs can in its enhanced form be applied to the contemporary task analysis, including studies of computer-based and computerized tasks.

We want to remind the reader that the focus here is not only on the nature of improvements but also on the method of determining their efficiency. It's very important to understand that this method allows conducting such an evaluation at the analytical stage of accessing the economic efficiency of the examined innovations. In our example, we analyzed three different versions of task performance that occur depending on the quantity and price of the received items. The same method of analysis can be applied when the difference in performance strategies is due to the individual style of performance. The suggested method allows to compare and access economic efficiency of each method of task performance.

References

Bedny, I. S. (2004). General characteristics of human reliability in system of human and computer. *Science and Education*, Odessa, Ukraine: # 8–9, 58–61.

Bedny, I. S. (2006). On systemic-structural analysis of reliability of computer based tasks. *Science and Education*, Ukraine, Odessa, Ukraine: 1-2, # 7–8, 58–60.

Bedny, G. Z. (2015). *Application of Systemic-Structural Activity Theory to Design and Training*. Boca Raton, FL and London, UK: CRC Press, Taylor & Francis.

Bedny, I. S. and Bedny, G. Z. (2011a). Abandoned actions reveal design flaws: An illustration by a web-survey task. In Bedny G. Z. and Karwowski W. (Eds.) *Human-Computer Interaction and Operators' Performance. Optimization of Work Design with Activity Theory* (pp. 149–185). London, UK and Boca Raton, FL: Taylor & Francis, CRC Press.

Bedny, I. S. and Bedny, G. (2011b). Analysis of abandoned actions in the email distributed tasks performance. In Kaber, D. B. and Boy, G. (Eds.). *Advances in Cognitive Ergonomics* (pp. 683–692). Boca Raton, FL, and London, UK: CRC Press, Taylor & Francis.

Bedny, G. Z. and Harris, S. (2013). Safety and reliability analysis methods based on systemic-structural activity theory. *Journal of Risk and Reliability*. Sage Publisher, V. 227, Number 5, pp. 549–556.

Bedny, G. and Meister, D. (1997). *The Russian Theory of Activity: Current Application to Design and Learning*. Mahwah, NJ: Lawrence Erlbaum Associates.

Bedny, I. S. and Sengupta, T. (2005). The study of computer based tasks, *Science and Education*, Odessa, Ukraine: 1-2, # 7-8, 82–84.

Bedny, I. S., Karwowski, W., and Bedny, G. Z. (2010). A method of human reliability assessment based on systemic-structural activity theory. *International Journal of Human-Computer Interaction*, 26, # 4, 377–402. Boca Raton, FL and London, UK: CRC Press, Taylor & Francis Group.

Bedny, I. S., Karwowski, W. and Bedny, G. (2012). Computer technology at the workplace and errors analysis. In Stanney K. M. and Hale K. S. (Eds.). *Advances in Cognitive Engineering and Neuroergonomics* (pp. 167–176). Boca Raton, FL, and London, UK: CRC Press, Taylor & Francis Group.

Jacko, J. A. (1997). An empirical assessment of task complexity for computerized menu systems. *International Journal of Cognitive Ergonomics*, 1 (2), 137–147.

Jacko, J. A. and Ward, K. G. (1996). Toward establishing a link between psychomotor task complexity and human information processing. *Computers and Industrial Engineering*, 31 (1–2), 533–536.

Jacko, J. A., Salvendy, G., and Koubek, R. J. (1971). Modeling of menu design in computerized work. *Interaction with Computer*. 7 (3), 304–330.

Kedrick, J. W. (1977). *Understanding Productivity: An Introduction to the Dynamics of Productivity Change*. Baltimore, MD: Johns Hopkins.

Kirwan B. (1994). *A Guide to Practical Human Reliability Assessment*. London, UK: Taylor & Francis.

Kotik, M. A. and Yemelyanov, A. M. (1993). *The Nature of Human-Operator's Errors. Examples Derived from Analysis Means of Transport*. New York, NY: Ronald Press

Smith, K. H. (2005). Measuring innovation. In: *The Oxford Handbook of Innovation* (pp. 148–177). New York, NY: Oxford University Press.

Conclusion

Systemic-structural activity theory (SSAT) is a relatively new and still-developing unified framework for the study of the efficiency of human performance and process, equipment, and software design. Several original scholarly books have been published in this field. However, technical and social changes in human activity require the continual improvement and advancement of the methods of studying human performance when interacting with continuously changing technology. This book presents new theoretical data and methods of perfecting human performance from the systemic perspective derived from it. The SSAT systemic approach implies that any complex sociotechnical system includes two basic subsystems. These are (1) a goal-directed human activity subsystem, which consists of various elements, the main of which are cognitive and behavioral actions; and (2) a technological subsystem that includes various technical components or elements. Each subsystem has a complex structure. Any changes in one subsystem immediately entail changes to the other subsystem. This leads to some basic principles of ergonomic design. One of the most important of them is that the efficiency of the equipment or software design can be evaluated based on the analysis of the structure of activity during execution of various tasks. Moreover, based on analysis of the structure of activity, it is possible to develop more efficient methods of task performance, evaluate the efficiency of training, increase productivity, and improve the user experience in general. Many specialists are involved in studies of productivity, including engineers, economists, work psychologists, and ergonomists. The key aspects of the relationship of performance and productivity are considered in this book. Special attention is paid here to the psychological aspects of time study, the complexity of work and the reliability of task performance, and to the evaluation of innovations and development of efficient strategies of task performance.

The main objective of this book was presenting the analytical methods of studying productivity. The existing methods of study outside of SSAT consist only of observation and experiments. However, observation and experiment are not always available or efficient, especially for new equipment or interfaces. Moreover, such methods are not sufficient for predicting the efficiency of performance and of the utilized technology. Enhancement of efficiency and/or productivity requires continuous improvement and innovation of the utilized processes. This means that people who are involved in the production process have to generate more output per units of time with less physical and mental effort, lower energy cost, enhanced reliability, enhanced usability, improved user experience, and so on.

Specific attention in this book was paid to the connection of work psychology and economics and the role of computerization in increasing productivity.

At present, the concept of design is substituted in ergonomics by experimentation. Such an approach reduces the external validity of task analysis. As a result, prediction of productivity becomes extremely difficult. Contemporary complex computerized and computer-based tasks include multiple decisions and are exceedingly variable. There are no experimental and especially analytical methods of task analysis and design adapted for such tasks outside of SSAT. These drawbacks are common specifically when considering psychological and economic aspects of design when cognitive components of activity are extremely important.

So there is a need for the consistent and clear description of the relationship between productivity and performance and for the methods of study that allow reduction of the number of the enhancement cycles of the designed objects. The book presented especially valuable information from this perspective. It described a powerful unified theory that offers new analytical procedures as well as qualitative and quantitative methods of analysis that can be used for design of complex sociotechnical systems.

SSAT offers several basic stages of design and redesign of such systems. It includes qualitative descriptive analysis, functional analysis, morphological analysis, and quantitative analysis of human activity. Experimental procedures of study can be utilized at any stage of analysis depending on the specificity of the designed or redesigned object.

When a totally new system is being designed, the options for utilizing experimental procedures are limited. The basic principle of design is development of models of activity that describe its structure.

Systemic description of the activity structure and evaluation of its efficiency are achieved via morphological analysis, which facilitates development of quantitative methods of analysis. The book presented new advanced principles of morphological analysis of activity and derived from them enhanced principles of task complexity evaluation. Task complexity determines the cognitive demands for its performance. The concept of task and job complexity is important in economic analysis of the relationship between productivity and human performance. The more complex the task is, the greater is its cost per unit of time. Hence, this should be a point of tight interaction between economics, psychology, and ergonomics.

The book presented a new and accurate method for estimating the probabilistic characteristics of activity during task performance and the method of reliability assessments of human performance derived from it. The quantitative analytical methods of analysis considered in the book are closely related to the qualitative methods. The functional analysis of activity is of particular importance, when a complex self-regulative system is under consideration.

The book presented the enhanced model of self-regulation of orienting activity, and the advanced models of self-regulation of all cognitive processes are presented for the first time. These models are beneficial for efficient prediction of the preferable strategies of task performance and for conducting advanced analytical methods of study. The analytical principles of task analysis suggested in the book allow a reduction in costly cycles of continuous enhancement and redesign solutions of production processes and repetitive improvement of software. In general, the book presented a new and advanced approach to the study of human performance from the SSAT perspective. It demonstrated application of this approach using extremely complex computer-based and computerized tasks.

Index